Transnational Frontiers of Asia and Latin America since 1800

Frontiers are "wild." The frontier is a zone of interaction between distinct polities, peoples, languages, ecosystems, and economies, but how do these frontier spaces develop? If the frontier is shaped by the policing of borders by the modern-nation state, then what kind of zones, regions, or cultural areas are created around borders?

This book provides 16 different case studies of frontiers in Asia and Latin America by interdisciplinary scholars, charting the first steps toward a transnational and transcontinental history of social development in the borderlands of two continents. Transnationalism provides a shared focus for the contributions, drawing upon diverse theoretical perspectives to examine the place-making projects of nation states. Through the lenses of different scales and time frames, the contributors examine the social processes of frontier life, and how the frontiers have been created through the exertions of nation-states to control marginal or borderland peoples. The most significant cases of industrialization, resource extraction, and colonization projects in Asia and Latin America examined in this book reveal the incompleteness of frontiers as modernist spatial projects, but also their creativity—as sources of new social patterns, new human adaptations, and new cultural outlooks and ways of confronting power and privilege. The incompleteness of frontiers does not detract from their power to move ideas, peoples, and practices across borders both territorial and conceptual.

In bringing together Asian and Latin American cases of frontier-making, this book points toward a comparativist and cosmopolitan approach in the study of statecraft and modernity. For scholars of Latin America and/or Asia, it brings together historical themes and geographic foci, providing studies accessible to researchers in anthropology, geography, history, politics, cultural studies, and other fields of the human sciences.

Jaime Moreno Tejada is a historical geographer based at Chulalongkorn University, Thailand.

Bradley Tatar is a cultural anthropologist based at UNIST, Ulsan, Korea.

Border Regions Series

Series Editor: Doris Wastl-Walter

University of Bern, Switzerland

In recent years, borders have taken on an immense significance. Throughout the world they have shifted, been constructed and dismantled, and become physical barriers between socio-political ideologies. They may separate societies with very different cultures, histories, national identities or economic power, or divide people of the same ethnic or cultural identity.

As manifestations of some of the world's key political, economic, societal and cultural issues, borders and border regions have received much academic attention over the past decade. This valuable series publishes high quality research monographs and edited comparative volumes that deal with all aspects of border regions, both empirically and theoretically. It will appeal to scholars interested in border regions and geopolitical issues across the whole range of social sciences.

For a full list of titles in this series, please visit www.routledge.com/geography/series/ASHSER-1224

Placing the Border in Everyday Life
Edited by Reece Jones and Corey Johnson

Borders, Fences and Walls
State of Insecurity?
Edited by Elisabeth Vallet

Israelis and Palestinians in the Shadows of the Wall
Spaces of Separation and Occupation
Edited by Stéphanie Latte Abdallah and Cédric Parizot

Informal Trade, Gender and the Border Experience
From Political Borders to Social Boundaries
Olga Sasunkevich

Russian Borderlands in Change
North Caucasian Youth and the Politics of Bordering and Citizenship
Tiina Sotkasiira

Transnational Frontiers of Asia and Latin America since 1800
Edited by Jaime Moreno Tejada and Bradley Tatar

Transnational Frontiers of Asia and Latin America since 1800

Edited by
Jaime Moreno Tejada and Bradley Tatar

LONDON AND NEW YORK

First published 2017
by Routledge
2 Park Square, Milton Park, Abingdon, Oxon OX14 4RN

and by Routledge
711 Third Avenue, New York, NY 10017

Routledge is an imprint of the Taylor & Francis Group, an informa business

British Library Cataloguing-in-Publication Data
A catalogue record for this book is available from the British Library

Library of Congress Cataloging in Publication Data
Names: Moreno Tejada, Jaime, editor. | Tatar, Bradley, editor.
Title: Transnational frontiers of Asia and Latin America since 1800 / edited by Jaime Moreno Tejada and Bradley Tatar.
Description: Abingdon, Oxon ; New York, NY : Routledge, 2017. Includes bibliographical references and index.
Identifiers: LCCN 2016015204 | ISBN 9781472470560 (hardback) | ISBN 9781315549866 (ebook)
Subjects: LCSH: Asia—Historical geography. | Latin America—Historical geography. | Geopolitics—Asia. | Geopolitics—Latin America. | Asia—Boundaries. | Latin America—Boundaries.
Classification: LCC DS5.9.T73 2017 | DDC 950/.3—dc23
LC record available at https://lccn.loc.gov/2016015204

ISBN: 978-1-4724-7056-0 (hbk)
ISBN: 978-1-315-54986-6 (ebk)

Typeset in Times New Roman
by Swales & Willis Ltd, Exeter, Devon, UK

To Álvaro, Jaime's godson, who was born as this book went into production.

Contents

Figures

Tables

Contributors

Greg Acciaioli lectures in Anthropology and Sociology at the University of Western Australia. His Ph.D. research dealt with Bugis migration in the Indonesian archipelago, concentrating upon the case study of Lindu in Central Sulawesi. Subsequent research has focused on the way Indigenous peoples such as the To Lindu have harnessed conservation idioms to further their local political aspirations. He has investigated conservation contestations in such contexts as the Tun Sakaran Marine Park in eastern Sabah, as well as conducting research on the social effects of rice intensification in South Sulawesi and the impact of oil palm plantations in Central Kalimantan.

Rosalia Avila-Tàpies is a human geographer. Rosalia graduated from the University of Barcelona and holds an M.A. and a Ph.D. in Letters (Geography) from Kyoto University (2000). She is currently a research associate of the Institute for the Study of Humanities and Social Sciences at Doshisha University in Kyoto. She has written extensively for geographical and academic reviews, and presently serves as an editor board member and article reviewer for reputable international geographical journals. Her research interests are in the areas of migration and ethnicity, and population issues in the Asia-Pacific region, including Japan.

Louise Cardoso de Mello is a Brazilian historian and anthropologist. She holds an M.A. with distinction in Indigenous History of Latin America from the Universidad Pablo de Olavide in Seville, Spain, where she continues her Ph.D. research in Amazonian Studies. Previous institutions she has collaborated with include the Université de Provence, La Sapienza, FUNDHAM in Brazil, the Universidad de San Carlos de Guatemala, the Universidad Nacional de Oriente, in Mexico, and the Universidad Nacional Amazónica de Madre de Dios, in Peru. She has further recently concluded an academic stay in the Division of Archaeology at the University of Cambridge.

Low Choo Chin is a lecturer in the History Section, School of Distance Education, Universiti Sains Malaysia (USM), Penang. She holds a Ph.D. from the University of Melbourne. Her research interests include comparative citizenship, migration, deportation, and diaspora. Her recent publications include:

"The Politics of Emigration and Expatriation: Ethnicisation of Citizenship in Imperial Germany and China," *Journal of Historical Sociology* (2015); "Debates over Liberalising Dual Citizenship: Prospects and Limits in Taiwan and the People's Republic of China," *International Journal of Asia Pacific Studies* (2015); "China's Citizenship Diplomacy at Bandung: An Evaluation from the British Perspective," *New Zealand Journal of Asian Studies* (2015); "The Repatriation of the Chinese as a Counter-insurgency Policy during the Malayan Emergency," *Journal of Southeast Asian Studies* (2014).

William T. Fischer received his B.A. in History from Carleton College in 2005 and his Ph.D. in Latin American History from the University of Florida in 2015. His research was supported by the University of Florida Center for Latin American Studies and by a Fulbright Award. He is Assistant Professor of History at Missouri Southern State University in Joplin, Missouri.

Dannelle Gutarra has earned the academic degrees of B.A. in Audiovisual Communications, M.A. in Communication Theory and Investigation, and Ph.D. in History from the University of Puerto Rico, Río Piedras Campus. Her academic performance won her the High Academic Distinction Medal for best academic performance from the Doctoral Program of History of the University of Puerto Rico and the Dr. Aída Caro Costas Award for best doctoral dissertation in the field of History from the University of Puerto Rico. Gutarra has taught undergraduate and graduate courses at the Inter American University of Puerto Rico, Metropolitan Campus, and Virginia Commonwealth University.

Christopher A. Howard is Lecturer in Cultural Anthropology at Chaminade University of Honolulu and Writing Fellow at Hiroshima University. He was previously Visiting Lecturer at Boston University, and has worked at academic institutions in New Zealand.

Pedro Iacobelli D. is Assistant Professor in History at the Pontificia Universidad Católica de Chile. He holds a B.A. in History from the Pontificia Universidad Católica de Chile, an M.A. in Asian Studies and a Ph.D. from the Australian National University. He is currently completing a manuscript about the role of the sending state in postwar migration movements from Japan to South America. His most recent publication is a co-edited volume titled *Transnational Japan as History: Empire, Migration and Social Movements* (2015). Among his academic interests, contemporary Japanese history, migration history of Asia, history of ideas, and Asia-Latin America relations are currently occupying most of his time.

Daniel S. Margolies is Professor of History and Chair of the History Department at Virginia Wesleyan College. He has been a Fulbright Senior Scholar at Sogang University in Seoul, Korea, a Faculty Fellow at the American Center for Mongolian Studies in Ulaanbaatar, and a Visiting Scholar at the Center for the Study of Law and Society at the University of California-Berkeley. His most recent book is *Spaces of Law in American Foreign Relations:*

Extradition and Extraterritoriality in the Borderlands and Beyond, 1877–1898 (2011). He is currently writing a book on jurisdictional assertions in U.S. foreign trade policy.

Graeme MacRae trained as an architect in Australia then as an anthropologist in Aotearoa/New Zealand. Since 1998, he has taught Anthropology at Massey University in Auckland. His recent research focuses on the ways in which human societies intersect with natural/ecological processes by way of technological interventions such as agriculture, waste management, disaster recovery, and architecture. While his primary research has been in Indonesia (mostly Bali), he has maintained a second stream of comparative research in India, most recently on basmati rice and mountain cultural ecology in the north.

Alastair McClure is a Ph.D. candidate from the University of Cambridge funded by the Arts and Humanities Research Council. His research examines colonial South Asia with a current focus on the history of criminal law and legal culture in the nineteenth and twentieth centuries. His chapter comes from recent fieldwork in India and London which was part funded by a grant from the Cambridge Smuts Memorial Fund.

Brent Metz received his Ph.D. from the University of Albany in 1995 and is Associate Professor in the Department of Anthropology at the University of Kansas. He has conducted research on the processes of ethnic transformations in and on the Ch'orti'-speaking region of eastern Guatemala since 1990, and the former Ch'orti'-speaking region of Guatemala, Honduras, and El Salvador since 1997. Among his publications are the books *Ch'orti' Maya Survival* (2006), *The Ch'orti' Area, Past and Present* (editor, 2009), and *Primero Dios* (2002).

Jaime Moreno Tejada received his Ph.D. from King's College London in 2009, and has since published a book and various academic papers on Upper Amazonian frontiers. He is interested in non-hegemonic modernity, mobility, and the everyday production of space. Recent publications include "Lazy labor, modernization and coloniality: mobile cultures between the Andes and the Amazon around 1900" (*Transfers*). Jaime is based at Chulalongkorn University (Thailand) where he investigates tropical and urban frontiers in transnational perspective.

Miki Namba is a doctoral candidate (Anthropology) at Hitotsubashi University. Her research focuses on the complex relation between infrastructural development, urbanization, and modernity in Laos. Her work is forthcoming in *Infrastructure and Social Complexity*, eds. P. Harvey, C. B. Jensen, and A. Morita (Routledge).

Jane M. Rausch earned a B.A. at DePauw University (1961) and her M.A. (1964) and Ph.D. (1969) at the University of Wisconsin in Madison. She joined the History Department at the University of Massachusetts-Amherst in 1969 where she taught Latin American History until her retirement in May 2010. A specialist in Colombian history and comparative frontier regions, she is the author of several books. Her most recent publications are *Colombia and the Transformation of the Llanos Orientales* (2013) and *Colombia and World War I: The Experience of a Neutral Latin American Nation during the Great War and Its Aftermath, 1914–1921* (2014).

Luis Roniger is Reynolds Professor of Latin American Studies and Political Science at Wake Forest University and a professor emeritus of Sociology and Anthropology at Hebrew University of Jerusalem. A comparative political sociologist, his work focuses on the interface between politics, society, and public culture. He is on the international board of academic journals published in Argentina, Colombia, Israel, Mexico, Spain, and the United Kindgom; and has published over 160 academic articles and 18 books, among them *Transnational Politics in Central America* (2011) and *La política del destierro y el exilio en América Latina* (2013, with Mario Sznajder), which has been awarded the 2014 Arthur Whitaker Prize of the Middle Atlantic Council of Latin American Studies in the USA.

Alka Sabharwal completed her Ph.D. in Environmental Anthropology at the University of Western Australia in 2015. Her doctoral research employed the framework of cultural politics to draw attention towards those complex cultural works embodied within the conservation contestations at the border region between India and China, that generally remains ignored in the traditional political ecology approach. Her research interests are centered upon the study of the state, bureaucracy, and the politics of conservation with an area interest in India and the Himalaya. At present she guest lectures and teaches in Social Anthropology at the University of Western Australia.

Bradley Tatar is a sociocultural anthropologist and Associate Professor of General Studies in Ulsan National Institute of Science and Technology (UNIST), in South Korea. His research includes many aspects of popular culture and politics in Latin America, the United States, and Korea. After researching post-Cold War political transition in Nicaragua, immigration politics in the U.S., and Mexican corrido songs, his current focus is on human/nature relations in the politics of whale conservation. Recent articles include "The Safety of Bycatch: South Korean Responses to the Moratorium on Commercial Whaling" (2014) and "Destroying Patriarchy: Struggle for Sexual Equality in Mexican-American Corridos and Anglo-American Ballads" (2015).

Diana Zhidan Duan, whose ancestry includes Tibetan, Muslim and Han Chinese migrants from Nanjing, Jiangsu, was raised in Dali, Yunnan Province of China. Her current focus is on the borderlands of China and Mainland Southeast Asia. She is actively engaged in the discussions of important themes that have been shaping the local history, including migration and immigration, cross-border communities, ethnic groups, the state ethnic and religious policies, indigenous economy, and environmental problems. She recently received her Ph.D. degree in Asian History from Arizona State University, and currently serves as Assistant Professor in History at Brigham Young University-Provo.

Acknowledgments

The authors thank the participants in *Frontier Regions, Modernity and Development: Borders and the Arts of Resistance*, a panel held in July 2015 at the 2015 IUAES Inter-Congress held in Thammasat University, Bangkok, Thailand. Participants included Attama Pocapanishwong, Xiaoping Sun, Greg Acciaioli, Miki Namba, Peter Kunstadter, Bradley Tatar, Yasmin Cho, and Jaime Moreno Tejada as the discussant. Special thanks to UNIST Division of General Studies for financial support of Bradley Tatar's participation in the panel. This book was partly funded by the Faculty of Arts at Chulalongkorn University; thanks go to CU's research section.

Introduction

Distance: modern transnational frontiers

Jaime Moreno Tejada

Modern frontiers and the case for transnational studies

For the purposes of this collection, frontiers may be defined as unfinished spatial projects. Despite their potential as gateways to wealth and regeneration, borderlands seem to stubbornly escape human mastery. But they do so over long periods of time, and the overall impression is that frontiers eventually close. The essays presented here explore a period commonly referred to as modernity. Modernity means many things, but we could narrow it down to three historically and locally rooted developments: industrialization, nation-state building, and imperialism. In the nineteenth century, with the rise of the world industrial economy, the modern frontier was intellectually recognized. Karl Marx and Friedrich Engels wrote about it in 1848 (n.p.): "The need of a constantly expanding market for its products chases the bourgeoisie over the entire surface of the globe. It must nestle everywhere, settle everywhere, establish connections everywhere." Modern capitalism goes hand in hand with the making of the nation-state, which is nothing if not a process of frontier making. In *Facundo: Civilization and Barbarism* (1845) Juan Domingo Sarmiento looked at the vast plains that lay beyond Buenos Aires and concluded, drawing from Johann Gottfried von Herder and James Fenimore Cooper in equal measure, that the Argentinian national character was the product of distance, solitude, and rugged *gaucho* (cowboy) livelihoods. Fueled by the profits and needs of industry, governments created new networks that reached beyond the nation-state. Imperialism was a long-distance form of border crossing that slowly turned the entire world into a gigantic space of uneven flows. Ideologically, nineteenth- and twentieth-century imperialism was justified on the grounds of sheer economic necessity and a moral obligation to civilize the innocent, barbarian hordes. It is worth recalling Rudyard Kipling's *The White Man's Burden* (1899)—"Your new-caught, sullen peoples/Half-devil and half-child"—but for balance's sake, let us quote from Mochiyi Rokusaburo, a member of the Japanese colonial administration in Taiwan, who in 1902 wrote:

> The problem of aboriginal land has yet to arrive at a successful solution. Yet if we do not solve this problem, our countrymen will likely fail to realize their great potential for overseas expansion. Occupying 50 to 60 percent of the

> entire islands, the aboriginal lands constitute a treasure trove rich in forest, agricultural, and mining resources. Unfortunately, we have not succeeded in unlocking this treasure trove because ferocious savages block our access to it . . . it is not a problem that one can hope to resolve by ethical means.
>
> (Rokusaburo quoted in Tierney 2010, 44)

It is useful to think of transnational frontiers as cultural networks that either surpass or ignore national borderlines. The prefix "trans" means "beyond" and "across." A transnational study should either be located *beyond* the nation or study movements that occur *across* lines dividing nations; at the very least, a transnational study must take into account the other side of the border. The second proposition seems straightforward—a transnational study is an international study—but it is not necessarily so. As historian Akira Iriye has explained (2013), transnational history differs from international history in that it brings the richness of context, cultural context, into an equation that traditionally contained only diplomacy and politics. Iriye's book *Power and Culture* (1981), a study of the Pacific War, identified the cultural turn taking place in academia after the publication of classics-to-be such as Clifford Geertz's "Thick Description" (1973) and Edward Said's *Orientalism* (1978). The accent has moved to *big* frontiers (e.g. imperial frontiers, as in the case of Said's Orient) and *small* frontiers (Geertz's cockfight in Bali being a case in point). A focus on culture as it really is, a shapeless thing that almost never coincides with the limits of any specific country, compels researchers to leave the nation in the background as a unit of analysis.

To be sure, as Jane Rausch writes in Chapter 1 in this volume, "the political significance of . . . international border[s] is a reality not readily dismissed," a point that is particularly emphasized by Daniel Margolies in Chapter 5. Other contributors to this book make similar observations. Miki Namba shows in Chapter 14 how governments build recreational infrastructures to block the natural porosity of borders. Culture itself has been a site of patriotic identification, akin to language and race, since at least the Romantic period. When combined with nationalism and radical politics, culture is a time bomb whose consequences range from mass deportation (Low, Chapter 15 this volume) to genocide. And, to paraphrase Fischer (Chapter 10 in this volume), a significant part of the frontier's history happens, i.e. is imagined, in the press and other media aimed at a general public that could not be more out of touch. It seems clear, though, that the tables were turning from the outset. DNA sequencing has sufficiently proven that human history is one of generalized migrations, never confined to the national geo-bodies (Winichakul 1994) concocted by antiquarians and map-makers in the nineteenth and twentieth centuries. Sociologists, from Georg Simmel (1902) to John Urry (2007), and everyone in between, have produced theories that explain how modern societies become deterritorialized—and in doing so, they spill *beyond* the nation. Modernity offers unprecedented opportunities and obligations to become mobile. People, things, capital, and ideas are on the move. Mobility, in turn, dislocates cultures. Think of the overseas migrant, of the peasant in the city and, in the terrain of fake documentary, of the African tribe thrown off balance by the chance discovery of a bottle of Coke (*The Gods Must Be Crazy* 1980). Finally, to be beyond the nation implies a certain kind of detachment that seems quintessentially modern: a refusal to be constrained by political borderlines, when traveling internationally for instance, and a refusal to stop consuming commodities regardless of their origin. Commodities essential to sustain middle-class lifestyles (oil, rubber, drugs, foodstuffs, and so on) are often imported from remote regions. Shopping and leisure habits signal a shift from Benedict Anderson's "imagined communities" (1991 [1983]) to Chua Beng Huat's "communities of consumers" (2006), and should be taken into account if the making of modern frontiers is to be fully understood.

At this point we must distinguish transnational practice from transnational theory, even if there is an obvious correspondence between the two. In the introduction to the *Transnational Studies Reader* (2008), Peggy Levitt and Sanjeev Khagram speak of five "intellectual foundations" of this program:

1 *Empirical transnationalism* aims to provide evidence of the transnationalization of modern society. Levitt's own work with migrants (2001) is an excellent example, but the hunt is on to document the cross-border experience of modernity on a global scale. In some instances the evidence confirms the Westernization of the world. In others, it points toward localization or indigenization. This wealth of data reveals unprecedented cultural processes, but it also supports David Harvey's statement (2001, 24) that "the contemporary form of globalization is nothing more than yet another round in the capitalist production and reproduction of space. It entails a further diminution in the friction of distance . . . through yet another round in innovation in the technologies of transport and communication." As Luis Roniger demonstrates in this volume (Chapter 13), history and geography still matter.
2 *Methodological transnationalism* involves the creation of new research methods. The political economy of *big* frontiers and the affective economies (Ahmed 2004) mediating the study of *small* frontiers have found ways to interact with each other. Multidisciplinary inquiry is the norm nowadays, although it has been decades in the making. Historians have long been in deep conversation with anthropologists, and anthropologists have long "moved to the city" in order to comprehend the fate of folk cultures. Cultural studies is perhaps the best example of a methodological melting pot designed to capture the complexity of our contemporary, transnationally networked society. On the downside, common themes in frontier research (non-urban spaces, historicism, unchosen livelihoods as opposed to liberally cherry-picked lifestyle choices, and political economy, among others) are underrepresented in the cultural studies literature.
3 *Theoretical transnationalism* is the lasting result of such surveys. A new discourse supported by an army of powerful metaphors has taken hold of academia in the past 30 years or so. In the 1990s, a decade of neoliberal frenzy surrounding the collapse of the USSR and the techno-utopian build up to the dot.com crash, some of the most popular assumptions overlooked the role of

resistance in the making and functioning of transnational links. In short, the image of the virtual network was too readily extrapolated to socioeconomic structures, at a time of financial and political optimism. A Marxian revival following the 911 attacks and the economic crisis of 2008 has corrected this oversight somewhat. Anna Lowenhaupt Tsing's *Friction* (2005) and Alison Hulme's *On the Commodity Trail* (2015) serve as a reminder that neither things nor people flow as easily as virtual information does.

4 *Philosophical transnationalism* is, in Levitt's own words (2), a "metaphysical assumption that social worlds and lives are inherently transnational." Even the most self-contained culture exists on the back of economic and non-economic exchange between peoples that see each other as foreigners. Tribal exogamy and the Kula ring (Malinowski 1922) provide "non-modern" evidence of this. Strangers, as the one that appears in Georg Simmel's same titled essay of 1902, are the rule and not the exception in virtually all human societies: "He is fixed within a particular spatial group, or within a group whose boundaries are similar to spatial boundaries. But his position in this group is determined, essentially, by the fact that he has not belonged to it from the beginning" (1950 [1902], 402). Strangers are frontier-makers, not only philosophically speaking but in actuality as well.

5 *Public transnationalism* is the official acknowledgment of modern mobility and the policies that result from it. Public transnationalism encompasses a proactive education in international brotherhood, shared historical experiences, and mutual interests. There is the UN and the EU after World War II, and more recently LAIA (Latin American Integration Association) and ASEAN (Association of Southeast Asian Nations), in Latin America and Asia respectively, but there have also been less noble efforts toward pan-Europeanism, pan-Americanism, and pan-Asianism, driven largely by imperialist design—the *Lebensraum*, the Monroe Doctrine, the Greater Asia Co-Prosperity Sphere. But again, grand discursive gestures do not convey (if anything, they hide) the complexity of cultural practice at the local and regional levels (see e.g. Ávila-Tàpies, Chapter 4 in this volume).

The "transnational turn" brings to mind Immanuel Kant's cosmopolitanism—"knowledge of the world . . . as a citizen of the world"—which the German philosopher understood as something of a moral obligation. On a more materialist level, there are at least two reasons why the nation-state ought to be considered in transnational terms. First, the state apparatus tends to be less effective, less impressive, as it moves away from the city. Frontiers are relatively distant regions, where the policeman and the tax collector are weak by comparison with their metropolitan counterparts (see McClure, Chapter 6, this volume). The importance of isolation in the making and sustaining of frontiers stands out in the work of James C. Scott (2009). For the nation-state, every village is a problem, a problem of distance to be dealt with. Substitute "village" with "colony" and the problem is not one of internal colonialism, but of colonialism as such. In *Domination and the Arts of Resistance* (1992), Scott finds inspiration in George Orwell's classic 1936 essay

"Shooting an Elephant," a masterful critique of imperialism and also a healthy reminder of the loneliness of the colonial agent. In the frontier too, nations overlap and transnational contacts occur, but by itself overlap does not produce a frontier. Rather, what does produce it is the tendency of places and people to escape, to a greater or lesser extent, passively or otherwise, the tight administration of the nation-state; hence, the frontiers of modernity may be defined as transnational in principle. Second, as explained above, transnationalism as a process has only intensified with the expansion of the world industrial economy. As the essays in this volume show, transnationalism does not invalidate the nation as a bounded, coherent self, although it makes it more and more difficult, more and more an exercise in territorial bigotry, to call oneself a nationalist. In many ways, transnationalism in the social sciences is a twenty-first-century version of Kant's cosmopolitanism, located somewhere along the line that separates the totalizing agenda of globalization from the close-up familiarity implied in cultural relativism.

Big frontiers (and everyday life)

Why Asia and Latin America? The editors of this book share a background in Latin American studies and a postdoctoral interest in Asia as a topic of both research and teaching. It was our goal to move not only beyond the nation but also beyond what is vaguely known as Area Studies. There has been ample debate about the validity of Area Studies at a time of all-pervading globalization (e.g. Sidaway 2013). In the chapter that closes this volume (Chapter 16), Pedro Iacobelli brings to light the linkages that exist between the birth of Area Studies and U.S. geopolitics during the Cold War. Argentinian semiologist Walter Mignolo (2009, 270) has gone as far as arguing that continents are colonial constructs, and has called for a "decolonization of narrated history and imperial historiographical thought." This is a compelling point, Saidian in spirit. The essays presented here sufficiently show that all frontiers are shape-shifting spaces, both in theory and in practice. However, continental borders *do* count. They count by virtue of physical geography. Dozens of countries and hundreds of cultures comprising each area of study offer more than enough material for one scholar to digest in a lifetime—albeit individual research is often limited to one country and/or one extra language. But we want to draw the readers' attention to the artificial barriers dividing university departments. Latin Americanists and Asianists know precious little of each other's work, and that represents a tremendous loss to the scholarly community. In the past decade or so, a proliferation of centers devoted to Global and Transnational Studies has begun to break down some of these barriers, but there is much to be done. *Transnational Frontiers of Asia and Latin America* makes a historico-geographical and anthropological contribution to this now popular trend (Geiger 2008; Readman et al. 2014).

Big frontiers and *small* frontiers come into play here. On the one hand, there is the vastness of scale. By global standards, and certainly by European standards, Asia and Latin America are enormous. Territorial size translates into an abundance of frontiers. In the strict sense, frontiers are relatively large, seemingly empty spaces. This statement may be misleading, because the pretended emptiness of the world has always been a convenient excuse for colonialist expansionism. Early twentieth-century political cartoonists often relied on the octopus as a visual metaphor to condemn—and to mock (see Figure I.1)—the slimy yet otherworldly ways in which states, capital, and empires wrapped themselves around the five continents. Distance is also a factor in Frederick Jackson Turner's "The Significance of the Frontier in American History" (1994 [1893]). This paper, first delivered as a speech, is rightly understood as a symbolic articulation of U.S. exceptionalism, but it should be noted that Turner defined the frontier objectively, in terms of low demographic density, highlighting the distinction between crowded Europe and open-spaced America. Throughout modern history, much of the "empty" Latin American subcontinent and large portions of Asia (let alone Africa) have been seen with telescopic, colonizing eyes.

On the other hand, there is everydayness. Colonizers are strangers who make the frontier come to life. They are settlers, soldiers, civil servants, and investors, but also tourists, journalists, explorers, and even conservationists (see below Howard, Chapter 9; MacRae, Chapter 11; and Acciaioli and Sabharwal, Chapter 2). The metaphor of the octopus does not express their real significance, or rather insignificance. Micro-analysis allows for intimate readings of cultural contact as it happens at ground level, beneath the overarching theme of the conqueror vs. the victim. Archival research complicates all attempts at ready-made sentimentalism, and in the field especially the world is a confusing place. Individual memory blends into local mythology, diluting historical facts in a maelstrom of circular times. Furthermore, all material leftovers of past modernizing efforts are indicative of loss and retreat, not of positive progress (Gordillo 2014). Through the microscope, in sum, the frontier has an uncanny air about it—unsettling yet familiar, disturbing yet candid. Take as example a photograph and caption from Lewis Cotlow's *Amazon Head-Hunters* (1954, 129). The image shows the explorer and filmmaker, whose professional life was devoted to a sort of spectacular salvage anthropology, sitting next to four native children in Amazonian Ecuador. They are all smoking. The caption reads: "Colorado Indian children liked the author's cigarettes, but the little girl next to him felt ill." Everydayness is a disarming experience, and in the face of thick description the researcher often finds himself bereft of moral and metaphorical support.

But, again, frontiers are the product of long-term, wide-ranging processes that transcend the deafening stillness of everyday life. The pattern of frontier expansion in Latin America has been the subject of an interesting debate (Hennessy 1978; Weber and Rausch 1994; Cecil 2012), especially in Brazil, a country that ignored its backlands until the mid-twentieth century, when the federal government was transferred from the old coastal city of São Paulo to the newly-built, high modernist experiment that is Brasilia (for a critique, see Berman 1982, and Scott 1999). Speaking of loss and retreat as constitutive features of teleological progress, in 1938, looking at the then still underpopulated hinterland of São Paulo, geographer Preston E. James wondered: "Why has the frontier of agricultural settlement swept over this territory, leaving derelict landscapes, shorn forest cover, with worn-out and eroded soils, but with little permanent settlement attached to

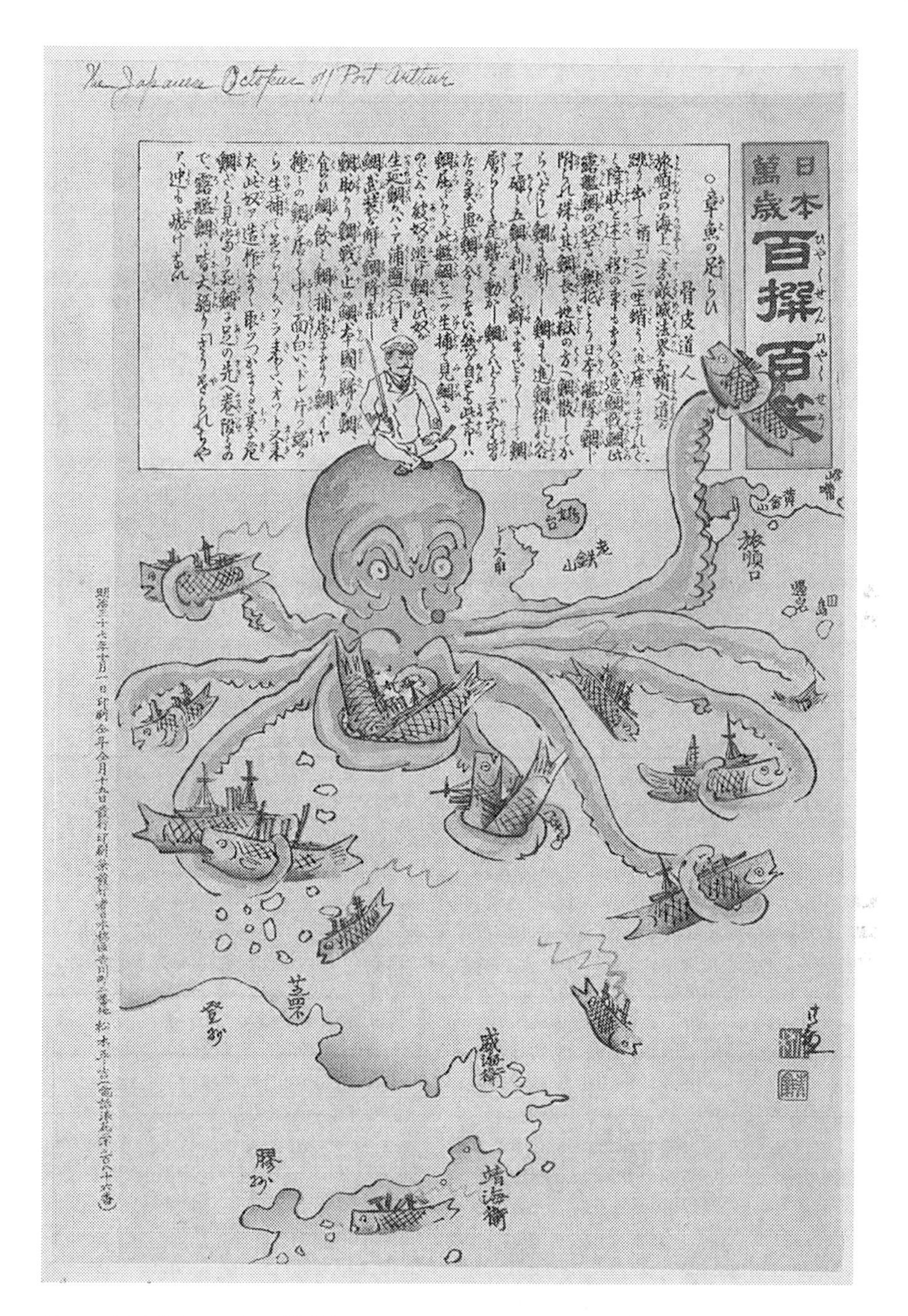

Figure I.1 "Tako no Asirai," "The Japanese Octopus of Port Arthur," 1904. Woodcut. Kobayashi Kiyochika.

Source: Library of Congress.

the land? Why is Brazil," Preston concluded (361), "still a land with a hollow frontier?" More generally, hollow frontiers are characterized by boom-and-bust cycles before they finally become depressed and stagnant.

The opposite of a hollow frontier would be a sustainably marginal border zone, which is how frontier ethnohistorian Ma Dazheng (2012, 68) envisions the role of China Proper in its cultural and territorial fringes. But this is policy-making, ideological thinking. In reality, a typical pattern of market and state expansion involves the growth of "nodes of prosperity" amid "waves of ruination" (Gordillo 2014, 126–127). Hollow frontiers are, in short, a feature of modernity. Feverish boomtowns grow "in the middle of nowhere," thin communication lines are laid down, and nothing is left behind (or along) when the rush comes to an end. The solitary pipelines that traverse Central Asia and even the party islands in the tropical South, where late-night hallucinatory excess turns into a wasteland of broken glass and vomit in the morning—what John Urry (2014) referred to as the offshoring of leisure—provide sensuous proof of the inner workings of hollow frontiers in transnational times.

Populist frontier mythology should not be overlooked, in part because it filters into the academic discourse too. The Turner Thesis created the myth that American democracy and prosperity were one with the frontier. The Hollow Frontier Thesis has gathered empirical evidence to suggest that Latin American borderlands—from Mexico (Redclift 2004) to Chile (Klubock 2014)—are harbingers of poverty and instability. Colonial prejudice and magical realism did the rest. Theodore Roosevelt, who believed that the frontier was indeed an integral part of the American psyche, and therefore supported his country's post-Wild West maritime adventures, traveled in South America in 1913, and reached the conclusion that the chief obstacles to the industrial exploitation of the tropical plains of the interior were two: mosquitoes and "the revolutionary habit" (Roosevelt 1914, 43). Historically, frontier underdevelopment has been blamed on both natural and cultural causes. In fiction literature, the climate turns the colonizer into a madman of Mr. Kurtz proportions. In government and company reports, such as the Japanese opinion on Taiwan quoted above, the uncooperative native, whether "lazy" or "unruly," makes any attempt at systematic use of the landscape futile. As Gutarra notes in this volume in relation to the slave-fueled French colonies in the Caribbean, transoceanic distance lends itself to vivid tropes of "difference" that function as a strategy of political, economic, and bodily exclusion.

In the twentieth century the idea was put forward that the world was the Great Frontier (Webb 1952) of European history. The successful tapping of distant regions became central to many academic explanations of European growth after 1500. Marxism turned this notion into a comprehensive critique of international capitalist exploitation. Dependency and world-systems theory (Cardoso and Faletto 1979 [1971]; Wallerstein 1976), still influential despite lacking some key elements such as thick description and native agency, were the result of this effort. Stephen Bunker (1984, 1017) spoke of Amazonian "modes of extraction" in an attempt to avoid Eurocentric accounts for the "ways in which the extraction and export of natural resources affect the subsequent developmental potential" of "extreme peripheries." Modes of extraction belong in places where the export of raw materials can sustain an economy with no developed modes of production—remote places, by and large, where the "wilderness" has not yet been domesticated.

Because of their relative isolation, extreme peripheries are potential conduits for extreme violence and repression. Economic demand in the metropolis, no matter how banal, can easily breed a "culture of death" (Taussig 1984) in the frontier. In the world that invented the telegraph and the Internet, news may still arrive late or be ignored at the centers of media production; rumors may not be corroborated and publicized for years. There is a long tradition of atrocities committed in the Great Frontiers of extraction, and in their scale and brutality they seem unreal, "pure" events as opposed to historical events (Lundborg 2012). Outside the parameters of this "space of terror" (again, Taussig 1984), there is the everyday violence that circulates in the socioeconomic margins of capitalism, where *longue durée* peonage is customary and local government corruption is rife.

Culture and nature, producers and consumers

The notion of wilderness-as-last-frontier is an enduring one. Its appeal is directly related to the history of modern markets, but it also belongs to the world of middle-class dreaming and consumption. Either way, these are loci of desire that continually shape the cultural and economic borders of the world-system. For some, the vanishing frontier has nostalgic overtones, reminiscent of Cooper's *Last of the Mohicans* (1826). Amazonian deforestation and the looming death of tribes "never before seen"—but surely on YouTube for everyone to see—are among the mass media's favorite apocalyptic scenarios, and rightly so. For others, the frontier is open for business, a last-chance or be-the-first-to-get-there type of invitation. Myanmar/Burma, comparatively rich in oil and gas resources, minerals, and timber, has been labeled "Asia's last frontier," and "the final frontier in South East Asia." The sense of urgency is palpable and speaks of a long process of transnationalization. On a broader scale, this is the story of how commodity chains became ubiquitous in contemporary life. Some of the most groundbreaking research themes in modern Asian and Latin American frontiers include precisely long-distance thing-tracking (e.g. Gootenberg 2015; Mizushima et al. 2014) and localized critiques of tropical tourism (Gmelch 2012; Hoefinger 2013) that resonate with what, by now, is a traditional idea: that the frontier is there to be used by someone else.

The polarization of the world between "the West and the Rest" is becoming increasingly problematic, but claims that "core" and "periphery" are no longer viable tenets may be an exaggeration at best, and a concession to those who equate flow with freedom at worst. The world's frontiers are still shrinking, following a Caspian Sea-type pattern that prevails since 1500 and that took on a new meaning at the onset of modernity. It may be the case that we are living post-natural times, in the sense that "there is no more nature that stands apart from human beings" (Purdy 2015, 3). We should, however, be wary of canceling the culture vs. nature divide altogether. The Amazon Basin, the forests of Burma and Central Africa, the Arctic Plains, and the colossal Coral Triangle are good examples of endangered primeval regions. While their *nature* has been altered by millennia of native labor and by two centuries of intensive industrial agency, these and a handful of other places are the last great *natural* frontiers of planet Earth. From 1800 there have been moments of rupture, stoppage, and acceleration, and yet the progressive loss of nature in the course of modern human history is unquestionable. We have not yet reached an endpoint.

The shift in global power toward multi-polarity will require an equal redistribution of blame and liability. This need is more acute given that contemporary modernity is far more advanced than, say, Cold War modernity. Economic and spatial optimization must be discussed, using *critical theory* as an interpretive frame. Producers and consumers are getting ever closer to each other and, as result, schizophrenic frontiers—for instance, Peruvian pineapples in Thai supermarkets—are taken-for-granted facts of transnational existence. The BRIC countries (Brazil, Russia, India, and China) appear to be confirming the main points of Francis Fukuyama's "End of History" prophecy (1992). The meaning of culture (whatever pertains to middle-class leisure) is more or less universally agreed upon. Let us not forget that Osama Bin Laden nurtured not only a burning hatred of the West but also a substantial stash of pornography. Present-day glocalized culture seems therefore less of a double-edged sword and more of a numbing shopping experience. Its highest philosophical expression is perhaps the English-language "pseudo-profound bullshit" (Pennycook et al. 2015) that swamps the social media in the form of supposedly inspirational quotes. In the "civilizing missions" of the twenty-first century, "civilization" is progressively deprived of non-economic ethical moorings—however cynical and murderous, catholic guilt and Victorian propriety did slow down European colonial expansion—and gradually confused with raw commercial profit and wishy-washy social virtues such as entertainment and fun. The global middle class has all the qualities of a drone operator, raping the wilderness in real time, while creating an insurmountable moral distance between killer and target, the pleasures of consumption and the perils of production. And if this is the Asian Century, as it seems to be, then it is sensible to think that Asia will be doing (is already doing) a lot of the drone operator's work in the deserts, seas, and forests of the world.

*

There is a rich literature on Asian historical frontiers, including the writings of Peter Perdue (e.g. 1987; 2005) on China. Great Frontiers of Asian modernity comprise vast swathes of the Chinese mainland (Billé et al. 2012), the Southeast Asian forests and deltas (Tagliacozzo 2005), the thousands of islands scattered between the Indian and Pacific Oceans (Shicun and Hong 2014), the sea frontier itself (Wang 2014), and the expanse between Northern India and Central Asia (e.g. Warikoo 2009). Much of this corresponds to Zomia, an area only recently conceptualized that, precisely because of its multinational vastness, offers all kinds of opportunities for contention (Scott, 2009; Jonsson, 2010; Michaud 2010; Zhidan Duan, Chapter 7 this volume). Zomia covers the rugged hills of northern

Thailand, Myanmar/Burma, and Indochina, the southwestern provinces of China, and for some authors the Himalayas and its western ramifications too, all the way to Afghanistan. Zomia has one distinctive feature: it is hard to reach. For that reason, the region remains not only enticing for speculators, but also embarrassingly ethnic (see Metz, Chapter 12, this volume) in relation to the pulling forces of the nation-state. Zomia, in James Scott's words (2009, 4), "represents the last great enclosure movement in South East Asia." It is a modern project of standardization—some would call it cultural annihilation—and, as a historic site of resistance, it is the source of age-old headaches among Asian policy makers.

The phrase "embarrassingly ethnic" seems appropriate here, as it defines a general attitude of the urban middle-classes in these rapidly modernizing countries toward the *unclean* peasantry and the *backward* countryside. Above all, the frontier has connotations of poverty. Material poverty in Asia and Latin America is remembered across generations and is also felt in the flesh through the senses—for example, in the sight of tan-skinned street vendors and in the smell of rotting food outside trendy shopping malls (and, inside the malls, in the self-effacing circulation of cleaners from upcountry; see Brody 2006). The nation-state and the capitalist economy are "closing in" in the city too. While this volume deals largely with out-of-the-way spaces—frontiers, both *big* and *small*, in the traditional sense—the investigation of transnational contact zones in Asia and Latin America asks for yet another kind of materialist approach: a study of gentrification (along the lines of Smith 1996) and bodily practices that will shed light on the metropolitan interaction between "civilization" and "barbarism." At any rate, when studying frontiers it seems logical to consider the different biographies of those in charge of the narrative. What looks fresh and quaint to Euro-American eyes, without a doubt because Westerners invented the Noble Savage, might appear old and tired to Asian and Latin American observers. British students have no trouble sympathizing with Margaret Mead's bucolic take on Samoan teenagers (1928), while Thai students will gaze at the PowerPoint slides with quiet skepticism.

At the level of everyday life the frontier offers endless angles for scholars to pursue, but the point remains that Zomia, Amazonia, and all other borderlands of Asia and Latin America are losing ground against the history of transnationalization. The frontier metaphor—the stranger in the wilderness—may well be a subconscious part of the human mind: to be touched from a distance, to be conquered and kept. But *actual* expansion into the next big empty land is a driving principle of modern life, of complex society, and of civilization. Hence, as Arjun Appadurai put it (2006, 9), "the anxiety of incompleteness." Geographical anxiety is neither metaphysical nor suspended in time; it is the result of knowing as a matter of fact that the land will be taken.

References

Ahmed, Sara. 2004. "Affective Economies." *Social Text* 79 22(2): 117–139.

Anderson, Benedict. 1991. *Imagined Communities: Reflections on the Origin and Spread of Nationalism* [1983]. London: Verso.

Appadurai, Arjun. 2006. *Fear of Small Numbers: An Essay on the Geography of Anger*. Durham and London: Duke University Press.
Berman, Marshall. 1982. *All That Is Solid Melts into Air: The Experience of Modernity*. London: Verso.
Billé, Franck, Grégory Delaplace, and Caroline Humphrey, eds. 2012. *Frontier Encounters: Knowledge and Practice at the Russian, Chinese and Mongolian Border*. Cambridge, UK: Open Book Publishers.
Brody, Alison. 2006. "The Cleaners You Aren't Meant to See: Order, Hygiene, and Everyday Politics in a Bangkok Shopping Mall." *Antipode* 38(3): 534–556.
Bunker, Stephen. 1984. "Modes of Extraction, Unequal Exchange, and the Progressive Underdevelopment of an Extreme Periphery: The Brazilian Amazon, 1600–1980." *American Journal of Sociology* 89(5): 1017–1064.
Cardoso, Fernando Henrique and Enzo Faletto. 1979. *Dependency and Development in Latin America [1971]*. Translated by Marjory Mattingly Urquidi. Berkeley, CA, London: University of California Press.
Cecil, Leslie G., ed. 2012. *New Frontiers in Latin American Borderlands*. Newcastle upon Tyne: Cambridge Scholars Publishing.
Chua, Beng Huat. 2006. "East Asian Pop Culture: Consumer Communities and Politics of the National." Paper presented at the International Conference *Cultural Space and Public Sphere in Asia*. Hallyim University, Seoul.
Cooper, James Fenimore. 1826. *The Last of the Mohicans: A Narrative of 1757*. [Online].
Cotlow, Lewis. 1954. *Amazon Head-Hunters*. Robert Hale: London.
Fukuyama, Francis. 1992. *The End of History and the Last Man*. New York: The Free Press.
Geertz, Clifford. 1973. "Thick Description: Toward an Interpretive Theory of Culture." In *The Interpretation of Cultures: Selected Essays*, 3–30. New York: Basic Books.
Geiger, Danilo. 2008. *Frontier Encounters: Indigenous Communities and Settlers in Asia and Latin America*. Copenhagen: IWGIA.
Gmelch, George. 2012. *Behind the Smile: The Working Lives of Caribbean Tourism*. Bloomington: Indiana University Press.
The Gods Must Be Crazy. 1980. Directed by Jamie Uys. South Africa, Botswana: Ster Kinekor, 20th Century Fox.
Gootenberg, Paul. 2015. "Toward a New Drug History of Latin America: A Research Frontier at the Centre of Debates." *Hispanic American Historical Review* 95(1): 1–35.
Gordillo, Gastón R. 2014. *Rubble: The Afterlife of Destruction*. Durham and London: Duke University Press.
Harvey, David. 2001. "Globalization and the 'Spatial Fix'." *Geographische Review* 3(2): 23–30.
Hennessy, Alistair. 1978. *The Frontier in Latin American History*. London: Edward Arnold.
Hoefinger, Heidi. 2013. *Sex, Love, and Money in Cambodia: Professional Girlfriends and Transnational Tourism*. London: Routledge.
Hulme, Alison. 2015. *On the Commodity Trail: The Journey of a Bargain Store Product from East to West*. London, New Delhi, New York, Sydney: Bloomsbury.
Iriye, Akira. 1981. *Power and Culture: The Japanese American War, 1941–1945*. Cambridge, MA: Harvard University Press.
——. 2013. *Global and Transnational History: Past, Present, and Future*. Basingstoke, Hampshire: Palgrave Macmillan.

James, Preston E. 1938. "The Changing Patterns of Population in São Paulo State, Brazil." *Geographical Review* 28(3): 353–362.

Jonsson, Hjorleifur. 2010. "Above and Beyond: Zomia and the Ethnographic Challenge of/for Regional History." *History and Anthropology* 21(2): 191–212.

Kipling, Rudyard. 1899. "The White Man's Burden." *McClure's Magazine*. [Online].

Kublock, Thomas Miller. 2014. *La Frontera: Forests and Ecological Conflict in Chile's Frontier Territory*. Durham and London: Duke University Press.

Levitt, Peggy. 2001. *The Transnational Villagers*. Berkeley and Los Angeles: University of California Press.

Levitt, Peggy and Sanjeet Khagram. 2008. "Constructing Transnational Studies." In *The Transnational Studies Reader: Intersections and Innovations*, edited by Shanjeev Khagram and Peggy Levitt, 1–22. New York and Abingdon: Routledge.

Lundborg, Tom. 2012. *Politics of the Event: Time, Movement, Becoming*. London and New York: Routledge.

Ma, Dazheng and Patrick Fuliang Shan. 2012. "Frontier History in China: A Scholarly Dialogue across the Pacific Ocean." *The Chinese Historical Review* 19(1): 65–78.

Malinowski, Bronislaw. 1922. *Argonauts of the Western Pacific: An Account of Native Enterprise and Adventure in the Archipelagos of Melanesian New Guinea*. London: Routledge & Kegan Paul.

Marx, Karl and Friedrich Engels. 1848. *Manifesto of the Communist Party*. [Online].

Mead, Margaret. 1928. *Coming of Age in Samoa: A Psychological Study of Primitive Youth for Western Civilization*. New York: William Morrow & Co.

Michaud, Jean. 2010. "Editorial: Zomia and Beyond." *The Journal of Global History* 5: 187–214.

Mignolo, Walter D. 2009. "La idea de América Latina (la derecha, la izquierda y la opción descolonial)." Edited by Horacio Tarcus, *Crítica y Emancipación* 1(2): 251–276. CLACSO (Consejo Latinoamericano de Ciencias Sociales).

Mizushima, Tsukasa, George Bryan Souza, and Dennis O. Flynn, eds. 2014. *Hinterlands and Commodities: Place, Space, Time, and the Political Economic Development of Asia in the Long Eighteenth Century*. Leiden and Boston: Brill.

Orwell, George. 1936. "Shooting an Elephant." [Online].

Pennycook, Gordon, James Cheyne, Nathaniel Barr, Derek Koehler, and Jonathan Fugelsang 2015. "On the Reception and Perception of Pseudo-Profound Bullshit." *Judgement and Decision Making* 10(6): 549–563.

Perdue, Peter C. 1987. *Exhausting the Earth: State and Peasant in Hunan, 1500–1850*. Cambridge, MA: Harvard University Press.

——. 2005. "Identifying China's Northwest: For Nation and Empire." In *Locating China: Space, Place, and Popular Culture*, edited by Jing Wang and David Goodman, 94–114. London: Routledge.

Purdy, Jedediah. 2015. *After Nature: A Politics for the Anthropocene*. Cambridge, MA and London: Harvard University Press.

Readman, Paul, Cynthia Radding, and Chat Bryant, eds. 2014. *Borderlands in World History 1700–1914*. London: Palgrave MacMillan.

Redclift, Michael. 2004. *Chewing Gum: The Fortunes of Taste*. New York: Routledge.

Roosevelt, Theodore. 1914. *Through the Brazilian Wilderness*. Scribner: New York.

Said, Edward. 1978. *Orientalism*. New York: Vintage Books.

Sarmiento, Juan Domingo. 2004. *Facundo: Civilization and Barbarism* [1845]. Translated by Kathleen Ross. Berkeley and Los Angeles: University of California Press.

Scott, James C. 1992. *Domination and the Arts of Resistance: Hidden Transcripts*. New Haven, CT: Yale University Press.

——. 1999. *Seeing Like a State: How Certain Schemes to Improve the Human Condition Have Failed*. New Haven, CT: Yale University Press.

——. 2009. *The Art of Not Being Governed: An Anarchist History of Upland South East Asia*. New Haven, CT: Yale University Press.

Shicun, Wu and Nong Hong. 2014. *Recent Developments in the South China Sea Dispute*. London: Routledge.

Sidaway, James D. 2013. "Geography, Globalization, and the Problematic of Area Studies." *Annals of the Association of American Geographers* 103(4): 984–1002.

Simmel, Georg. 1950. "The Stranger [1902]." [Online].

Smith, Neil. 1996. *The New Urban Frontier: Gentrification and the Revanchist City*. London: Routledge.

Tagliacozzo, Eric. 2005. *Secret Trades, Porous Borders: Smuggling and States along a Southeast Asian Frontier, 1865–1915*. New Haven, CT: Yale University Press.

Taussig, Michael. 1984. "Culture of Terror—Space of Death: Roger Casement's Putumayo Report and the Explanation of Torture." *Comparative Studies in Society and History* 26(3): 467–497.

Tierney, Robert Thomas. 2010. *Tropics of Savagery: The Culture of Japanese Empire in Comparative Perspective*. Berkeley and Los Angeles: University of California Press.

Tsing, Anna L. 2005. *Friction: An Ethnography of Global Connection*. Princeton, NJ: Princeton University Press.

Turner, Frederick Jackson. 1994. "The Significance of the Frontier in American history." In *Rereading Frederick Jackson Turner*, ed. John Mack Faragher. New York: Henry Holt. Pp. 31–60.

Urry, John. 2007. *Mobilities*. Cambridge: Polity Press.

——. 2014. *Offshoring*. Cambridge: Polity Press.

Wallerstein, Immanuel. 1976. *The Modern World-System: Capitalist Agriculture and the Origins of the European World-Economy in the Sixteenth Century*. New York: Academic Press.

Wang, Wensheng. 2014. *White Lotus Rebels and South China Pirates: Crisis and Reform in the Qing Empire*. Cambridge, MA, and London: Harvard University Press.

Warikoo, K., ed. 2009. *Himalayan Frontiers of India: Historical, Geo-Political and Strategic Perspectives*. Abingdon: Routledge.

Webb, Walter Prescot. 1952. *The Great Frontier*. Cambridge, MA: The Riverside Press.

Weber, David J. and Jane M. Rausch, eds. 1994. *Where Cultures Meet: Frontiers in Latin American History*. Wilmington, DE: Scholarly Resources.

Winichakul, Thongchai. 1994. *Siam Mapped: A History of the Geo-Body of a Nation*. Honolulu: University of Hawaii Press.

Part I

Theories

1 Globalization and changing conceptions of Colombia's Llanos frontier since 1980[1]

Jane M. Rausch

The Llanos are an extensive region of seasonally-flooded tropical plains consisting of grasslands and forests shared by Venezuela and Colombia. Artificially divided by the international boundary between the two countries, they in fact comprise a single ecosystem that stretches in a northeast direction from the foothills of the Colombian Andean Mountains and along the course of the Orinoco River nearly to its delta in the Atlantic Ocean. Limited by the Andes in the west, the Venezuelan Coastal Range in the north, and the Amazonian wilderness in the south, the region covers some 451,474 square kilometers. Around 39 percent (176,359 sq. km.) lies within Colombia accounting for 17 percent of that county's territory, while 61 percent (275,115 sq. km.) make up 31.2 percent of Venezuela's territory (Rivas et al. 2002, 265). During the dry season (November to April) the land is parched and the grass brown, brittle, and inedible while during the rainy season much of the area is inundated. Notwithstanding this harsh climate, the region teems with wildlife, harboring more than 100 species of mammals and over 700 species of birds. The Llanos also harbors the Orinoco crocodile, one of the most critically endangered reptiles on earth, along with other rare species including the Orinoco turtle, giant armadillo, giant otter, and several species of catfish.

Until recently geographers and historians who studied the Colombian portion of this region, known as the Llanos Orientales, have regarded it as a frontier territory, for isolated as they were by the Eastern Andean Cordillera, these tropical plains appeared to be a peripheral area of little consequence in the evolution of national history, despite the fact that towns, ranches, and missions established there have been interacting with the region's aboriginal inhabitants since the sixteenth century. Since the 1980s, however, the rapid exploitation of petroleum in the plains by multinational companies, the arrival of thousands of migrants, and consequent urbanization have brought significant changes to the region's demographic composition, social structure, and political dynamics. This impact of globalization has prompted some scholars to modify the concept of "frontier" when referring to the Llanos, while others prefer to regard the plains as "Orinoquia" (a trans-national region including the Llanos of Colombia and Venezuela), and still others have suggested that the area should be analyzed as an international borderland. Since each of the three approaches provides new insight, the object of this chapter is to suggest some ways twentieth-century social scientists have employed these models to analyze the role of the Llanos frontier in the shaping of the Colombian nation.

The frontier in U.S. historiography

In 1973 when I first began to study the history of the Llanos Orientales, it is perhaps not surprising that, as a young North American *gringa*, I chose to adopt a variation of a North American model, the Frontier Thesis proposed by Frederic Jackson Turner, as a framework interpreting my data. In 1893 Turner delivered a seminal speech entitled "The Significance of the Frontier in American History" which was to have an enormous impact on the study of U.S. history. In this address Turner suggested that a major theme in the history of the United States was a "moving frontier" which he defined as the line between "civilization and barbarism" (Billington 1973, 18). He argued first, that this line moved westward across the continent offering "free land" to North Americans seeking to make a new life, and second, that their struggle for survival required them to rebuild their societies afresh. In short Turner maintained that the two social processes of settling and surviving on free frontier lands shaped American character and institutions, and that "to the frontier, the American intellect owes its striking characteristics of inventiveness, practicality, inquisitiveness, restlessness, optimism and individualism" (Turner cited by Weber and Rausch 1994).

For much of the twentieth century U.S. scholars regarded the Turner Thesis as the single most useful concept for understanding the distinctive features of North American civilization, and it was not until the 1980s, that a new generation of historians emerged who were willing to challenge the idea of "frontier" as a useful category of analysis (Weber and Rausch 1994, xxxii). Taking into account their earlier numerous (and valid) objections, John Mack Faragher in 1992 concluded that "Turner's thesis long ago found its way onto the trash heap of historical interpretations" (Faragher 1992, 30), but others such as Martin Ridge and Richard Slatta argued that "reports of the trash heap may be premature." They pointed out that Turner's thesis continues to attract those who find it useful in modified form, and that the frontier remains a compelling heuristic device used even by Turner's detractors (Ridge 1991, 76). In his book, *Comparing Cowboys and Frontiers: New Perspectives on the History of the Americas* (1997), Richard Slatta suggests that, revisionist historians notwithstanding, the frontier as an analytical construct remains important in framing the history of the Western Hemisphere. He writes, "I believe it is counterproductive to bury the concept of frontier simply because Turner's formulation has proved incorrect." In fact, the study of frontiers "has been enriched by framing them within the process of incorporation . . . Frontier social changes and interaction need more probing. Even if we acquiesce to burying Turner, we should demur from burying the concept of the frontier with him" (Slatta 1997, 131).

Frontier historiography and the Colombian Llanos

Before 1970 the scholars who attempted to apply Turner's view of the frontier to the Colombian Llanos were North American geographers. They noted that while the Spanish conquistadors quickly incorporated into their New World Empire the Amerindians living in the high Andes and along the coast, their impetus was checked by the barrier posed by the Eastern Andean Cordillera, the inhospitable jungles of the Amazon Basin, and the equally unattractive tropical plains broken up by the Orinoco River. Blocked by a combination of geographic obstacles, deadly climate, native resistance, and lack of material incentives, the Spanish contented themselves with extending nominal control over thousands of miles of unexplored tropical wilderness. These geographers observed that although settlers did move into the tropical plains as early as the seventeenth century, they quickly found themselves isolated from the *altiplano* (highlands) by the rugged Andes and were forced to resign themselves to a self-contained existence based on cattle ranching and subsistence crops. Unlike the situation in the U.S. West, in the Llanos during the next three centuries there was little expansion eastward, thus giving rise to the notion of a "permanent" as opposed to a moving frontier (Bowman 1931, 84; Brunnschweiler 1972, 88).

To test the concept of a "permanent frontier," between 1984 and 2013 I wrote four books that traced the history of the Colombian Llanos (a region which includes the present day departments of Meta, Casanare, Arauca, and Vichada) from the Spanish conquest to the present. Covering developments up to 1950, the first three volumes largely substantiated that the frontier line established by the seventeenth century along the edge of the Eastern Andean Cordillera, expanded eastward only slightly despite improved health conditions and technology that made the topical lowlands potentially more accessible by the early twentieth century.[2] The settlers of plains might rightfully be called pioneers even though they had lived on farms or in villages occupied by their families for generations. The Llanos frontier with regard to Colombia's population center in the Andes was characterized by its immobility.

This immobility, I argued, did not keep the region from participating in the gradual evolution of the nation. Although traditional histories ignore its role, it is certain that, in the colonial period, Casanare was a major supplier of cattle to Boyacá and that some of its inhabitants took an active part in the Comunero Revolt of 1781. Llaneros were a decisive force in the War of Independence and the War of the Thousand Days. In the 1930s, President Alfonso López Pumarejo signaled out the Intendencia del Meta as a key area for development during his so-called *Revolución en Marcha*. Much of the fighting during La Violencia (1948–1964) took place in Meta, Casanare, and Arauca, and one needs only to read José Eustacio Rivera's *La vorágine* or Eduardo Carranza's poems to appreciate the contribution of the Llanos to Colombian literature (Rausch 1988, 32–40).[3]

The transformation of the Llanos in the twentieth century

The last decades of the twentieth century witnessed a sharp rise in national awareness of the significance of the Llanos. The Department of Meta was of special interest for, as a result of improvements in the Bogotá–Villavicencio highway, it had become a major supplier of cattle and agricultural commodities to the vast market in Cundinamarca. Although the Violencia compelled at least 6,000 people to abandon the plains between 1948 and 1953, they were soon replaced by the arrival of 16,000 new emigrants fleeing violence in other parts of Colombia (Ojeda Ojeda 2000, 187, 205). During the National Front (1958–1974) improvements in health measures and communication, the availability of public lands and the organization of colonization programs prompted thousands of settlers to move out to the Llanos with the hope of beginning a new life for their families. By 1993 the population of the four departments was 1,077,711 or nearly double the number counted in 1973. Although this figure was less than one percent of Colombia's total population of 40 million, in demography as well as economy, the region had emerged as one of the fastest growing segments of the country (Iriarte Núñez 1999, 134).

Frontier historiography and the Llanos since 1980

After 1980 the increasing importance of the Llanos thanks to the petroleum boom, and the activities of drug cartels, guerrilla groups, and paramilitaries has prompted some Colombian scholars (all either born in the Llanos or with close ties to the plains) to offer new conceptual models of the region. Five of them concentrating primarily on Meta have employed variations of the Turnerian definition of frontier to interpret the recent history of the department. Others have suggested that all four departments should be regarded as Orinoquia i.e. as a "region" that transcends national boundaries to encompass the Llanos of Venezuela as well as those of Colombia, while a third group, focusing on Arauca and Vichada propose exploring Llanos history from the standpoint of an international frontier or borderland with Venezuela (see Figure 1.1).

Neo-Turnerians

The first five scholars (who I regard as "neo-Turnerians") agree on essential points but diverge in their emphasis. In his 1997 book, *Un pueblo de frontera: Villavicencio 1840–1940*, historian Miguel García Bustamante challenged the traditional view that the Llanos formed a permanent frontier by pointing out that, notwithstanding the geographical difficulty of intercourse between them, the proximity of Villavicencio to Bogotá had from the seventeenth century involved a reciprocity characterized by unequal relations since Villavicencio was dependent politically on Bogotá, and its export economy of cattle and food commodities was directed almost exclusively to the national capital (García Bustamante 1997, 11). In 2003 García Bustamente refined his analysis further to suggest that the Llanos region per se consisted of two different frontiers. The piedmont area or *Llanos arriba* was a *frontera provisoria* or temporary frontier characterized by constant interaction with the highlands, while the grasslands east and north of the piedmont or the *Llanos abajo* remained a permanent frontier where development took place much more slowly (García Bustamante 2003, 40–41).

Anthropologist Nancy Espinel Riveros in her books has emphasized that Villavicencio is a frontier city from the standpoint of the east as well as the west, for just as it receives a constant stream of migrants from Colombia's highland departments, it continues to be the western terminal for cattle regularly driven

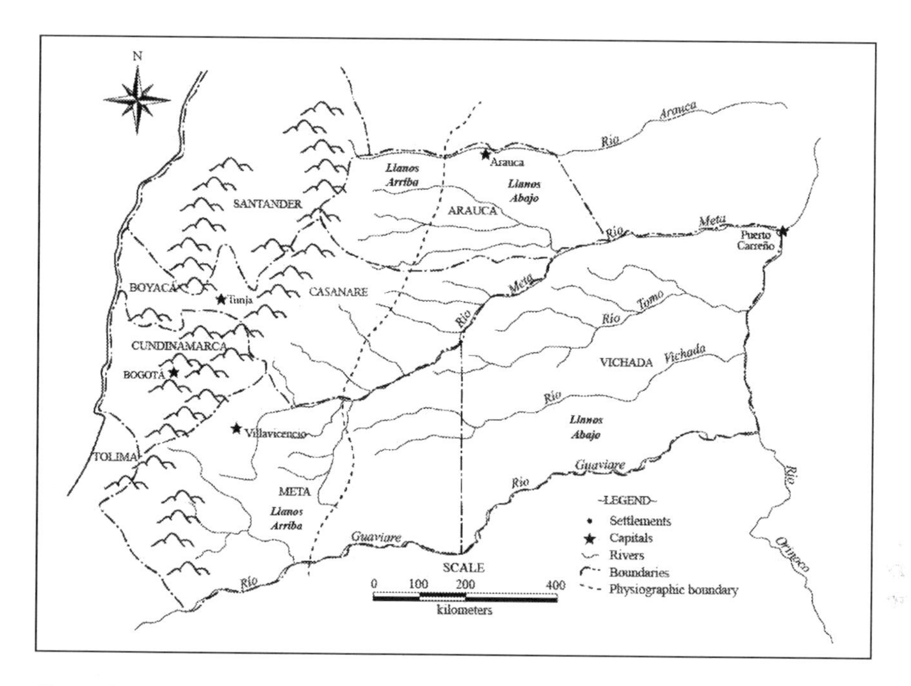

Figure 1.1 Map of the Llanos of Colombia showing the Llanos Arriba and the Llanos Abajo.

from the northern and far eastern plains of Casanare and Arauca to be sold eventually in Bogotá. Since its founding Villavicencio has been a mixture of two distinct cultures—Andean and Llanero—a duality that persists to this day. In her view the city has been converted into a human "crucible" in which are mixed both these distinctive traditions (Espinel Riveros 1997, 201–202).

Describing the same phenomenon of a dual plains frontier, economist Alberto Baquero Nariño takes a dependency point of view in *El caso llanero: Villavicencio*, but he substitutes the term *frontera interior* for Bustamante's *frontera provisora.* He suggests that Villavicencio exhibits the characteristics of a *frontera interior* because "immigrants pass through the city with little desire to settle in it; residency is impermanent; there is little substantial financial investment in the region, and the inhabitants do not display a sense of civic concern" (Baquero Nariño, 1990, 32–34). Still worse, from Baquero's point of view, is that, although Villavicencio is the center of Llanero economy, that economy since 1950 has been characterized by "savage capitalism" by which he means that it is dominated by "continual export of economic products and the absence of a genuine agro-industrial economy that could generate wealth for the region." The transport of cattle on the hoof for Bogotá and the predominance of rice and African palm oil produced by large corporations are characteristics of an extractive economy of very little regional value and minimal generation of employment (Baquero Nariño 1990, 69).

Writing in 1990 Baquero Nariño does not include the impact of the petroleum boom in his analysis, but it is evident from publications by historian Reinaldo Barbosa Estera that, first in Arauca and then in Casanare and Meta, oil exploitation has been similarly skewed to profit multinational corporations rather than the economic growth of the Llanos. The plains have a proven reserve of 500 million barrels of oil, but in those locations where it has been extracted, petroleum taxes have not fulfilled the promise of generating expected local revenues. Barbosa further asserts that the decision by the Ministerio de Minas y Energía to hand over the pipelines to multinational companies for 99 years "can be seen as a rebirth of colonialism" (Barbosa Estera 1998, 164–168).

In contrast to the previous four scholars who focus attention on the Department of Meta, economist Fredy Preciado has been exploring the reasons for Casanare's arrested development—a lack of growth that is especially striking when compared with the rapid change occurring in nearby Meta, and he offers a solution. Conceding that the exploitation of oil in Casanare has only reinforced its dependency, Preciado suggests "endogenous development" as a strategy to combat this situation, by which he means the strengthening of local resources and local community empowerment in order to counteract the economic model based on extraction and exploitation (Preciado n.d., 4:37). To enable "endogenous development" in Casanare Preciado lays out a five-point plan: First, local micro-business people must use their expertise to produce products that can compete in regional and national markets. Second, universities, institutes and *colegios* (schools) must adopt programs that will endow local entrepreneurs with the knowledge to make necessary innovations. Third, it is essential to foster local values and institutions to support this development. Fourth, local administrators should create enterprise zones favorable to production and business development. Finally, to sustain local development attention must be paid to Casanare's specific environmental potentials and limitations (Preciado n.d. 4:37).

The Llanos as a region

While many scholars in the 1990s found some variation of the Turnerian concept of frontier still relevant, others increasingly employed the term Orinoquia when referring to the Llanos Orientales (see Figure 1.2). The name Orinoquia suggests a region, rather than a frontier, and the concept is frequently expanded to encompass the entire Orinoco River basin to include the Llanos of Venezuela. Within Colombian historiography, the study of regions is a tried and true approach for it is a methodology fostered by the broken geography of the country. Historian John Lynch describes New Granada on the eve of the war of independence as a "conglomeration of regions, isolated from each other by mountains, jungles, plains and rivers" (Lynch 1989, 229), and David Bushnell's survey of modern Colombian history emphasizes that because of the survival of these regions, Colombia still lacks "a true national identity or proper spirit of nationalism, at least compared to most of its Latin American neighbors" (Bushnell 1993, viii). It is likely however, that Lynch and Bushnell would agree that not all Colombian regions are equal. Cundinamarca, Antioquia, Cauca, and Atlántico have dominated the country both

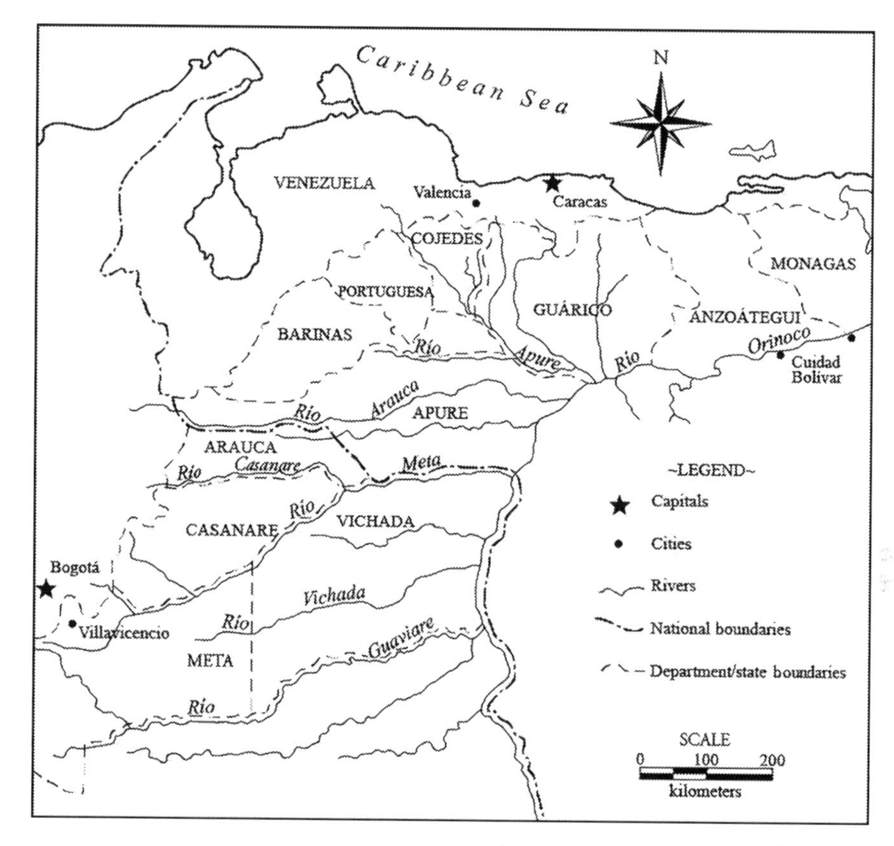

Figure 1.2 Map of the region of Orinoquia.

in terms of political power and historiography. Compared to the vast amount of research concerning these departments, studies of Orinoquia lag far behind.

It was during the 1st Simposio de Historia de los Llanos Colombo-Venezolanos in Villavicencio on August 10–13, 1988 that Omar Baquero and Luis Yesid Andoval issued a call for a regional history of the Llanos. They pointed out that, with some notable exceptions, previous studies had been either "global descriptions" drawn largely from secondary sources or biographies of exalted heroes, especially those associated with the Wars of Independence. What was lacking, they argued, to provide the building blocks of a true regional history, was systematic research that addressed such essential themes as the development of capitalist agriculture, the oil boom, guerrilla warfare, the presence of drugs, colonization, and the relationship with neighboring nations—Brazil and Venezuela (Baquero and Yesid Sandoval 1988, 455). Bogotá officials were already reinforcing this call by employing the term "Orinoquia" when they referred to the Llanos, a change underscored by President Virgilio Barco's decision to create the Corpes Orinoquia in 1989 to serve as an agency to promote development not only in the plains but also in Amazonia.

The existence of Corpes promoted common affinities between the Llanos departments and thus contributed to the developing sense of regionalism. As a result, in 2003 Leonel Pérez Bareño, a former director of both DAINCO (Departamento Administrativo de Intendencias y Comisarías) and Corpes Orinoquia, made a strong case not just for writing a history of Orinoquia but for the actual consolidation of the region as a political entity. In a paper presented at the VIII Simposio Internacional de la Historia de los Llanos Colombo-Venezolanos, Pérez argued that five factors supported the dismantling of the four Llanos departments and their amalgamation into a single administrative unit. First, the region had cultural, sociological and geographical homogeneity. Second, its division into departments being very recent was not firmly established. Third, there existed an indisputable capital city—Villavicencio. Fourth, there was an abundance of royalties from petroleum to support the change. And fifth, during the last 60 years, Orinoquia was the only Colombian region that had had experience with supra-departmental forms of government under DAINCO and Corpes Orinoquia (Pérez Bareño 2003, 357).

Pérez's plan does not seem to have garnered support from congress, but the goal of promoting the historical study of Orinoquia both as a Colombian region and as one that includes the Venezuelan plains has consistently gained more appeal. At the same 2003 Simposio Alberto Baquero noted that, once the artificial international border was set aside, there was no natural division between the Colombian and Venezuelan portions of Orinoquia since they shared the same environment, culture, and economy. He called for solidarity between Llanero historians of both countries as they sought to create the image (*ideario*) of the *País del Orinoco* (Baquero Nariño 2004, 8–13). Since that year four more international symposia have been held: Villavicencio (2006 and 2012); Barinas (2008), and Támara (2010).[4] These and other like meetings have allowed scholars to exchange ideas, affirm academic ties, and forge friendships as they study what might be termed Greater-Orinoquia, and there is evidence that at least in Colombia the term "Orinoquia" is slowly supplanting the traditional name, *los Llanos Orientales*.[5] For example, in 2014 both the Universidad de los Andes and the Universidad Nacional in Bogotá were offering undergraduate courses on the history of Orinoquia, and plans were underway for a similar course at the Universidad de los Llanos in Villavicencio.

The Llanos as an international frontier

In ordinary conversation the Spanish word, "frontera," means "boundary, border, or limit." Academics like Baquero Nariño and Pérez Bareño may dream of a "greater-Orinoquia" that transcends national boundaries, but the political significance of the international border between Colombia and Venezuela is a reality not readily dismissed (see Figure 1.3). Since the 1990s renewed recognition of its presence has opened up another way to conceptualize the Llanos—as an international frontier.[6]

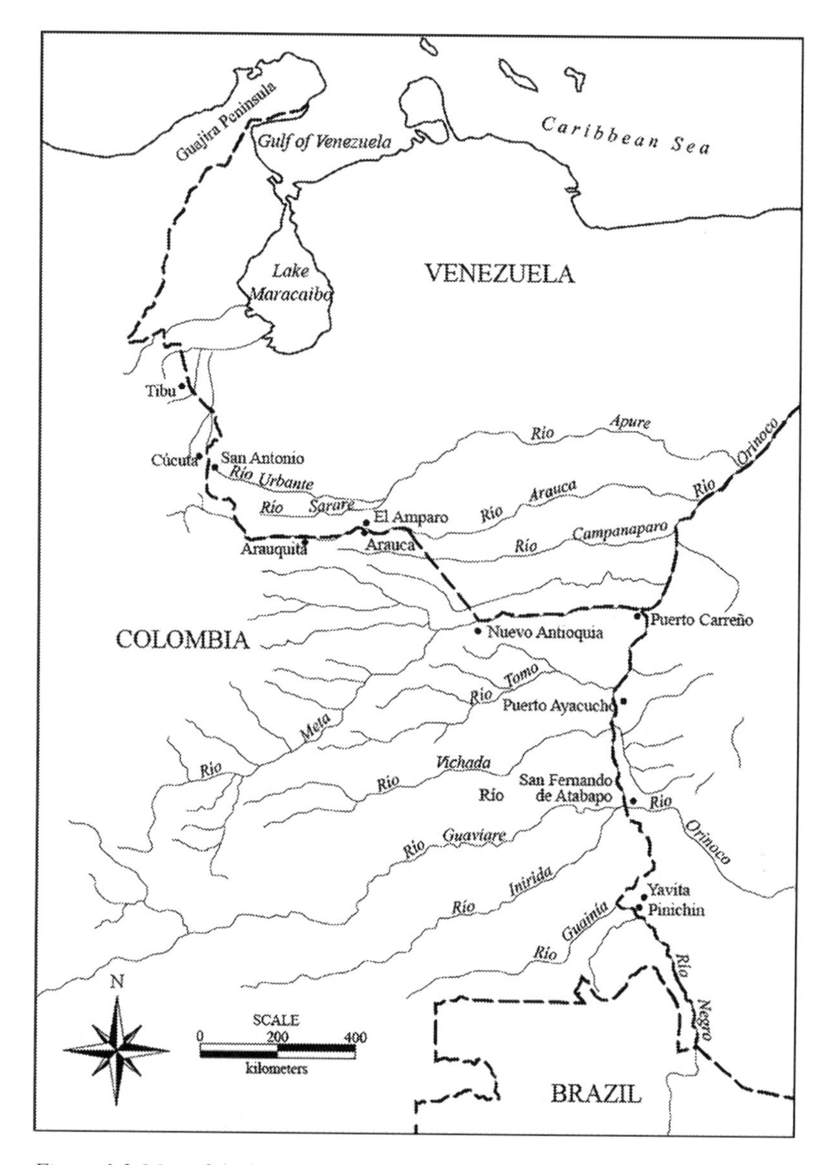

Figure 1.3 Map of the border between Colombia and Venezuela as established by the treaty of 1941.

Thanks to its location in the northwest corner of South America, Colombia borders four other countries besides Venezuela: Brazil, Peru, Ecuador, and Panama. Yet, since winning independence from Spain, Bogotá in the high Andes has paid scant attention to the regions lying along these lines, even after the secession of Panama in 1902 and the near loss of Leticia to Peru in 1932. The Constitution of 1991's elevation of the remaining peripheral national territories to departments reflected a belated recognition of their strategic importance. On June 23, 1995 Congress attempted to redress past neglect by decreeing Ley 191 known as the Ley de Fronteras. The object was "to establish a special rule for frontier zones in order to promote and facilitate their economic, social, scientific, technological and cultural development."[7] More specifically, the law required state action in these areas to protect human rights, strengthen processes of integration and cooperation between Colombia and its neighbors, create conditions to support economic development and the merger of frontier zones into the national and international economy, and joint action with neighboring states to provide security and fight international crime. To achieve these objectives, the law permitted governors and mayors of frontier departments and *municipios* to negotiate treaties with corresponding Venezuelan authorities on the other side of the border to foster cooperation.

Ley 191 directly affects Arauca and Vichada for they abut the Venezuelan states of Táchira, Barinas, Apure, and Bolívar and the Federal Territory of Amazonas. Because the border is really an artificial division that cuts through the center of Orinoquia, there has always been great affinity between the mestizos and Amerindians that live on either side of it. Over decades social, economic, cultural and family ties have created a common identity, and lacking strong state presence, the border itself has had little meaning. Given the absence of roads to the east, Venezuela was the natural market for cattle and other products produced in Arauca and Vichada, and Venezuelans accounted for the bulk of the population in Arauca City and Puerto Carreño. At the inauguration on September 6, 1967 of the José Antonio Páez Bridge that spanned the Arauca River to link Arauca City with El Amparo, Luis Serrano Reyes wrote in *El Tiempo,* "Nationality is blurred in these two cities that complement, support and trade with each another" (*El Tiempo,* September 6, 1967).[8]

In the 1990s the expansion of oil exploration and the tendency of Colombian guerrillas to seek safe haven in Venezuela alerted Bogotá to the need to pay more attention to its eastern boundary. Responding to a joint initiative by the Universidad Nacional and the Universidad Central of Venezuela, Congress ordered the Escuela Superior de Administración Pública (ESAP) to carry out a study of the "international frontier reality" and "to offer recommendations that might improve the quality of life of the people resident on Colombia's international frontier" (Ladino Orjuela and Duarte Moreno 2010, 9). ESAP chose to focus on Vichada, and the result was the publication in 2010 of *Apropiación de instrumentos fronterizos por parte de actores locales y su proceso de internacionalización: El caso del departamento del Vichada* by Wilson Ladino Orjuela and Pedro Duarte Moreno.

The purpose of the Ladino-Duarte study was to "deepen knowledge of one of the sections of the extensive and complex international frontier between Colombian and Venezuela," and to investigate one of its least known segments—the Department of Vichada and its interactions with the Venezuelan states of Apure, Bolívar and Amazonas. After tracing the history of Vichada since the

Spanish conquest, Ladino and Duarte show that until the 1970s inhabitants of the territory were overwhelmingly native Americans with a small mestizo element concentrated in the principal city, Puerto Carreño. Economic activities included extraction of rubber and other natural products, ranching, and contraband. Between 1993 and 2005 Vichada's population increased from 36,336 to 44,592 with most of the growth coming from migration of mestizos from other parts of Colombia. In 2005 the department was divided into four *municipios*, and there were 46 legally recognized Indian *resguardos* (reservations). Economic activities continued to be agriculture and ranching, but coca cultivation was extending along the Guaviare River. Across the border, Apure had a population of 337,756; Bolívar 1,214,464, and Amazonas 70,464. The latter, with its capital Puerto Ayacucho on the Orinoco River, had an economy similar to Vichada's, based on agriculture, ranching and mining (Ladino Orjuela and Duarte Moreno 2010, 41–62).

Throughout the twentieth and twenty-first centuries Colombian ranchers have owned *hatos* (ranches) in Apure, Bolívar or Amazonas and vice versa. The population of Puerto Carreño includes citizens of both nationalities and is characterized by a low level of education. In the 1990s when the bolívar, the national currency of Venezuela, was strong in relation to the peso, its Colombian counterpart, Venezuelans crossed the border to shop in Puerto Carreño, but since 2000, because of programs established by Hugo Chávez, the situation has been reversed. Food, fuel, and construction materials are cheaper in Venezuela, and Colombians visit Puerto Ayacucho with the object of buying these items.

The expansion of drug trafficking prompted both Colombia and Venezuela to strengthen their military presence on the border, increasing tensions between the two nationalities. Television programs broadcast from Caracas and Bogotá contribute to hardening stereotypes. For the Venezuelans, Colombians are the cause of violence, drug trafficking, and para-militarism, while Colombians believe that royalties from petroleum permit Venezuelans a more comfortable life style (Ladino Orjuela and Duarte Moreno 2010, 61).

Chávez's government proposed several projects to tie Vichada more closely to Venezuela including building a railroad, constructing bridges, establishing cement factories, and developing the petrochemical industry. These proposals reflected the asymmetrical relationship between the Venezuela and Colombian segments of Orinoquia. Although treaties have been made and signed, they remain formalities, for Vichada lacks the political institutions necessary to implement them and to function effectively as a department. Native Americans, without leadership and protection, are slowly migrating to the Venezuelan side of the border, and many mestizos are eager to acquire Venezuelan citizenship.

To offset these disadvantages Ladino and Duarte recommend that Colombia adopt a long-term development policy to maintain its hold on Vichada. Among other steps the government should establish schools in the *municipios*, support the formation of competent administrators, and promote agreements between Colombian and Venezuelan entities to establish a bi-national agenda that would fortify the ties between the local authorities (Ladino Orjuela and Duarte Moreno 2010, 124–125).

Apropriación de instrumentos fronterizos is a pioneering study of the Llanos as an "international frontier borderland," and it demonstrates that this approach will be a valuable one for future historians. Given the spread of globalization and the not always tranquil diplomacy between the Santos regime and the Bolivarian state, controlled since 2013 by Chávez's successor Nicolás Maduro, a clearer understanding of the history and current situation of the people who live along the Colombo–Venezolano border appears to be one of the most pressing projects awaiting academic study.

Summary and conclusion

To summarize, since the mid-twentieth century modernization has transformed the piedmont areas of the Llanos Orientales almost beyond recognition while the *Llanos abajo* continue to languish in their traditional state. Academic study of the region has made great strides. The frontier model has diversified, and each of its manifestations—neo-Turnerian, region, and international borderland—offers valid insights about the plains and suggest comparisons that might be made with other South American peripheral regions. It is to be hoped that during the next decade the proliferation of universities in Orinoquia will produce newly-minted historians and graduates in a variety of disciplines who will take on the challenge of increasing and deepening understanding of the history and culture of this region that is destined to have such a significant impact on Colombia's future.

Notes

1 This chapter is reprinted with permission from the University of Florida Press. See Jane M. Rausch, "Changing Concepts of the Llanos Frontier in the Last Half of the Twentieth Century," in *Colombia: Territorial Rule and the Transformation of the Llanos Orientales*, by Jane M. Rausch, 132–246 (Gainesville: University Press of Florida, 1999). An earlier, Spanish-language version of this chapter was published in Alexander Betancourt Mendieta y José Guadalupe Rivera González (eds.), *Territorios y fronteras: miradas desde las Ciencias Sociales y las Humanidades* (Barcelona: Anthropos Editorial/Universidad Autónoma de San Luis Potosí, 2015).

2 These books are: *A Tropical Plains Frontier: The Llanos of Colombia, 1531–1831* (Albuquerque: University of New Mexico Press, 1984); *The Llanos Frontier in Colombian History, 1830–1930* (Albuquerque: University of New Mexico Press, 1993); *Colombia: Territorial Rule and the Llanos Frontier* (Gainesville: University Press of Florida, 1999); and *Colombia and the Transformation of the Llanos Orientales* (Gainesville: University Press of Florida, 2013).

3 The era of La Violencia was a period of undeclared civil war fueled by political hatreds that lasted from 1946 to 1964 and claimed between 100,000 and 250,000 lives. Although it affected nearly every Colombian region, much of the fighting that occurred between 1947 and 1953 was localized in the Llanos.

4 A meeting planned for San Luís de los Morros in Venezuela in 2014 failed to take place.

5 In addition to social scientists and historians, Colombian and Venezuelan Llanero poets and novelists have met regularly. See the papers of the 1st Encuentro Colombo-Venezolano de Escritores Llaneros published as *Sobre Los Llanos* (Occidental de Colombia Inc.,1988).

6 In 1978 Alistair Hennessy included the "political frontier" in his frontier typologies pointing out that throughout Latin America international borders precipitated conflicts over territorial sovereignty, cut through sedentary Indian communities, and generated contraband trading. He predicted that, as internal frontiers of settlement moved closer to the international borders, disputes over territorial sovereignty would come to assume greater importance underlining "the urgent need for international cooperation and regional integration before it is too late" (Hennessy 1978, 106–109).
7 Law 191 defined "Frontier Zones" as "those municipios and corregimientos especiales of the frontier departments located on the international border of the Republic of Colombia and those in which economic and social activities are directly influenced by their location." See www.secretariasenado.gov.co (accessed September 9, 2011).
8 On August 14, 2004 *El Tiempo* reported that, due to lack of maintenance, the bridge was in such poor condition that it might collapse into the Arauca River.

References

Baquero, Omar and Luis Yesid Sandoval. 1988. "La 'Historia' de la Historia Regional: Hitos y Perspectivas." In *Los Llanos: una historia sin Fronteras*, edited by María Eugenia Romero Moreno and Jane M. Rausch, 445–455. Villavicencio: Academia de Historia del Meta.

Barbosa Estera, Reinaldo. 1998. "Frontera agrícola orinoquense: de la precariedad estatal a la crisis de derechos humanos." In *Conflictos Regionales: Amazonia y Orinoquia*, edited by José Jairo González, 155–195. Bogotá: Tercer Mundo.

Barquero Nariño, Alberto. 1990. *El caso llanero: Villavicencio*. Villavicencio: Editorial Siglo XX.

——. 2004. "Derecho y Deber al Ideario del País del Orinoco." In *VIII Simposio Internacional de historia de los llanos Colombo-Venezolanos*, edited by Alberto Baquero Nariño, 8–13. Villavicencio: Editorial Juan XXIII.

Billington, Ray Allen. 1973. *The American Frontier Thesis: Attack and Defense*. Washington, D.C.: American Historical Association.

Bowman, Isaiah. 1931. *The Pioneer Fringe*. Special Publication No. 12. New York: American Geographical Society.

Brunnsschweiler, Dieter. 1972. *The Llanos Frontier of Colombia: Environment and Changing Land Use in Meta*. East Lansing: Michigan State University.

Bushnell, David. 1993. *The Making of Modern Colombia: A Nation in Spite of Itself*. Berkeley: University of California Press.

Carranza, Eduardo. 1975. *Los pasos cantados: El corazón escrito*. Bogotá: Instituto Colombiana de Cultura.

Espinel Riveros, Nancy. 1997. *Villavicencio, dos siglos de historia comunera: 1740–1940*. Villavicencio: Editorial Juan XXIII.

Faragher, John Mack. 1992. "Gunslingers and Bureaucrats: Some Unexpected Thoughts About the American West." *The New Republic* 30, December 14: 29–36.

García Bustamante, Miguel. 1997. *Un pueblo de frontera: Villavicencio 1840–1940*. Bogotá: Universidad de los Llanos.

——. 2003. *Persistencia y cambio en la frontera oriental de Colombia: El piedemonte del Meta 1840–1950*. Medellín: Fondo Editorial Universidad EAFIT.

Hennesey, Alistair. 1978. *The Frontier in Latin American History*. Albuquerque: University of New Mexico Press.

Iriate Núñez, Gabriel. 1999. *Colombia a su alcance*. Bogotá: Planeta.

Ladino Orjuelo, Wilson and Pedro Duarte Moreno. 2010. *Apropiación de instrumentos fronterizos por parte de actores locales y su proceso de internacionalización: El caso del departamento del Vichada.* Bogotá: ESAP.

Ley 19. (1995). http://www.secretariasenado.gov.co. Accessed October 11, 2011.

Lynch, John. 1989. *The Spanish American Revolutions 1808–1826.* New York: Norton.

Mantilla Trejos, Eduardo. 1988. *Sobre los llanos.* Bogotá: Fotomecánica Industrial.

Ojeda Ojeda, Tomás. 2000. *Villavicencio entre la documentalidad y la oralidad.* Villavicencio: Oscar Giraldo Durán Ediciones.

Pérez Bareño, Leonel. 2003. "Experiencias de integración regional en la Orinoquia Colombiana." In *VIII Simposio Internacional de Historia de los Llanos Colombo-venezolanos*, edited by Alberto Baquero Nariño, 350–361. Villavicencio: Editorial Juan XXIII.

Preciado, Fredy. n.d.. "Desarrollo endógeno de una región de frontera: los Llanos Orientales, departamentos de Casanare y Meta (Colombia)." Yopal: unpublished document.

Rausch, Jane M. 1984. *A Tropical Plains Frontier: The Llanos of Colombia, 1531–1831.* Albuquerque: University of New Mexico Press.

——. 1988. "Región olvidada: Los Llanos en la historia de Colombia." *Revista de la Academia de Historia del Meta* 2(1): 32–40.

——. 1993. *The Llanos Frontier in Colombia History, 1830–1930.* Albuquerque: University of New Mexico Press.

——. 1999. *Colombia: Territorial Rule and the Llanos Frontier.* Gainesville: University Press of Florida.

——. 2013. *Colombia and the Transformation of the Llanos Orientales.* Gainesville: University Press of Florida.

Ridge, Martin. 1991. "Frederick Jackson Turner and His Ghost." In *Writing the History of the American West*, edited by George Miles, 65–76. Worcester: American Antiquarian Society.

Rivas, J. A., J. V. Rodríguez, and C. G. Mittermeier. 2002. "The Llanos." In *Wildernesses*, edited by R. A. Mittermeier, 265–273. Mexico: CEMEX.

Rivera, José Eustasio. *La Vorágine.* 1935. Translated by Earle K. James. New York: G. P. Putnam's Sons.

Slatta, Richard. 1997. *Comparing Cowboys and Frontiers: New Perspectives on the History of the Americas.* Norman: University of Oklahoma Press.

Sobre los Llanos. 1988. Bogotá: Occidental de Colombia Inc.

Weber, David J. and Jane M. Rausch. 1994. *Where Cultures Meet: Frontiers in Latin American History.* Wilmington: Scholarly Resources.

2 Frontierization and defrontierization

Reconceptualizing frontier frames in Indonesia and India[1]

Greg Acciaioli and Alka Sabharwal

Introduction: reconceptualizing frontiers as trajectories of frontierization and defrontierization

The concept of the frontier has recently become re-energized and extended as a frame for studying human-environment interactions in a global context. Frederick Jackson Turner first brought the notion of the frontier into academic discussion with his 1893 address, "The Significance of the Frontier in American History" (1994), spawning an ongoing debate concerning the significance of the frontier for constituting the distinctively American brand of free enterprise and the *laissez-faire* polity that sustained it. Although some of Turner's insights, such as the succession of different sorts of frontiers in any one region, have remained valuable, his emphasis upon American exceptionalism (Geiger 2008a, 77)—the uniqueness of the American experience of the frontier—has hindered attempts to generalize this notion to other regions, as did his concentration upon the "heroic" plainsmen with the consequent neglect of the indigenous peoples whose land was occupied by them.

The historian Walter Prescott Webb (1952) did extend Turner's framework to the entire periphery of the colonial expansion of western Europe:

> Once we conceive of western Europe as a unified, densely, populated small region with a common culture and civilization—which it has always had basically—and the frontier also as a unit, as a vast and vacant land without culture, we are in a position to view the interaction between the two as a gigantic operation extending over more than four centuries, an operation which may well appear as the drama of modern civilization.
>
> (Webb 1952, 7–8)

Webb's contrast of the Great Frontier and the Metropolis, as he termed the European hub, anticipated some of the distinctions and processes later developed in World Systems theory (Wallerstein 1974), as did his analysis of economic "booms" occasioned by frontiers. However, as had Turner, Webb restricted his perspective to the expanding colonialists, largely erasing the peoples into whose lands the Westerners expanded in what was basically a triumphalist narrative of progress—the Metropolis gained wealth, while the Great Frontier received "culture." Webb's treatment remained an archetypical exemplar of accounts of "Europe and the people without history" (Wolf 1982).

However, almost concurrently Owen Lattimore began expanding the usefulness of the notion of the frontier beyond the context of Euro-American expansion. Pursuing the project of developing a "scientific" model of the way human societies form, evolve, grow, decline, mutate and interact with one another along frontiers (Rowe 2007, 259), Lattimore concentrated on historical transformations along imperial China's borders, which he analyzed as sites of contestation; in certain periods the peoples from beyond, such as the Mongols and Manchurians, emerged as the conquerors of the heartland. The expansion of civilization and barbarian incursion were mutually constitutive processes in an ebb and flow of interacting populations guided by ecological parameters inflected by the relative levels of technology of those societies. The frontier was thus a fluctuating social zone rather than being defined by a distinct geographic boundary, emerging as occupying communities began to concentrate on new forms of subsistence enabled by new technologies. Lattimore thus emphasized the interaction between socially and economically differentiated communities, both the occupying and the occupied, and the diverse trajectories that could ensue, including amalgamation or expulsion when the occupying community was of a different (more complex) kind and (larger) scale than the occupied or even the synthesis of qualitatively new communities distinguished from both original communities when they were of roughly of the same "social vigor and institutional strength" (Lattimore 1968, 374). Where local leaders in frontier regions were transformed into "chiefs" who served as intercultural brokers with the expanding society (Lattimore 1968, 377–378), the frontier could function as the very origin of such social formations as tribalism and feudalism. However, despite this processual outlook, Lattimore still concentrated more upon the expanding society than those whose lands the former occupied, characterizing his work as written "from the perspective of a Chinese frontier administrator" (Lattimore 1968, 382).

More recently, the transition to an anthropological outlook more closely attuned to balancing the perspectives of the occupiers with those of the occupied has been advanced by the work of Anna Tsing (2005). Tsing emphasizes the impacts of the transformation of the montane homeland of the Meratus of South Kalimantan, Indonesia, into a frontier marked by the contradictory co-existence of both the most orderly of economic impositions—vast expanses of mono-crop plantations of acacia, oil palms and other crops—and the wildness of individuals advancing into the area in order to "make a killing" through logging, mining, and other enterprises. Indonesians themselves have noted the parallels of this situation to the American frontier conceptualized by Turner; as an Indonesian Minister of the Environment declared: "Kalimantan at this time is part of the Wild West . . . like parts of America in the nineteenth century" (quoted in Tsing 2005, 31). Tsing notes the continuing relevance for political figures and others of Turner's (and, one might add, Webb's) conceptualization of the frontier as a meeting place of savagery and civilization. However, Tsing views the frontier as neither a place nor a process, but a project; a particular region is envisaged as a set of resources that entrepreneurs and their workers then occupy to exploit. This resource frontier is thus a specifically capitalist frontier, defined by harvesting products destined as commodities. Tsing (2005, 31–32) does note other types of frontiers—the nation-making frontier, the techno-frontier, and the salvage frontier—characterizing the current situation in Kalimantan as an unstable mix of both the resource frontier and the salvage frontier, with destruction, production and conservation all operating simultaneously. Yet, such emphasis upon simultaneity does not allow a more nuanced tracing of how various types of frontiers may be operating in different temporal trajectories, such that while one type of frontier is gaining greater salience in a region, another may be reducing in importance. That is, her Foucauldian emphasis on the frontier as project, defined by discursive formations moulding certain types of human subjects (Tsing 2005, 30), backgrounds the processual rise and decline of frontiers in a region—processes of frontierization and defrontierization.

In order to foreground this processual dimension, we turn here to a different contemporary typology of the frontier, that of Danilo Geiger (2008b), who defines frontiers as "*areas remote from political centres which hold strategic significance or economic potentials for human exploitation, and are contested by social formations of unequal power*" (Geiger 2008a, 78, italics in the original).[2] Geiger's inclusion of "economic potentials" echoes Tsing's emphasis on the construction of resources in her resource frontier, but his complementary emphasis upon "strategic significance" also highlights the political dimension of state control. His ending allusion to "social formations of unequal power," using more explicitly Marxian terminology, reiterates and clarifies Lattimore's focus upon the economic and social differentials of communities interacting in frontier contexts. However, defining a frontier as an "area" conflates the notion of a region with the frame that is imposed upon it. We consider it more productive to view a frontier as a framing of a region by the state or other "unequal social formation" that redefines a region with consequences for patterns of settlement, exploitation, and securitization.

Following from his definition, Geiger proposes a tripartite typology of frontiers. First is the "frontier of settlement," which builds most overtly upon Turner's conceptualization of the frontier. Settlers may range from those explicitly sponsored by the state, through those who make their own way but are facilitated by governments through loans, provision of tools and other supplies in the new settlement, to fully spontaneous migrants, though even the latter often further state aims by bringing with them and imposing in their region of settlement such state-sanctioned notions of development as intensified agriculture, cash crops, and other innovations. "Frontiers of extraction," Geiger's second type, aligning with Tsing's "resource frontier," are catalyzed by the discovery or creation of "resources" in an area, a process that depends upon a social formation coming up with the technology to harvest or harness what it identifies as resources in the region to be frontierized. Enterprises, either state-owned or private, and increasingly transnational corporations working together with domestic companies and agents of the state either work to extract what have been identified as resources (e.g. the minerals identified by mining companies) or create the economic constellations (e.g. large-scale mono-crop plantations) that can make use of identified resources (e.g. particular types of land or other habitat suitable for such cultivation). Geiger's third type, "frontiers of control," prioritizes the securing of the sovereignty of the nation-state or other political forms (e.g. rajahdom, Empire, and so on.).[3] Though such zones are predominantly found along national borders, these "frontiers of control" may also occur well within the borders of a state in instances of insurrection by an ethnic or religious constituency. However, even zones of domestic unrest are often found near the borders of a state (e.g. the Chittagong Hill Tracts of Bangladesh, Kashmir in India, and Papua in Indonesia), as they are often sites of resistance because, in addition to being politically peripheral, they are frontiers of intensive extraction. Conceptualizing frontiers of control requires considering not only the operations of the direct agents of state organs such as the administrative bureaucracy and military, but also those who are members of organizations that are not, strictly speaking, arms of the state, but, nevertheless, further state interests in controlling a periphery. Missionaries can be seen as furthering state aims of domesticating peripheral populations and socializing them in standards of orderliness and civilization envisaged and enforced by the state, despite seeing themselves as mainly operating in their own kind of spiritual frontier. Hence, the spiritual frontier requiring cultivation by Christian evangelization, Islamic *dakwah*, and analogous techniques of other religions to create properly disciplined religious subjects may be seen as part of the "frontier of control" largely operated by the state.

In addition to Geiger's three categories, we suggest a fourth type: frontiers of conservation. Tsing adumbrated this fourth type in her notion of the salvage frontier, which she saw as ironically co-existing with the resource frontier, as "protected areas" often become the preferred region for the operation of military companies and other enterprises run by cronies of the state. Yet, in theory a frontier of conservation is more the obverse than the complement of the frontier of extraction, proscribing what the other prescribes as resources to be exploited. Conservation creates new types of boundaries around protected areas—reserves, wetlands, national parks and others—and new zones at these boundaries—buffer areas, production zones, and others—that allow different types of extraction and use, and thus generates the conditions to transform relations between Indigenes and settlers, the occupied and the occupiers. This new frontier also attracts a new community of occupiers: conservationists, research scientists, and the bureaucrats with whom they work in securing these new boundaries. Frontiers of conservation, like other frontiers, may come into being gradually, with protected areas often long suspended in a candidate status, but are activated as a relevant frame more intensively when certain thresholds are passed, such as the gazetting and notification of a protected area.

While Tsing has noted the simultaneity of more than one type of frontier, it is important to note that across historical eras these frontier frames assume differential relevance. At one historical juncture, for example following a boundary war, a region may be most importantly framed as a frontier of control, while temporarily declining as a frontier of extraction, though perhaps even intensified as a frontier of settlement to provide a buffer population as a further warrant for asserting a national boundary. Each of these four frontier frames has its own trajectory, responding to technological advances, local political tensions, global pressures, state imperatives and a host of other conditions. Turner and Webb emphasized the succession of frontiers in the expansion of America and the West, Lattimore noted how the Chinese frontier was one of oscillation, and Geiger has labeled frontiers as cyclical rather than linear. Such characterizations warrant a focus not simply upon frontiers as imposed frames for regions, but also upon the rise and demise of the salience or relevance of a particular frontier frame: processes of frontierization and defrontierization. Even those who discussed the "closing of the frontier" in the American West have acknowledged (e.g. Webb 1952, 4) that such "closing" was a gradual process, most apparent from 1880 to 1910, rather than a discrete event, just as was the "opening" of the frontier. We reconceptualize the "closing" of frontiers by suggesting that processes of defrontierization occur when the occupied peoples are able successfully to exercise their agency in diminishing the political and economic power of occupiers through such means as the revitalization of local customary institutions, social movements that can result in reclamation of land and other resources or similar (re)assertions of local control.

Having established the background of this reconceptualization of frontiers as trajectories of frontierization and defrontierization along four dimensions, we seek to demonstrate the usefulness of this framework by briefly exemplifying the differential relevance of various frontier frames at different historical periods in two Asian contexts: the highland valley Lindu in Central Sulawesi (Indonesia) and the highland plains of Changthang in Ladakh (India). We seek to conceptualize the convergences and divergences of the historical trajectories of these two locations in terms of the differential rise and fall of our four frontier frames across various periods of their history from pre-colonial times to the present. While this overview can only be schematic given constraints of space, hopefully such an approach will reveal how these two regions have responded to various imperatives of succeeding states, from their pre-colonial conditions as peripheries of minor rajahdoms to their current status as foci of internationally recognized biodiversity hotspots (Himalaya and Wallacea respectively), in ways that reveal larger trajectories of state development in Asia.

Throughout their histories Lindu and Changthang have been subjected to the incursions of others that have framed these regions as frontiers. However, simply stating that fact does not allow for a nuanced comparison of the two areas nor any elaboration of the theoretical notion of frontier based upon analyzing the similarities and differences of the histories of these two regions. We present here a more articulate comparison by tracing how the various trajectories of frontierization and defrontierization in regard to control, extraction, settlement and, in more recent times, conservation have intersected to constitute the histories of these regions as successions of types of place defined by different intermeshings of intensifying and weakening frontier frames. The very logic of interrelation of these frontier frames does lead to some convergences in the histories of these two regions, but each of them has also displayed different conjunctures of these trajectories due to their contrasting geopolitical positions as peripheries and the ways in which opportunities for the exercise of agency among the constituent players in the frontiers, the occupiers and the occupied, have been deployed.

As a graphical representation for which we make no claims other than providing a crude heuristic summary of our qualitative historical assessments, Figure 2.1 presents the trajectories of frontierization, represented by upward slopes, and defrontierization, represented by downward slopes, in all four aspects—control, extraction, settlement, and conservation—across the five historical periods we have isolated for the two regions Lindu and Changthang. A cursory glance at the two sets of trajectories reveals significant similarities, as would be expected from their embedding as peripheries in two Asian contexts that have experienced somewhat analogous histories of colonization, decolonization, and development as independent nation-states. However, certain differences are evident in the intersections and divergences of these lines as well, formally representing the divergent histories of the two regions as differently situated peripheries.

Both regions function as relatively remote peripheries in the pre-colonial context, but Changthang draws more attention from the powerful local pre-colonial states, particularly Ladakh, due to its role in the lucrative trade in pashm connected to the famous silk trade and Kashmir's shawl industry (Drew 1875; Rizvi 1999). It thus begins at higher levels of control and especially extraction, due to levies on that trade, than does Lindu, where the only mode of control exerted by the lowland rajahdom of Sigi is through exaction of minor tribute, largely smoked fish (Valentijn v.1, pt.b 1724, 74; Kruyt 1938), thus also comprising the only mode of extraction by outsiders (hence the coincidence of those lines for the two types of frontier trajectory in this period). Neither region is subjected to substantial settlement from the outside in the pre-colonial period, and resource conservation imposed by outsiders is not yet a relevant factor.

In the colonial period, the Buddhist Changthang region experiences a peak in extraction from outside, as it is subjected to overt government by the Hindu Rajput dynasty of Dogra and more systematic pashm taxation by the British administration (Drew 1875; Beek 1999), though with the lessening of the pashm trade in the build-up to India's independence this levels off and then precipitously declines. The 1846 Battle of Chushul between the Sino-Tibetan army and the Dogra rulers not only redefines a new frontier of extraction in Ladakh through post-war treaties, but also opens it to the new foreign traders entitled to local free labor services in order to intensify Dogra control in the region (Drew 1875; Sheikh 1999), demonstrating how trajectories of frontiers of control and extraction often (but not always—see below) move hand in hand. Lindu is transformed quickly into a frontier of control at the beginning of direct colonial penetration in the Ethical Period beginning in the first decade of the twentieth century; administrators, missionaries and others sponsored by the Dutch regime enter the area and

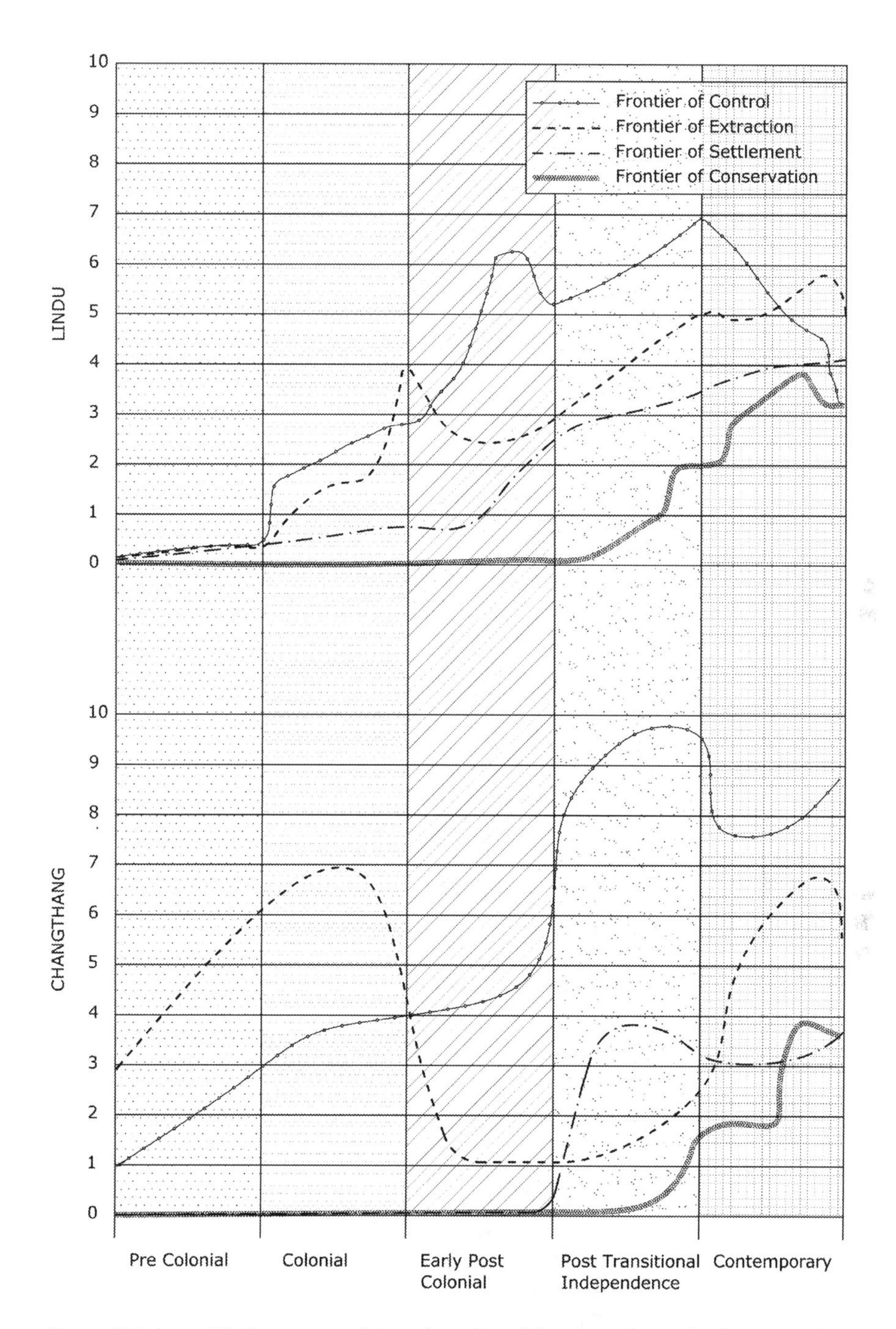

Figure 2.1 A graphical summary of the trajectories of frontierization and defrontierization in all four aspects—control, extraction, settlement, and conservation—across the five historical periods isolated for the two regions Lindu and Changthang.

begin imposing programs of development—improvement of wet rice irrigation, introduction of cash crops such as coffee, forced resettlement from the scattered settlements in the surrounding hills into nucleated villages in proximity to Lake Lindu, and other programs (Davis n.d., 3; Kaudern 1940, 35; Kruyt v.1 1938, 14). The extension of control brings with it intensification of extraction, as the introduced cash crop of coffee is used for payment of taxes, and corvée labor is also extracted to improve paths up to the highland plain (Vorstman n.d., 10–11), facilitating more settlement by outsiders, as does the extension of missionization to Lindu by the Salvation Army (Brouwer 1977), but frontierization of settlement initially increases quite slowly. Extraction spikes precipitously at the end of the colonial period with the punitive exactions of the Japanese during World War II, intensifying the trajectories of frontierization of extraction and (to a lesser extent) control at Lindu, though not those of settlement and conservation.

At the beginning of India's early independence period, despite a leveling off of extraction, there is at first some increase in control exerted from outside Changthang. Activities of mapping, measuring, and counting the inhabitants and the products of the area—techniques to render the populace and its resources visible to the new state (Scott 1998; Beek 1999)—soon yield an increase in control, whose trajectory spikes toward the end of this period due to concerns about the intentions of neighboring China. In fact, the Sino-Indian War of 1962 becomes the event that marks the end of this period for Changthang, so the trajectory of administrative control increases more steeply at the end of this period and reaches unprecedented levels with the intensification of the presence of the Indian army in the region in the ensuing period (Aggarwal 2004). Intensification of Changthang as a frontier of control far exceeds the control ever exercised at Lindu due to the former's geopolitical position at an international border, as compared to Lindu's geopolitical situation near the geographic (though not political) centre of the nation-state of Indonesia, only lying along a border of state control during the period of regional rebellions (1950–1965) of the early Republic (Harvey 1977, n.d.). Those rebellions, and the increased presence of the national military they elicited in the vicinity of Lindu, are evidenced in the steep rise of the trajectory of control for Lindu, which eases off with the quelling of those rebellions, as Lindu once more becomes a relatively neglected periphery within the nation-state, though never returning to the low levels of outsider control of the pre-colonial and colonial eras.

Concurrent with the sharp intensification of India's control in Ladakh in this period is a substantial decline in extraction, as the region is increasingly cast as economically backward, offering little to the Indian state as a whole (Beek 1999; Aggarwal 2004). In contrast, extraction at Lindu, after an initial dip at the beginning of the period of engulfing regional rebellions begins a steady upward rise that will be continued throughout the next era due to the increasing exploitation of resources of the highland plain, especially the fish earlier introduced by the Fisheries department into Lake Lindu, by migrants, paramount among them the Bugis from South Sulawesi (Acciaioli 2000). Their arrival and proliferation at Lindu is marked by the increase in the trajectory of settlement, as the Bugis, as well as other migrants, first arrive as Internally Displaced Persons (IDPs) from the regional rebellions and then initiate a process of chain migration that increases their presence as occupiers into the ensuing periods. In contrast, settlement of Changthang by outsiders remains negligible, as its image as economically backward and physically inhospitable provides no lure for in-migration or even policy attention from the state (Sabharwal n.d.; Beek 1999; Aggarwal 2004).

The subsequent political phase in each of these regions, labeled post-transitional independence in Figure 2.1, is marked by the onset of the New Order in Indonesia and the intensification of Changthang as India's national border with China. Lindu witnesses the rise in importance of all its aspects as a frontier. Intensive implementation of development by the New Order state increases extraction and population movement into the region and is accompanied by increasing administrative control. The New Order, pursuing its master plan for national "improvement" (Li 2007), homogenizes the administrative structure of all regions within its borders, including exclusion of local customary councils from participation in village governance, thus reducing opportunities for the exercise of agency by Indonesian indigenous peoples such as the To Lindu (Acciaioli 2009). Lindu first begins to emerge in this period as a frontier of conservation with the declaration of several protected areas throughout the nation, including in the region encompassing Lindu (Watling and Mulyana 1981), as a strategy to consolidate and expand bureaucratic control and enhance governmental legitimacy in the face of economic liberalization (Cribb n.d., 26, 36). The declaration of the Lore Lindu National Park at first intensifies conservation frontierization, as well as the concomitant intensification of the frontier framework of administrative control inevitably accompanying that trajectory, but eventually there is a leveling off as the park languishes for a decade in candidate status (Acciaioli 2008, 2015).

In contrast, the intersection of trajectories of frontierization and defrontierization is much more complex and expansive in Changthang during this period. As a contested national border of India in the post Sino-Indian War period, Changthang is subjected to ever-increasing military control; the Changthang border is declared an Inner Line Area restricting civilian entry from other parts of India (Aggarwal and Bhan 2009). However, despite this restriction, outsiders do enter the region from a different direction, as Changthang experiences a substantial influx of refugees from Tibet, thus being opened as a "frontier of settlement" from outside India. Changthang, adversely affected by the post-war demise of the cross-border pashm trade and the substantial loss of local winter pastures to China, remains one of the most remote and underdeveloped regions of India (Dawa 1999) early in this period. However, extraction begins to rise again due to both the influx of Tibetan refugees and deployments of the Indian military and para-military forces (Goodall n.d.; Dollfus 2013; Sabharwal n.d.). Near the period's end Changthang begins, somewhat precipitously, to emerge as a frontier of conservation; adopting a wilderness approach, the Indian state, in a coalition with staunch conservationists and the scientific community, declares Changthang a protected area to conserve some of the globally recognized Trans-Himalayan avifaunal species, eventually succeeding in imposing a hunting ban (Bhatnagar et al. 2007).

Our final period, beginning in the early 1990s in both regions, is characterized by the re-empowerment of local peoples in these frontier regions and hence the intensified exercise of agency on their part that defines defrontierization (at least in some trajectories), as facilitated by the indigenous peoples' movement in Indonesia (Li 2000; Acciaioli 2007) and the achievement of Scheduled Tribe status in Ladakh. With the weakening and then demise of the New Order regime, succeeded by Reformation and its policy of Regional Autonomy (Erb et al. 2005), control from the centre declines. Indigenous To Lindu succeed in blocking the erection of a hydroelectric dam (Sangaji 2000), and eventually their campaign to regain political power in their own region finds success with the establishment of Lindu as a Conservation Subdistrict, with all political offices occupied by To Lindu and the Lindu customary council once more incorporated as part of the governmental apparatus (Acciaioli 2009). This success marks a substantial defrontierization of Lindu in regard to central state control; in effect, regional autonomy allows the indigenous To Lindu to themselves become the state within their own region, realizing in some respects the goal of the "sovereignty of indigenous people" in their own domain, as trumpeted by the national Indigenous peoples' movement (Acciaioli 2007). Extraction by outsiders continues, however, as Central Sulawesi experiences a cocoa boom, with many of its profits accruing to outsiders who have been savvy enough to acquire land for garden cultivation from Indigenes during the regional monetary crisis that toppled the New Order (Li 2002). Thus, throughout much of this period Lindu remains a frontier of settlement, though in-migration levels off when the newly re-empowered customary council declares the right of the To Lindu to control movement into their own domain (Acciaioli 2008). The transformation of Lindu from an open access region back to a commons regulated by customary law levels the trajectory of Lindu as a frontier of settlement, just as customary restrictions on the amount of land that can be worked by residents in the plain begins to level some processes of extraction. The To Lindu are able to begin reversing the balance of economic and political relations between themselves and previous migrants, both spontaneous (e.g. Bugis) and state-sponsored (i.e. local transmigrants from elsewhere in the subdistrict), by using the rationale of conservation, eagerly signing on to a community conservation agreement and codifying their customary law (*hukum adat*) as primarily a community-based resource management system along guidelines set by The Nature Conservancy (TNC) (Haller et al. 2015). So, while experiencing the intensification of a frontier of conservation, the To Lindu are able to increase their own participation and stature in this process, thus reducing its impact as a trajectory of frontierization controlled by outside conservation interests (Acciaioli 2008).

The partial lifting of Changthang's status as an Inner Line Area at the beginning of the analogous period in India marks a sharp decline in its position as a frontier of control, but a new "cosmetic federalism" through the course of development and an escalation of military infrastructure intensify administrative control again in this era (Baruah 2003). The establishment of nature tourism enterprises at the beginning of this period by outside tour operators based in the district capital Leh intensifies Changthang's status as a frontier of extraction, but the local inhabitants eventually succeed in assuming control of tourism enterprises themselves through locally enforced regulations (Sabharwal n.d.), providing a salient example of defrontierization in the region. Outside settlement of the area by refugees levels off, but toward the end of the period other types of settlement, such as that of migrant labor from within India, begin to rise with the increased role of the military in local development and administration. Changthang is intensified as a frontier of conservation even more saliently than Lindu with the Indian Supreme Court's notification of the Changthang Cold Desert Wildlife Sanctuary and increasing intervention by the Worldwide Fund for Nature (WWF) in response to descriptions of the pressures upon wildlife, especially wild herbivores, due to "overstocking" of domestic animals by local pastoralists (Mishra et al. 2001, 279). However, near the end of this era local pastoralists reverse this intensification to some degree, as evidenced by their demolition of the conservation fence earlier erected to prevent access to Lake Tsomoriri, as well as their opportunistic coalitions with the Indian military (Sabharwal n.d.) Yet, in comparison to the To Lindu's successful harnessing of the conservation idiom to advance their own local interests within their enclave, the pastoralists of Changthang have had little success in recovering their pasturelands from the sanctuary (except indirectly through the military's denotification of parts of the sanctuary), although accommodating to the situation in the short term by orienting to associated tourism opportunities.

Conclusions: frontierization, defrontierization and recovering histories

In conclusion, we want to emphasize again some of the ways in which our reconceptualization of the frontier advances the study of relations between the occupiers and the occupied in frontier regions. Although we have built upon Geiger's (2008) typology positing frontiers of control, settlement and extraction, we have added a fourth type, the frontier of conservation, paralleling the salvage frontier adumbrated by Tsing (2005), in order to account for frontierization effected by the incursions of the agents of conservation in protected areas such as sanctuaries and national parks. As do control, settlement, and extraction, conservation sets new boundaries, creates new zones, alters the mix of these three other dimensions, and recasts the relations of occupiers and the occupied. The frontier frame of conservation will become increasingly important, as the implementation of such programs as carbon marketing, including REDD+ and other compensation schemes for local stewards of forests, further alters patterns of control, settlement and extraction in peripheral regions subject to frontierization. We have also emphasized that placement of frontier frames, as well as their retraction, is a continuing process rather than a decisive alteration from one state to another. We believe our notion of trajectories of frontierization and defrontierization, which are even amenable to rough graphical representation, captures this processual dimension more adequately. Finally, in our characterization of the trajectory of defrontierization, we have not regarded as criterial the attainment of some threshold of population density among settlers or some similar index, as earlier posited for the closing of the frontier in the American West. Rather, assertions of agency by local peoples in their efforts to reverse the asymmetry in the relations of (dominating) occupiers and the (subordinated) occupied are criterial for our characterization of trajectories of defrontierization. This emphasis restores to the frontier as shaping actors those peoples who had been erased in its earlier conceptualization, allowing them to emerge once more in accounts of frontiers as people with a history.

Notes

1 Many thanks to Paramjeet Singh for the digital drawing of Figure 2.1.
2 Geiger reiterates this definition on p. 99 of the same work, and also provides the more succinct gloss "frontiers are loosely-administered spaces rich in resources, coveted by non-residents" (Geiger 2008a, 78). He acknowledges that his definition derives from the fuller specification of Hvalkof (2008) in an essay later in the same volume.
3 We will simply use the state as a term to cover all these political formation in the following argumentation.

References

Acciaioli, Greg. 2000. "Kinship and Debt: The Social Organization of Bugis Migration and Fish Marketing at Lake Lindu, Central Sulawesi." In *Authority and Enterprise among the Peoples of South Sulawesi* (Verhandelingen 188), ed. Greg Acciaioli, Kees van Dijk, and Roger Tol. Leiden: KITLV [Koninklijk Instituut voor Taal-, Land- en Volkenkunde] Press. Pp. 210–239.

——. 2007. "From Customary Law to Indigenous Sovereignty: Reconceptualizing the Scope and Significance of Masyarakat Adat in Contemporary Indonesia." In *The Revival of Tradition in Indonesian Politics: The Deployment of Adat from Colonialism to Indigenism*, ed. Jamie S. Davidson and David Henley (Routledge Contemporary Southeast Asia Series). Abingdon, Oxon: Routledge. Pp. 295–318.

——. 2008. "Strategy and Subjectivity in Co-management of Lore Lindu National Park (Central Sulawesi, Indonesia)." In *Biodiversity and Human Livelihoods in Protected Areas: Case Studies from the Malay Archipelago,* ed. N. Sodhi, G. Acciaioli, M. Erb, and A. K.-J. Tan. Cambridge: Cambridge University Press. Pp. 266–288.

——. 2009. "Conservation and Community in the Lore Lindu National Park: Customary Custodianship, Multi-Ethnic Participation, and Resource Entitlement." In *Community, Environment and Local Governance in Indonesia: Locating the Commonweal*, ed. Carol Warren and John McCarthy. Abingdon, Oxon: Routledge. Pp. 88–118.

Aggarwal, Ravina. 2004. *Beyond Lines of Control: Performance and Politics on the Disputed Borders of Ladakh, India*. Durham and London: Duke University Press.

Aggarwal, Ravina and Mona Bhan. 2009. "'Disarming Violence': Development, Democracy, and Security on the Borders of India." *The Journal of Asian Studies* 68(2): 519–542.

Baruah, Sanjib. 2003. "Nationalising Space: Cosmetic Federalism and the Politics of Development in Northeast India." *Development and Change* 34(5): 915–939.

Beek, Martijn van. 1999. "Hill Councils, Development and Democracy: Assumptions and Experiences from Ladakh." *Alternatives* 24(4): 435–439.

Bhatnagar, Yash Veer, C. M. Seth, Jigmet Takpa, Saleem Ul-haq, Tsewang Namgail, Sumanta Bagchi, and Charudutt Mishra. 2007. "A Strategy for Conservation of the

Tibetan gazelle *procapra picticaudata* in Ladakh." *Conservation and Society* 5: 262–276.

Brouwer, Melattie. 1977. *60 tahun Bala Keselematan di Sulawesi Tengah*. Bandung: Bala Keselematan.

Cribb, Robert. n.d. [c. 1990] *The Politics of Environmental Protection in Indonesia* (Working Paper No. 48). Clayton, VIC: The Centre of Southeast Asian Studies, Monash University.

Davis, Gloria. n.d. [c. 1973] "The People and Legends of Lake Lindu, Central Sulawesi, Indonesia." Unpublished manuscript produced for Department Pendidikan dan Kebudayaan, Tingkat Propinsi, Palu.

Dawa, Sonam. 1999. "Economic Development of Ladakh: Need for a New Strategy." In *Ladakh: Culture, History and Development between Himalayan and Karakoram*, ed. Martijn van Beek, Kristoffer Brix Berielsen, and Poul Pedersen. Aarhus N: Aarhus University Press. Pp. 369–378.

Dollfus, Pascale. 2013. "Transformation Process in Nomadic Pastoralism in Ladakh." *Himalaya, the Journal of the Association for Nepal and Himalayan Studies* 32: 61–72.

Drew, Frederick. 1875. *The Jummoo and Kashmir Territories*. Delhi: Oriental Publishers.

Erb, Maribeth, Priyambudi Sulistiyanto, and Carol Faucher (eds). 2005. *Regionalism in Post-Suharto Indonesia*. New York: RoutledgeCurzon.

Geiger, Danilo. 2008a. "Turner in the Tropics: The Frontier Concept Revisited." In *Frontier Encounters: Indigenous Communities and Settlers in Asia and Latin America* (IWGIA Document No. 120), ed. Danilo Geiger. Copenhagen: International Work Group for Indigenous Affairs, Swiss National Centre of Competence in Research North–South. Pp. 77–215.

——. (ed.) 2008b. *Frontier Encounters: Indigenous Communities and Settlers in Asia and Latin America* (IWGIA Document No. 120). Copenhagen: International Work Group for Indigenous Affairs, Swiss National Centre of Competence in Research North–South.

Goodall, Sarah K. n.d. *From Plateau Pastures to Urban Fringe: Sedentarisation of Nomadic Pastoralists in Ladakh, Northwest India*. PhD thesis, University of Adelaide, 2007. http://digital.library.adelaide.edu.au/dspace/bitstream/2440/46332/2/01front.pdf (accessed May 23, 2013).

Haller, Tobias, Greg Acciaioli, and Stephan Rist. 2015. "Constitutionality: Conditions for Crafting Local Ownership of the Institution-Building Processes." *Society and Natural Resources: An International Journal*. http://www.tandfonline.com/doi/full/10.1080/08941920.2015.1041661.

Harvey, Barbara Sillars. 1977. *Permesta: Half a Rebellion*. Ithaca, NY: Cornell Modern Indonesia Project, Southeast Asia Program, Cornell University (Monograph Series, Publication No. 57).

——. n.d. "Tradition, Islam, and Rebellion: South Sulawesi 1950–1965." Unpublished PhD thesis, Department of Government, Cornell University, 1974. Ann Arbor: University Microfilms.

Hvalkof, Søren. 2008. "Colonization and Conflict on the Amazon Frontier: Dimensions of Interethnic Relations in the Peruvian Montana." In *Frontier Encounters: Indigenous Communities and Settlers in Asia and Latin America*, ed. Danilo Geiger (IWGIA Document No. 120). Copenhagen: International Work Group for Indigenous Affairs, Swiss National Centre of Competence in Research North–South. Pp. 219–286.

Kaudern, Walter. 1940. "The Noble Families or Maradika of Koelawi, Central Celebes." *Etnologiska Studier* 11: 30–124.

Kruyt, Albertus Christiaan. 1938. *De West-Toradjas op Midden-Celebes*. 4 vols. Amsterdam: De N.V. Noord-Hollandsche Uitgevers-Maatschappij. (Verhandelingen Der Koninklijke Nederlandsche Akademie van Wetenschappen te Amsterdam, Afdeeling Letterkunde, Nieuwe Reeks 40.)

Lattimore, Owen. 1968. "The Frontier in History." In *Theory in Anthropology: A Sourcebook*, ed. Robert A. Manners and David Kaplan. Chicago: Aldine Publishing. Pp. 374–385. (Originally published in *Relazioni del X Congresso Internazionale di Scienze Storiche* (Florence, 1956).)

Li, Tania Murray. 2000. "Articulating Indigenous Identity in Indonesia: Resource Politics and the Tribal Slot." *Comparative Studies in Society and History* 42(1): 149–179.

——. 2002. "Local Histories, Global Markets: Cocoa and Class in Upland Sulawesi." *Development and Change* 33(3): 415–437.

——. 2007. *The Will to Improve: Governmentality, Development, and the Practice of Politics*. Durham and London: Duke University Press.

Mishra, Charudutt, Herbert H. T. Prins, and Sipke E. van Wieren. 2001. "Overstocking in the Trans-Himalayan rangelands of India." *Environmental Conservation* 28: 279–283.

Rizvi, Janet. 1999. *Trans-Himalayan Caravans: Merchant Princes and Peasant Traders in Ladakh*. New Delhi: Oxford University of Press.

Rowe, William T. 2007. "Owen Lattimore, Asia, and Comparative History." *The Journal of Asian Studies* 66(3): 759–786.

Sabharwal, Alka. n.d. "Barren Frontiers Pristine Myths: The Cultural Politics of Wildlife Conservation in the Indian Trans-Himalaya." PhD thesis, University of Western Australia, 2015.

Sangaji, Arianto. 2000. *PLTA Lore Lindu: Orang Lindu Menolak Pindah*. Yogyakarta: Pustaka Pelajar kerjasama dengan Yayasan Tanah Merdeka, ED Walhi Sulawesi Tengah.

Scott, James Campbell. 1998. *Seeing Like a State: How Certain Schemes to Improve the Human Condition Have Failed*. New Haven: Yale University Press.

Sheikh, Abdul Ghani. 1999. "Economic Conditions in Ladakh during the Dogra Period." In *Ladakh: Culture, History and Development between Himalayan and Karakoram*, ed. Martijn van Beek, Kristoffer Brix Berielsen, and Poul Pedersen Aarhus: Aarhus University Press. Pp. 339–349.

Turner, Frederick Jackson. 1994. "The Significance of the Frontier in American history." In *Rereading Frederick Jackson Turner*, ed. John Mack Faragher. New York: Henry Holt. Pp. 31–60.

Tsing, Anna Lowenhaupt. 2005. *Friction: An Ethnography of Global Connection*. Princeton: Princeton University Press.

Valentijn, Francois. 1724–1726. *Oud en Nieuw Oost-Indiën, Vervattende een Naaukerige en Uitvoerige Verhandelinge van Nederlands Mogentheyd In die Gewesten, Benevens Eene Wydlustige Beschryvinge Der Moluccos, Amboina, Banda, Timor, en Solor, Java, en alle de Eylanden onder de zelve Landbestieringen behoorende; het Nederlands Comptoir op Suratte, en de Levens der Groote Mogols . . .* 1st ed. 5 vols. Dordrecht & Amsterdam: Joannes Van Braam En Gerard Onder De Linden, Boekverkoopers.

Vorstman, J. A. n.d. "Aanvullende Memorie van Overgave van den Controleur van Paloe. Unpublished Memorie van Overgave, Algemeen Rijksarchief, Tweede Afdeling – Koloniale Aanwinsten," Memories van Overgave Nederlands-Indië KIT Nr. 690, 28 February 1935.

Wallerstein, Immanuel. 1974. "The Rise and Future Demise of the World Capitalist System: Concepts for Comparative Analysis." *Comparative Studies in Society and History* 16: 387–415.

Watling, Dick and Y. Mulyana. 1981. *Lore Lindu National Park Management Plan 1981–1986* (Report prepared by the World Wildlife Fund – Indonesia Programme [Bogor] for the Directorate of Nature Conservation, Directorate General of Forestry, Republic of Indonesia).

Webb, Walter Prescott. 1952. *The Great Frontier*. Cambridge, MA: The Riverside Press.

Wolf, Eric R. 1982. *Europe and the People without History*. Berkeley: University of California Press.

Part II

Empires

3 Nation and race in the historical juncture of the Haitian Revolution

Dannelle Gutarra

Introduction

In the year 1791, a group of slaves and former slaves defied the limitations attributed to their condition and organized the most successful slave revolt of all time. The tumultuous tale of the Haitian Revolution underlines how the diverse social groups of the colony of Saint-Domingue seized the context of the French Revolution in order to voice their long-lived desperate claims: the white colonists' demand for political autonomy, the longing for French citizenship by the free men of color,

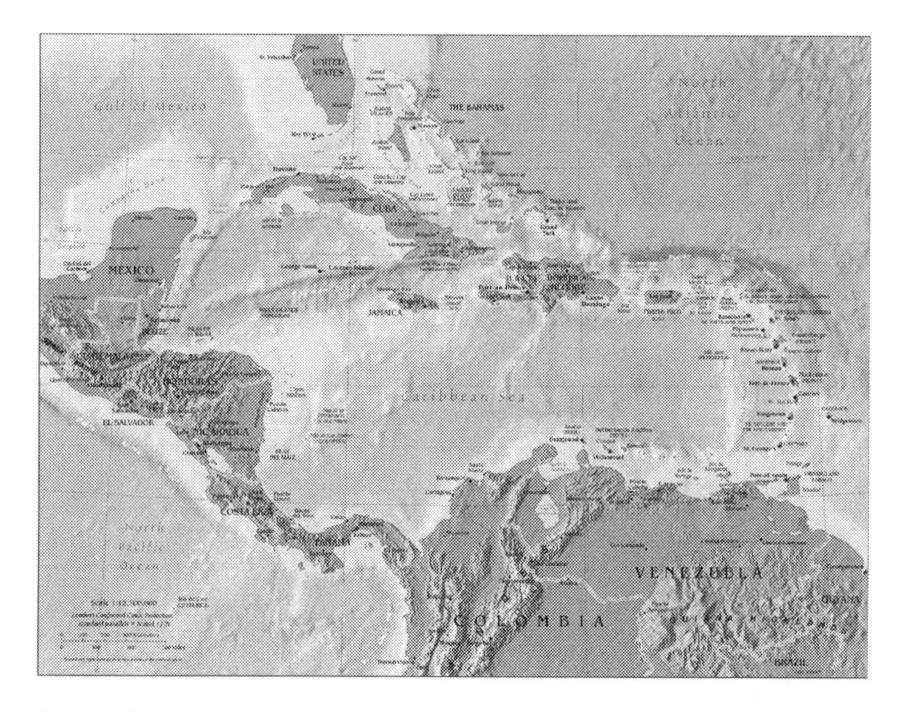

Figure 3.1 Map of the Caribbean.

Source: United States Central Intelligence Agency's World Fact book.

and the struggle of slaves for general liberty. Meanwhile, the French government strove to guarantee its sovereignty in the colony; the negotiations of power with the different social-racial groupings of Saint-Domingue and the persistence of the slave insurrections led to the pioneering abolition of slavery in 1794.

One of the leaders of the Haitian Revolution, Toussaint L'Ouverture, even became governor of the colony as a method for the pacification of the slave populations. Nevertheless, L'Ouverture transgressed the negotiations of power with the French metropole when he implemented the Constitution of 1801, which permanently abolished slavery in the colony, innovatively prohibited racial discrimination, established a level of political autonomy for the territory, and named L'Ouverture governor for life. The Napoleonic administration responded to this challenge with an aggressive expedition, the overthrow of L'Ouverture, and the acceleration of the restoration of slavery. The declared war was won by the determination of the former slave population to protect their freedom and culminated with the independence of Haiti in 1804.

During the historical episode of the Haitian Revolution, the diverse social-racial groupings of the colonial population of Saint-Domingue adhered to distinct conceptions of nation, frontier, and citizenship, which were simultaneously mediated by conflicting ideas of race. The power relations between the colonial population and the French metropolitan government show evidence of the escalation of pseudo-scientific theories that would gain even more prominence during the nineteenth century. The main purpose of this chapter is to explore the connections between ideas of frontier and race in colonial relations in Saint-Domingue during the historical juncture of the Haitian Revolution. This investigation is largely based on primary sources available in the Nemours Collection of Haitian History and the Bibliothèque Nationale de France.

This chapter will first identify the constructions of nation and frontier articulated by the different social groups of the colony of Saint-Domingue in the transatlantic debate about general liberty. Second, it will explore the idea of nation set forth after the abolition of slavery in 1794 and the influence of the notion of national regeneration in French revolutionary discourse. Third, it will examine the oscillations in ideas about the frontier during the fall of Toussaint L'Ouverture's government. Lastly, it will analyze the historical pertinence of the Haitian Revolution in the studies of the tie between "borders" and racialist thought.

Conflictive discourses of nation and frontier in the dawn of the Haitian Revolution

According to *Imagined Communities* by Benedict Anderson, nationalism is a modern construction that builds a sense of limited and sovereign community and that was generated during the historical juncture of the Enlightenment and the French Revolution (Anderson 1983, 7). Meanwhile, Immanuel Wallerstein argues in "The French Revolution as a World-Historical Event" that the French Revolution transformed the "cultural apparatus" of world-systems, even though it did not substantially modify the political, social, and economic condition of France (Wallerstein

1990, 130). The historical juncture of the French Revolution has been analyzed by multiple scholars as an episode that is both distinguished by constant transformation (Berman 1982, 17) and the celebration of grand narratives (Lyotard 1979, xxiii), like the modern nation; it is therefore deemed a historical juncture in which the notion of frontier is constantly negotiated and mutated. It is relevant to explore the oscillations in the ideas of the frontier during the most radical revolution of this historical juncture, the Haitian Revolution.

Lynette Russell (2001) defines the frontier as an idea and proposes that people involved in the building of frontiers do not necessarily have an awareness of its symbolic components; Russell highlights the opportunism in this phenomenon and describes the generation of frontiers in "settler-colonies" as "premised on the 'logic of elimination'" of the eminent domain of indigenous populations, even though it simultaneously leads to problematic "spatial coexistences" (Russell 2001, 2). In the case of the Americas, various scholars have contended that current tensions within frontiers of the Americas have arisen because of the tendency of oppression toward "ethnic minorities" that is inherited from colonial times (Doyle and Villela Pamplona 2006, 8). Frontiers and the racial relations that occur within them are not only based on economic value, but also acquire "strategic or symbolic importance to the state" (Wilson and Donnan 1998, 9).

In the case of colonialism, the symbolic implications of frontiers are inexorably tied to the implementation of politics of difference. In their volume *Empire in World History: Power and the Politics of Difference*, Jane Burbank and Frederick Cooper affirm that the "concept of empire presumes that different peoples within the polity will be governed differently" (Burbank and Cooper 2010, 8). According to the historians, empires are preserved by the core element of their "politics of difference"; the politics of difference are enforced through a calculated racialization of colonial hierarchies. It was of essential value for imperial rules to stimulate loyalty from colonial populations, even though this persuasion would be evocative of metropolitan control and not of resemblance. The politics of difference were vital not only for colonial exploitation, but for the transnational projection of the advancement of the empire and its authority, materializing a historical geography (Guha 2002, 12).

Within the historical geography designed by modern thought, colonies were in limbo. "The White man's burden" justified many colonizing enterprises; elaborate justifications of colonialism would have to confront the following paradox: a colonial space had to be "less inferior" just by the cleansing force of the colonizer's agency, which means that colonies had to be perpetually projected as spaces that were simultaneously evolving and paralyzed in time. Colonies had to be emblems of both discipline and negligence in order to validate metropolitan authority. This paradox generated hesitations in ideas of "colonial frontiers," since ideological frontiers could be symbolically framed within the metropole or could be alternatively emphasized by utilizing "distance" as a euphemism for "difference." The primary sources of the Haitian Revolution provide evidence of the centrality of the concept of "frontier" in negotiations of power during the diverse phases of the revolution that was analogous to French revolutionary history in the colonial actuality.

The French Revolution functioned as an impulse for the claim made by the *grands blancs* (white colonists) for the political autonomy for the colony. The longing of the white colonists for self-government already had antecedents before the outbreak of 1791, and had been concretized in brief rebellions against the French monarchy in the 1670s, 1720s, and 1760s (Geggus 2009, 12). The *grands blancs* resented the colonial structure that favored metropolitan profit; they pursued greater sovereignty for Saint-Domingue in order to control the vibrant colonial economy. The colonizers distanced themselves from metropolitan agency and verbalized their own subjectivities among the multiplicity of subjectivities that emerged in colonial actualities (Stoler 1989, 137). The vocal *grands blancs* proposed in the transatlantic debate about general liberty that the French Revolution had motivated their urge to reevaluate the limits of colonial relations, referring to the "reverberations" of the Revolution.[1] The colonizers recognized a double discourse from French revolutionary thought, or an incongruity in the development of democratic procedures and the simultaneous obstinacy in the persistence of despotic colonial relations. The *grands blancs* therefore insinuated that politics of difference were due to the distance of the colony from France.

The colonizers of Saint-Domingue decided then to use the argument of the distance between the colony and the metropole against the French government to support their yearning for self-determination, emphasizing the fact that "an immense sea separated us from France."[2] The Atlantic Ocean became a signifier of the distinction between the commercial interests of continental Frenchmen and the needs of the colonizers of Saint-Domingue. This "distance" is employed as the principal argument for the right of *grands blancs* to political autonomy; this argument is also intertwined with allusions to racialist theories of geographical determinism, referencing not only colonial frontiers but also racial and behavioral "frontiers" (Curtin 1964).

Due to the negotiations of power, discursive hesitations predictably arose. A speech given by deputies to the National Assembly on September 30, 1790 depicted Saint-Domingue as "a great part of this Empire," celebrating their "attachment to the Motherland."[3] This speech also presented two discursive paradigms of dissertations by the *grands blancs* of Saint-Domingue: allusions to the oath of the colony to "the Nation, the Law, and the King" and the statement "no commerce, no Colonies; likewise no Colonies, no commerce." The document employed diverse terms to describe the relationship between France and the colony: "attachment," "union," and "family"; when cultivating the metaphor of being a "family," the speech affirmed that France is the "mother" of the colonial population of Saint-Domingue, not a mere "godmother," which is why the *grands blancs* allegedly did not pursue political independence. The speech summarized its claim when it stated: "Our needs give us a right; our wishes entitle us to participate in the regeneration of this beautiful empire." In this case, the *grands blancs* stress on the "union" along with an implicit threat to finalize it if their claims are not recognized, which would have a resulting impact upon an economy based on colonial productivity; they contextualized the actuality of Saint-Domingue within the "regeneration of the empire." These discursive paradigms were to be recreated time after time by the *grands blancs* until the independence of Haiti in 1804.

The outburst of the French Revolution provided the occasion for the growing population of free men of color to request French citizenship and to battle the racial legislation that marginalized them from the economy of Saint-Domingue (Trouillot 1992, 154). Their claim was grounded on the guarantee of equality in first article of the Declaration of the Rights of Man and of the Citizen, which conceded both the rights of liberty and equality (Gouvernement Français 1789, 22). The discourse of the vocal free men of color of Saint-Marc frequently mentioned their self-identification as Frenchmen, their devotion to the French metropole, and their predilection for their white ancestry. "We suggest that, in all the French Empire, no class of men is more faithfully tied to the motherland than the citizens of color. The half of their origin rendered all its treasure, all its glory."[4] Just like the *grands blancs*, the free men of color referred to their "attachment" in the articulation of their claim; they also contextualized it with an appropriation of racialist theories of biological determinism.

The struggle for general liberty of the slave population of Saint-Domingue propelled the publication in France of reports, speeches, and books about "solutions" to the discords of the colony. A decree by the French National Assembly responded to colonial claims by asserting, "Reciprocal attachment, common advantages, inalterable fidelity: there it is, People of the Colonies, that which she (the Nation) promises you and that which she demands of you." The National Assembly's decree emphasized the "reciprocity of the attachment" to minimize the revolutionary claims of the social groups of Saint-Domingue.[5] Armand Guy-Kersaint advised the French government to take a proactive stance on independence movements and to focus on the protection of the "integrity of borders," intensifying maritime forces and surveillance of the frontiers of the colony.[6] The Committee of Public Safety of France concluded that colonial conflicts had emerged because of the "distance" from France, employing racialist theories of geographical determinism to portray "the effervescence of passions, under a burning sun" (Laurent 1953, 245).[7] French Admiral Louise Thomas Villaret de Joyeuse proclaimed "Saint-Domingue is also a Vendée to reconquer," establishing an analogy between the colony and a counterrevolutionary department of France.[8]

The French government responded to revolutionary claims with the expedition of three new civil commissioners from France: Mirbeck, Roume, and Saint-Léger. On December 3, 1791, the civil commissioners gave their respective speeches in front of the General Assembly of Saint-Domingue.[9] Mirbeck referred to the relationship between the colony of Saint-Domingue and the French government as an "indissoluble attachment" and as a "union that must be eternal." Roume defined Saint-Domingue as one of the "most distant parts from the center of the empire." Saint-Léger stated: "Saint-Domingue is a section too important for the French empire, and the French in the metropole and in the Colonies are now connected with very indissoluble knots." The strategic usage of concepts like "eternal union" and "indissoluble knots" connotes the purpose of the mission to operate as pacification and protection of metropolitan authority over the colony after the initiation of revolutionary turmoil. Throughout the official documents of the civil commissioners, their discourse celebrates the attachment to the "name of the Nation, the Law, and the King,"[10] constructs French nationalism as obedience to the law,[11] and employs the metaphor of anti-abolitionists belonging to "the empire of law and reason."[12]

On December 21, 1791, the civil commissioners released a document that described colonial relations and justified the perpetuation of slavery after the Declaration:

> Our colonies, in the actual sense of that word in France, are integral parts of the empire; but, who says colony means also a part separated from the center of the empire by a vast ocean; a part populated by whites, blacks, the free, the enslaved, and the mixes of whites and blacks; a part that, because of the nature of its population, needs a local constitution for the state of the existence of slaves.[13]

This last statement encapsulates the oscillations in ideas of frontier during the dawn of the Haitian Revolution. The pronouncement defines Saint-Domingue as an "integral part of the empire" and simultaneously highlights the "vast ocean" as its frontier. Not only does it refer to the sea to generate "frontiers" between the metropole and the colony, but it also rationalizes politics of difference based on the "nature of the population," establishing "racial frontiers." These "frontiers" serve as the structure for the "borders of slavery" that frame the colonial space in which the Declaration does not apply. The civil commissioners promoted submission to the law as something that bound France and Saint-Domingue and emphasized on the "indissoluble" quality of colonial relations as a mode to pacify the revolutionary territory. As Homi Bhabha argues:

> the postcolonial space is now "supplementary" to the metropolitan centre; it stands in a subaltern, adjunct relation that doesn't aggrandize the presence of the West but redraws its frontiers in the menacing, agonistic boundary of cultural difference that never quite adds up, always less than one nation and double.
>
> (Bhabha 1994, 241)

Territoriality in the idea of nation set forth after the abolition of slavery

Nations and Nationalism since 1780 by Eric Hobsbawm classifies the historical juncture of the French Revolution as the era of the tie between nation and frontier. "The equation nation=state=people, and especially sovereign people, undoubtedly linked nation to territory, since structure and definition of states were now essentially territorial" (Hobsbawm 1990, 19). Hobsbawm deems an illustrative contradiction the fact that modern French nationalism was intertwined with imperialism: the revolutionary period was one of much volatility of the frontiers. More importantly, the revolutionary accent on the "people" did not clarify what the concept of the "people" meant and its relation with frontiers. The ambiguity in the delimitation of the "people" and the volatile nature of frontiers are exemplified in the episode of the abolition of slavery in the colony of Saint-Domingue by Civil Commissioner Léger-Félicité Sonthonax, in response to slave insurrections.

Sonthonax's proclamation of the abolition of slavery in Saint-Domingue on August 29, 1793, which was ratified by the National Convention on February 4, 1794, is an official document that encapsulated the discursive paradigms of his mission in the Caribbean colony. In this text, Sonthonax contended that the age of French racism had concluded in both the colony and the metropole, and he attributed the termination of this mode of thought to republicanism and revolutionary principles:

> The French Republic wants the freedom and equality of all men regardless of color; the Kings are only happy among slaves: they are the ones that, on the coasts of Africa, sold you to whites; they are the tyrants of Europe that would want to perpetuate this infamous traffic. The Republic adopts you to the number of its children; the Kings only aspire to cover you with chains or destroy you.[14]

The civil commissioner distinctly established a parallelism between abolitionism and the republican cause, connoting the "frontiers of slavery." Sonthonax argued that the French Nation was the one that could guarantee the citizenship and freedom of emancipated slaves; he concurrently associated the institution of slavery with the Old Regime. Therefore, the declaration aimed to trigger anti-royalist sentiments and ally the revolutionary slave population with the republican cause. The civil commissioner urged emancipated slaves to fight counterrevolutionary movements and to reject the concept of independence in response to the concession of French citizenship. Sonthonax also antagonized other European nations as slave-driven markets that would eliminate the triumph of the end of slavery, in order to prevent pacts with the English or the Spanish. Nonetheless, emancipated slaves would have to labor for their former masters unless they were judicially allowed to leave their previous dwellings. This proclamation substantiated multiple discursive paradigms of Sonthonax's mission in Saint-Domingue: an insistent reference of racialism as related to the Old Regime, a moral conception of work, and the praise of revolutionary discourse, anti-royalist ideas, and republicanism.

Throughout his mission in Saint-Domingue, Sonthonax functionally decided to emphasize in his discourse the "frontiers of slavery" in order to protect colonial frontiers. In order to ally both the population of the free men of color and the rebel slaves that were victorious in multiple insurrections, the civil commissioner not only cultivated "frontiers" that separated France from other European powers, but also insisted that the narrative of Saint-Domingue fell within those "frontiers." The power relations between the civil commissioner and the social-racial groupings of Saint-Domingue persistently displayed a parallelism between factions during the French Revolution. While free men of color and former slaves were regarded as French citizens and supporters added to the Nation, *grands blancs* were vilified as the "aristocracy of colors."[15]

In *Politics, Culture and Class in the French Revolution*, Lynn Hunt proposes that the French Revolution was fundamentally a political episode of "national regeneration," since it marked the advent of a new political rhetoric that aimed to generate a "new nation" with no reference to the past (Hunt 1984, 27). Sonthonax's proclamations indeed depicted the suppression of slavery as a rupture with the royalist past and a commemoration of the revolutionary notion of national regeneration. The primary sources of Sonthonax's mission in Saint-Domingue portrayed the colony as one permanently tied to the destiny of the French metropole. "The Colony of Saint-Domingue is a precious part of the French Empire; it was populated by the French, and its destiny is inseparable from the fate of the motherland."[16] The revolutionary concept of national regeneration therefore altered French colonial history, since colonial territories were abstracted as figurative manifestations of metropolitan yearnings (Nelson 2013, 74). John Breuilly defines nationalism as a "pseudo-solution to the problem of the relationship between state and society" (Breuilly 1985, 69). With his French nationalist contextualization of colonial affairs, Sonthonax intended to precisely annihilate manifest discords between the colonial state and the revolutionary society of Saint-Domingue. There was therefore a transposition of sense in which the colony of Saint-Domingue became a complex allegory of the victory of French revolutionary thought. In official documents of Sonthonax's mission in the colony, the proclamation of the abolition of slavery was predominantly concerned with the formation of a new *French Frontier*.

Oscillations in ideas of frontier during Toussaint L'Ouverture's narrative

Toussaint L'Ouverture encapsulated his projection in his official correspondence with the French government when he described the instruction he desired for his sons in Paris. "I hope they are studious in the sciences which will adorn their spirits; hope they become reputable citizens, attached to the Republic, to their duties, and to the religion in which they were born!" (Laurent 1953, 449).[17] C. L. R. James would spark the historiographical interpretation of this historical character as one molded by revolutionary thought (James 1963, 290). Similarly, Aimé Césaire would affirm that it was L'Ouverture who authentically "vivified" French revolutionary concepts (Césaire 1960, 417). Doris Garraway argues that L'Ouverture employed French nationalist rhetoric as the foundation of the discourse of the struggle for general liberty (Garraway 2008, 75).

In his correspondence, his proclamations, and his memoirs, the revolutionary leader recurrently mentioned his "attachment" to France. He also depicted his temporary alliance with Spain as an unintentional mistake (Laurent 1953, 104) and the idea of independence as too much of a burden for the colony of Saint-Domingue (Laurent 1953, 446). He recreated the allegory of analogical territoriality in "both hemispheres," referred to the concept of "reverberations" of the Revolution, and intertwined them with the promotion of the applicability of laws in both sides of the Atlantic. "It is presumable that the victories of the French, which reverberate in both hemispheres, contribute much to imprint in the hearts of its inhabitants the principles of liberty, equality, and submission to the laws of the Republic" (Laurent 1953, 203).[18] Césaire would distinguish L'Ouverture's hesitancy to pursue independence as a defect of his rule; he proposed his mandate was exempt of a political discourse that could promote his control over the colonial mass that yearned for radical social transformation (Césaire 1960, 371). James was even more categorical in his disparagement of L'Ouverture's attachment to the French government, portraying his perseverance in this political tie as the germ of his downfall and a great historical tragedy (James 1963, 289). L'Ouverture would reaffirm his "attachment" to France even in his memoirs written shortly before his death in captivity (Saint-Rémy 1853, 100). Interestingly, the governor consistently used the word "attachment" differently from how it was typically employed during the dawn of the Revolution; instead of connoting an attachment to the frontiers of France, L'Ouverture mostly referred to a personal emotional attachment to the metropole, whether in relation to himself or his longing for his children and his "people." This emotional attachment could easily culminate in resistance if France restored slavery.

On January 28, 1801, Toussaint L'Ouverture as governor conquered Spanish Santo Domingo, extending the frontiers of Saint-Domingue to all coasts of the former Hispaniola; Napoleon Bonaparte received the news as a challenge, since, even though the Treaty of Basel of 1795 allowed this takeover, the First Consul had not given orders to formally alter the borders of Saint-Domingue (Dubois 2004, 253). The Constitution of 1801, signed by Toussaint L'Ouverture and implemented since July 3 of that year without the consent of the French government, is the definitive manuscript that contested the preexistent colonial relations and the restrictions of his post as colonial governor. The Constitution dictated the permanent abolition of slavery and the prohibition of racial discrimination, declaring the colonial territory a "free land."

> Art. 3.—There cannot be any slaves in this territory; slavery is abolished forever. Every man is born, lives, and dies free and French. Art. 4.—Every man, whichever his color, is eligible for any job. Art. 5.—There is not any other distinction that the virtues, talents, or other superiority that the law provides in the exercise of public service. The law is the same for everyone, be it as punishment, be it as protection.
>
> (Janvier 1886, 8)[19]

This regulation aimed to safeguard the permanency of the abolition of slavery and to concretize French citizenship for emancipated slaves of Saint-Domingue, rights already granted in the awaited decree of February 4, 1794 (Hunt 1996, 115). The Constitution of the Year III (1795) of France then outlawed slavery in its Declaration of Rights, the declaration that was subsequently removed in the first

Napoleonic constitution, the Constitution of the Year VIII (1799) (Gouvernement Français 2010, 105). Neither the decree of February of 1794 nor the Constitution of the Year III (1795) emphasized the perpetuity of the rights of freedom and citizenship for the freed slaves of French dependencies. Meanwhile, the Constitution of the Year VIII (1799) eliminated the preliminary declaration of rights, did not warrant the abolition of slavery, and dictated that colonies would be ruled by special laws. The language of the Constitution of 1801 is much more categorical in the perennial illegality of slavery and racial discrimination, accentuating the territoriality of the abolition of slavery; the implementation of the constitution without the authorization of the French metropole implies suspicion of the Napoleonic motives.

Toussaint L'Ouverture dared to move forward with a constitution that had not been approved by the French government and extended the frontiers of the colony without metropolitan consent. These political actions represented an overt statement to the effect that the redefinition of colonial relations was envisioned with or without the approval of the French government. As Sybille Fischer contends, even though it retained political ties with France, the constitution did not elaborate on "what France's authority actually amounted to" (Fischer 2004, 229). It was the instant in which Toussaint L'Ouverture challenged metropolitan power, surpassed the limits of his expected obedient service to the French Republic, and situated himself as a revolutionary leader with political agency despite his racial and colonial circumstance that he crossed contemporary ideological frontiers.

The official revelation to Toussaint L'Ouverture of General Charles Leclerc's expedition to overthrow him was concretized in a missive from the First Consul. The main argument of the letter is that L'Ouverture's response to the expedition would be the indicator of the legitimacy of his "attachment" and subservience to the French metropole. According to the missive, the reasoning behind the expedition was the challenge posed by the Constitution of 1801, particularly the articles that granted autonomy to the colonial government. "The constitution you have done, containing many good things, contains things that are contrary to dignity and the sovereignty of the French people, from which Saint-Domingue is no more than a portion" (Leclerc 1937, 307).[20] The First Consul conceived Saint-Domingue as a "portion," a territory subordinated to France that did not possess the right to implement a constitution and that should have been instead regulated by special laws determined by the French government.

On November 13, 1801, Napoleon submitted a letter to the Minister of Foreign Affairs; this document articulated the urgency of the expedition to Saint-Domingue. Bonaparte envisioned that, if the French government failed to coordinate an expedition by the year 1802, he would be forced to acknowledge Toussaint L'Ouverture's authority, to surrender his sovereignty over the colony, and to tolerate the advent of a black state in the Caribbean. The First Consul asseverated that the "liberty of the blacks of Saint-Domingue" was vital and strategic to ensure the permanence of France in the New World. The most significant statement in this document is the affirmation that the mission of Toussaint L'Ouverture's fall was a symbolic enterprise of the historicist discourse of white supremacy. "That, in the part that moved me to annihilate the government of the blacks in Saint-Domingue, I was less guided by considerations of commerce and finances than because of the necessity of suppressing, in all parts of the world, any kind of seed of restlessness or disturbance" (Bonaparte 2006, 851).[21]

General Leclerc's expedition was an international spectacle that included the deployment of 19,000 soldiers. In his correspondence with the First Consul, Leclerc refers to Saint-Domingue as "a fourth of France" (Leclerc 1937, 55)[22] and insistently alludes to the "ignorance" in France of matters related to the colony (Leclerc 1937, 245). A letter written by Leclerc on August 6, 1802 reveals Napoleon's plan to ultimately restore slavery in Saint-Domingue. Leclerc recriminates the First Consul for enabling the inhabitants of Saint-Domingue to detect the dispositions of the French government for their colonies in America (Leclerc 1937, 202). After the Peace of Amiens, a law had been ratified, which transformed France into a slave-holding republic once again (Gouvernement Français 1802, 1). Although the law of May 20, 1802 did not pertain to Saint-Domingue, it originated a sense of greater uncertainty, especially because this law revoked the decree of February 4, 1794, which had abolished slavery in all French colonies.

In a missive written to the Minister of the Navy on August 7, 1802, Napoleon Bonaparte articulated his rationalization of the restoration of slavery in French Guyane. His argumentation demonstrates that the First Consul deemed the institution of slavery a vital factor for the perpetuity of colonialism, since he believed it pacified the fear of the colonizer that the potential energy of the "blacks" could be converted from obedience to insurrection.

> A man destined to spend his life in the colonies must feel that, if the blacks were capable of maintaining the colonies against the Englishman, they would direct their fury against us, they would kill the whites, they would forever threaten to burn down our properties, and they would not present any guarantee to commerce, which would not offer any capital and remain without confidence.
>
> (Bonaparte 2006, 1059).[23]

In Napoleon's worldview, slavery signified a means to perpetuate control over black colonial populations, to guarantee sufficient French residents in the colonies, and to incite a transnational sense of trust in colonial efficiency and the projection of triumph of white supremacy; the "frontiers of slavery" had to recreate colonial borders and to guarantee economic structures based on racial hierarchies. Therefore, the frontiers of Saint-Domingue, the richest colony in the Americas, had to be protected by the French metropole, and slavery had to be restored as a way to inspire transnational admiration in the narrative of the French Empire; the symbolic significance of colonial frontiers had much to do with the building of the Napoleonic Empire and the sophistication of its racialist discourse. Toussaint L'Ouverture challenged Napoleonic sovereignty and the racialist foundations of French colonialism, since he reversed the emphasis in the territoriality of slavery when he established his "frontiers of liberty."

The ideological emphasis on frontiers as related to "slavery" and "race" is not only exemplified by the enterprise of L'Ouverture's public demise, but also by the narrative of genocide that followed. On November 2, 1802, General Rochambeau took command of Saint-Domingue after Leclerc's death by yellow fever, and initiated a "war tactic of genocide," purchasing bulldogs from Cuba that had been trained to kill; his intention was to rid the territory of slaves that lacked docility (Fick 1990, 229). After the mobilization of the revolutionary discourse to the cause of independence, Jean-Jacques Dessalines became the first ruler of independent Haiti on January 1, 1804; his discourse proclaimed Haitians as the ones who had "avenged America" and he rationalized the genocide of approximately 4,000 white colonizers (Jenson 2011, 81). Hence, in the decade of the 1800s, two genocidal enterprises were carried out, both intending to purge Saint-Domingue of a racial agent from within its frontiers.

Conclusions

Both the essay "Race and Nationalism" by Étienne Balibar (Balibar 1991, 57) and the text *Genealogy of Racism* by Michel Foucault (Foucault 1993, 65) define nationalism as a philosophy of history that is grounded on a war of races. In the context of the Haitian Revolution, this war of races was not only justified by a metropolitan enterprise of protection for the physical frontiers of imperial power, but also by the intellectual formulations that tied these borders to the limits of racial diversity, slavery, and colonial subordination. This accent on frontiers was intended to alleviate the paradoxes in the projection of the colonizing enterprise.

Throughout the primary sources regarding the negotiations of power between the French government and the colony of Saint-Domingue, there are oscillations in the projection of the frontiers that separated the metropole and its overseas territory. In some instances, the discourses emphasize distance, using the Atlantic Ocean as a trope for the politics of difference. In other instances, the actuality of the colony is represented as either analogous to the reality of the metropole, or alternatively, the distance is negated by underlining a symbolic proximity with terms like "union," "attachment," and "indissoluble knots." The structures of power and colonial populations functionally moved back and forth from these constructions of "frontier" to frame either their revolutionary claims or their calls for submission to the law. Furthermore, the colony was sometimes constructed as an analogical presence of French affairs. The Haitian Revolution was contemporarily minimized by metropolitan authorities through their focus on the project of national regeneration and its "reverberations" in Saint-Domingue, but nevertheless portraying the project as still being a strictly French enterprise. Ironically, this was vital to sustain the ideological frontiers for the un-applicability of revolutionary discourse in the colony. The French Revolution belongs to the historical juncture of the escalation of racialist thought and the intensification of the sophistication of "capitalist enslavement" (Buck-Morss 2009, 21), and this is exemplified by problematic colonial relations in the midst of a slave revolutionary movement.

The enterprise of the defeat of Haitian revolutionaries and the protection of colonial frontiers was expected to function as a symbolic spectacle of the vindication of white supremacy. The premeditated discursive accent on borders evidences the racialist constructions of frontiers in modern empires; the politics of control of racialized bodies within borders tragically led to ethnic cleansings. The devastating repercussions of racialism in the History of the Atlantic World during the nineteenth and twentieth centuries are precisely tied to the employment of metaphors of "frontiers" in the service of the accumulation of power in the Age of Revolutions.

Notes

1 "L'Assemblée Générale de la Partie Françoise de Saint-Domingue aux François à 13 septembre 1790," Biblioteca José M. Lázaro (BJML), Nemours Collection, box FR.
2 "Discours prononcé à l'Assemblée Nationale, le 2 octobre 1790, au nom de l'Assemblée general de la partie françoise de Saint-Domingue," BJML, Nemours Collection, box FR. Original text: "une immensité de mer nous séparoit de la France."
3 "Adresse prononcée à l'Assemblée Nationale, séance du 30 septembre, au soir, par les députés des Paroisses du Port-au-Prince et de la Croix-des-Bouquets (1790)," BJML, Nemours Collection, box FR. Original text: "une grande partie de cet Empire"; "attachement à la Mère-Patrie"; "à la Nation, à la Loi & au Roi"; "point de commerce, point de Colonies: aussi point de Colonies, point de commerce"; "Nos besoins nous donnoient un droit; nos vœux nous donnoient un titre pour participer à la régénération de ce bel empire."
4 "Réflexions politiques sur les troubles et la situation de la partie françoise de Saint-Domingue, publiées par les commissaires des citoyens de couleur de Saint-Marc et de plusieurs paroisses de cette colonie, auprès de l'Assemblée Nationale et du Roi, à Paris, ce 5 Juin 1792," BJML, Nemours Collection, box FR. Original text: "Nous pouvons avancer que, dans tout l'empire français, nulle classe d'hommes n'est plus fidèlement attachée à la mère-patrie que celle des citoyens de couleur. La moitié de leur origine fait tout leur trésor, toute leur glorie."
5 "Décrets de l'Assemblée Nationale concernant les Colonies; suivis d'une instruction pour les Isles de Saint-Domingue, la Tortue, la Gonave & l'Isle-à-Vaches ets (1791)," BJML, Nemours Collection, box FR. Original text: "Attachement réciproque, avantages communs, inaltérable fidélité: voilà, Peuple des Colonies, ce qu'elle vous promet & ce qu'elle vous demande."
6 "Moyens proposés à l'Assemblée Nationale, pour rétablir, la paix et l'ordre dans les colonies par Armand Guy-Kersaint, député suppléant, administrateur au département de Paris, chef de division des armées navales, à Paris, 1792," BJML, Nemours Collection, box FR.
7 Original text: "l'effervescence des passions, sous un soleil brûlant."
8 "Conseil des Cinq-Cents, Discours de Villaret-Joyeuse, Député du Morbihan, sur l'importance des Colonies & les moyens de les pacifer, Séance du 12 Prairial an 5," BJML, Nemours Collection, box FR. Original text: "Saint-Domingue est aussi une Vendée à reconquérir."
9 "Discours prononcés dans la séance de l'Assemblée générale de la partie française de Saint-Domingue, le 3 décembre 1791," BJML, Nemours Collection, box CM. Original text: "liens indissolubles"; "union qui doit être éternelle"; "parties les plus distantes du centre de l'empire"; "Saint-Domingue est une section trop importante de l'empire français, et les Français de la métropole et des Colonies sont maintenant liés par des nœuds trop indissolubles."

10 "Lettre des Commissaires nationaux-civils, aux braves Hommes de couleur de Sainte-Suzanne, Cap-Français, le 1 février 1792," BJML, Nemours Collection, box CM. Original text: "nom de la Nation, de la Loi et du Roi."
11 "Réponse des Commissaires-Nationaux-Civils, à des personnes des paroisses du Fond-des Nègres, de la Croix-des Bouquets et de l'Arcahaie, au Cap-Français, le 8 Janvier 1792," BJML, Nemours Collection, box CM.
12 "Lettre des Commissaires-Nationaux-Civils, à la Municipalité du Port-au-Prince, au Cap-Français, le 8 Janvier 1792," BJML, Nemours Collection, box CM. Original text: "l'empire des lois et de la raison."
13 "Réponse des Commissaires Nationaux-Civils, aux personnes réunies à la Croix-des-Bouquets, au Cap, le 21 décembre 1791," BJML, Nemours Collection, box CM. Original text: "Nos colonies, dans le sens actuel de ce mot en France, sont des parties intégrantes de l'empire; mais, qui dit colonie entend aussi une partie séparée du centre de l'empire par le vaste océan, partie peuplée de blancs, de noirs, de libres, d'esclaves et du mélange des blancs et des noirs; partie qui, par la nature de sa population, nécessite une constitution locale pour l'état d'existence des esclaves."
14 "Au nom de la République, Proclamation, Nous, Léger-Félicité Sonthonax, Commissaire Civil de la République, délégué aux Isles Françaises de l'Amérique sous le vent, pour y rétablir l'ordre et la tranquillité publique," Bibliothèque Nationale de France, Lk12-28. Original text: "La République Française veut la liberté et l'égalité entre tous les hommes sans distinction de couleur; les rois ne se plaisent qu'au milieu des esclaves: ce sont eux qui, sur les côtes d'Afrique, vous ont vendus aux blancs: ce sont les tyrans d'Europe qui voudraient perpétuer cet infâme trafic. La République vous adopte au nombre de ses enfans; les rois n'aspirent qu'à vous couvrir de chaînes ou à vous anéantir."
15 "Au nom du peuple français, Nous Étienne Polverel et Léger-Félicité Sonthonax, Commissaires civils de la république, délégués aux îles françaises de l'Amérique sous le vent, pour y rétablir l'ordre & la tranquillité (3 mai 1794)," BJML, Nemours Collection, box CS.
16 "Proclamation au nom de la Nation, Nous, Étienne Polverel, Léger-Félicité Sonthonax & Jean-Antoine Ailhaud, Commissaires Nationaux-Civils, délégués aux îles Françaises de l'Amérique sous le vent, pour y rétablir l'ordre et la tranquillité publique, Aux Hommes Libres de la partie Française de Saint-Domingue; à tous les Volontaires Nationaux, Soldats de la Garde-Nationale, Troupes de Ligne, et Matelots employés dans l'expédition," BJML, Nemours Collection, box CSA. Original text: "La Colonie de Saint-Domingue est une partie précieuse de l'Empire Français; elle est peuplée de Français; et sa destinée est inséparable du sort de la mère patrie."
17 Original text: "Qu'ils soient bien appliqués aux sciences dont on va orner leurs esprits; qu'ils deviennent des citoyens estimables, attachés à la République, à leurs devoirs et à la religion dans laquelle ils sont nés!"
18 Original text: "Ce qui est présumable, c'est que les victoires des français, qui retentissent dans les deux hémisphères, contribuent beaucoup à imprimer dans les cœurs de ces habitants les principes de liberté, et d'égalité et la soumission aux lois de la République."
19 Original text: "Art. 3.—Il ne peut exister d'esclaves sur ce territoire, la servitude y est à jamais abolie. Tous les hommes y naissent, vivent et meurent libres et Français. Art. 4.—Tout homme, quelle que soit sa couleur, y est admissible à tous les emplois. Art 5.—Il n'y existe d'autre distinction que celle des vertus et des talents, et d'autre supériorité que celle que la loi donne dans l'exercice d'une fonction publique. La loi est la même pour tous, soit qu'elle punisse, soit qu'elle protège."
20 Original text: "La constitution que vous avez faite, en renfermant beaucoup de bonne choses, en contient qui sont contraires à la dignité et à la souveraineté du peuple Français, dont Saint-Domingue ne forme qu'une portion."
21 Original text: "Que dans le parti que j'ai pris d'anéantir à Saint-Domingue le gouvernement des Noirs, j'ai moins été guidé pas des considérations de commerce et des finances que par la nécessité d'étouffer, dans toutes les parties du monde, toute espèce de germe d'inquiétude et des troubles."

22 Original text: "un quart de la France."
23 Original text: "Un homme destiné à passer sa vie dans les colonies doit sentir que, si les Noirs ont pu maintenir dans les colonies contre les Anglais, ils tourneraient leur rage contre nous, égorgeraient les Blancs, menaceraient sans cesse d'incendier nos propriétés, et ne présenteraient aucune garantie au commerce, qui n'offrirait plus de capitaux et resterait sans confiance."

References

Primary sources

Archives

Biblioteca José M. Lázaro, Colección Josefina del Toro Fulladosa, Colección Nemours.
Bibliothèque Nationale de France.

Printed works

Bonaparte, Napoleon. 2006. *Correspondance générale III: Pacifications 1800–1802*. Paris: Fayard.
Gouvernement Français. 1789. *La Déclaration des droits de l'homme et du citoyen*. Toulouse: N. Étienne Sens.
——. 1802. *Loi relative à la traite des Noirs et au régime des Colonies; du 30 Floréal, an X de la République une et indivisible*. Paris: L'Imprimerie de la République.
——. 2010. "Constitution de l'an III (1795)." In *Constitutions of the World from the Late 18th Century to the Middle of the 19th Century*, edited by Herausgegeben von Horst Dippel. Gottingen: Gruyter.
Janvier, Louis Joseph, ed. 1886. *Les Constitutions d'Haïti (1801–1885)*. Paris: C. Marpon et E. Flammarion.
Laurent, Gérard, ed. 1953. *Toussaint L'Ouverture à travers sa correspondance, 1794–1798*. Port-au-Prince: H. Deschamps.
Leclerc, Charles Victoire Emmanuel. 1937. *Lettres du General Leclerc*. Paris: Leroux.
Saint-Rémy, M., ed. 1853. *Mémoires du général Toussaint L'Ouverture, écrits par lui-même, pouvant servir a l'histoire de sa vie – Précédés d'un etude historique et critique*. Paris: Moquet.

Secondary sources

Anderson, Benedict. 1983. *Imagined Communities*. London: Verso.
Balibar, Étienne. 1991. "Racism and Nationalism." In *Race, Nation, Class: Ambiguous Identities*, edited by Étienne Balibar and Immanuel Wallerstein. London: Verso.
Berman, Marshall. 1982. *All That Is Solid Melts Into Air: The Experience of Modernity*. New York: Verso.
Bhabha, Homi K. 1994. *The Location of Culture*. New York: Routledge.
Breuilly, John. 1985. *Nationalism and the State*. Chicago: University of Chicago Press.
Buck-Morss, Susan. 2009. *Hegel, Haiti, and Universal History*. Pittsburgh: University of Pittsburgh Press.
Burbank, Jane and Frederick Cooper. 2010. *Empire in World History: Power and the Politics of Difference*. Princeton: Princeton University Press.
Césaire, Aimé. 1960. *Toussaint Louverture: la Revolución Francesa y el problema colonial*. La Habana: Instituto del Libro.

Curtin, Philip D. 1964. *The Image of Africa: British Ideas and Action, 1780–1850.* Madison: University of Wisconsin Press.

Doyle, Don H. and Marco A. Villela Pamplona. 2006. *Nationalism in the New World.* Athens: University of Georgia Press.

Dubois, Laurent. 2004. *Avengers of the New World: The Story of the Haitian Revolution.* Cambridge, MA: Harvard University Press.

Fick, Carolyn E. 1990. *The Making of Haiti: The Saint Domingue Revolution from Below.* Knoxville: University of Tennessee Press.

Fischer, Sibylle. 2004. *Modernity Disavowed: Haiti and the Cultures of Slavery in the Age of Revolution.* Durham: Duke University Press.

Foucault, Michel. 1993. *Genealogía del racismo: la guerra de las razas al racismo del Estado.* Buenos Aires: Altamira.

Garraway, Doris L. 2008. "'Légitime Défense': Universalism and Nationalism in the Discourse of the Haitian Revolution." In *Tree of Liberty: Cultural Legacies of the Haitian Revolution in the Atlantic World*, edited by Doris Garraway. Charlottesville: University of Virginia Press.

Geggus, David Patrick. 2009. "Saint-Domingue on the Eve of the Haitian Revolution." In *The World of the Haitian Revolution*, edited by David Patrick Geggus and Norman Fiering. Bloomington: Indiana University Press.

Guha, Ranajit. 2002. *History at the Limit of World-History.* New York: Columbia University Press.

Hobsbawm, Eric. 1990. *Nations and Nationalism since 1780, Second Edition.* Cambridge: Cambridge University Press.

Hunt, Lynn. 1984. *Politics, Culture and Class in the French Revolution.* Berkeley: University of California Press.

——. 1996. *The French Revolution and Human Rights: A Brief Documentary History.* Boston: Bedford.

James, C.L.R. 1963. *The Black Jacobins: Toussaint L'Ouverture and the San Domingo Revolution.* New York: Vintage Books.

Jenson, Deborah. 2011. *Beyond the Slave Narrative: Politics, Sex, and Manuscripts in the Haitian Revolution.* Liverpool: Liverpool University Press.

Lyotard, Jean-François. 1979. *The Postmodern Condition: A Report on Knowledge.* Minneapolis: University of Minnessota Press.

Nelson, William M. 2013. "Colonizing France: Revolutionary Regeneration and the First French Empire." In *The French Revolution in Global Perspective*, edited by Suzanne Desan, Lynn Hunt, and William M. Nelson. Ithaca: Cornell University Press.

Russell, Lynette. 2001. *Colonial Frontiers: Indigenous-European Encounters in Settler Societies.* Manchester: Manchester University Press.

Stoler, Ann Laura. 1989. "Rethinking Colonial Categories: European Communities and the Boundaries of Rule." *Comparative Studies in Society and History* 31: 134–161.

Trouillot, Michel-Rolph. 1995. *Silencing the Past: Power and the Production of History.* Boston: Beacon Press.

Wallerstein, Immanuel. 1990. "The French Revolution as a World-Historical Event." In *The French Revolution and the Birth of Modernity*, edited by Ferenc Fehér. Berkeley: University of California Press.

Wilson, Thomas M. and Hastings Donnan. 1998. *Border Identities: Nation and State at International Frontiers.* Cambridge: Cambridge University Press.

4 Expanding the Japanese empire to the Manchurian frontier

Immigration and ethnicity in the South Manchuria Railway towns

Rosalia Avila-Tàpies

Introduction

For centuries, the northeast region beyond China's Great Wall was a vast frontier dominion of the Chinese Empire, a peripheral dependency that had been populated from very early times by the ancestors of the Manchus, so in historical foreign usage, it was commonly referred to as "Manchuria."[1] After having conquered the China of the Mings and establishing the Manchu's Qing dynasty in 1644, the Manchu officials in Beijing identified Manchuria as their ancestral homeland and adopted a policy of closing it to the world, isolating it from the rest of China. As a consequence, a large part of the region still remained sparsely settled and undeveloped at the end of the nineteenth century, when it captured the imperial imaginations of the surrounding powers due to its economic possibilities and the new relevance of its geographical position. Since then, and for the next half a century, Manchuria became a zone of great migrations and it was a highly contested space for what came to be known as "the cradle of conflict," "the cockpit of Asia," or "the tinder box of Asia" (Lattimore 1931; Stewart 1936). There, the two dominant East Asian imperial powers at the time—Russia and Japan—engaged in competition for its control, while a threatened China struggled to maintain its extensive territorial boundaries.

This chapter examines the territoriality of imperial Japan in Manchuria and the morphology of its inter-ethnic relations. Specifically, the focus is on the South Manchuria Railway towns, which were planned and administered by Japan's South Manchuria Railway Company from 1907 to 1937. They were the main residential districts of the Japanese colonists until the demise of Japanese-ruled Manchukuo in 1945. Along with the Japanese, Chinese, Koreans, and citizens of other nationalities migrated to these towns, which became segregated spaces characterized by ethnic difference. In this study, the nature of the inter-ethnic contact is examined through the analysis of the levels of residential segregation or coexistence, and through the determination and analysis of the causes of segregation. The results show significant levels of ethnic residential segregation, and the existence of barriers and socio-spatial separations, both before and after the creation of Manchukuo (1932), despite the fact that the new state claimed to be based on ideologies of inter-ethnic harmony (*Minzoku kyôwa*), equality and

Oriental civilization, in its declared attempt to build an ideal and moral state in Asia and create a kind of pan-Asian utopia (Yamamuro 2006, 4).

The contested Manchurian borderland

Since the late nineteenth century, Russia and Japan competed to affect, influence or control the actions and interactions in Manchuria, as well as access to the region, through the use of territoriality; that is, by asserting and attempting to enforce control over specific geographic areas, transforming places, cultures, and economic resources into "bounded spaces" (Sack 1983; Sack 1986; Paasi 2011). The competing powers created bounded spaces by forcing imperialist treaties and agreements—the unequal treaties—on the Qing dynasty. Thus, Russia and Japan bordered and re-bordered their areas of influence and jurisdictional territories, the boundaries marking the degrees of control they exercised over people, resources, and relationships, while they constructed new spaces of belonging or exclusion (Wastl-Walter 2011, 5). These modern encroachments started in the mid-nineteenth century, when Czarist Russia infiltrated and later obtained the northernmost territories of Manchuria (also known as Outer Manchuria) in two treaties—the Treaty of Aigun (1858) and the Treaty of Peking (1860)—changing the Manchu-Russian border on the Amur and Ussuri rivers in favor of Russia. China's Qing government saw a foreign threat against the Manchus' homeland. Thus, fearing further Russian southward advances and because of the population pressure from China proper, it modified its policy of closing Manchuria (Lee 1970, 4, 103). The exclusion policy was reversed, and the strict border crossing prohibitions that had lasted for 200 years ended with opening of the Manchurian frontier to the Han Chinese, along with the sponsorship of actively planned colonization in central and north Manchuria. However, despite this countermeasure, Russia continued to expand into the region, followed by Japan which, after the Sino-Japanese War of 1894–1895, became a new contender in the struggle for the control of Manchurian territory at the expense of China and Russia.

The making of the Manchurian homeland

At the turn of the twentieth century, Manchuria was still an isolated and sparsely populated frontier region with a population estimated at 15 million, according to a preliminary enumeration made by the imperial Chinese government in 1907 for the three provinces. This low population and low density—particularly in central and northern Manchuria—was the consequence of an important population loss of Manchus in the mid-seventeenth century, when they left Manchuria in large numbers to undertake the task of ruling the empire after their conquest of China proper (Sakatani 1932/1980, 11); another contributory factor was the official closure of the region in the early Qing dynasty. The latter occurred despite an initial encouragement of Han colonization just after the conquest to relieve pressure on North China and promote the agricultural development of

Manchuria (Reardon-Anderson 2005), when its estimated population was just three million—one million Manchus and two million Hans (Taeuber 1945, 260)—which was considered insufficient for development.

The closing of Manchuria (*fengjin*) at the end of the seventeenth century and the enforcement of restrictions on emigration to Manchuria would have important demographic, economic, and geopolitical effects in the long term that ultimately favored Russian and Japanese encroachments in the region two centuries later (Rhoads 2000, 41). The decision had been made for political motives by the Manchu rulers in Beijing. For them, Manchuria symbolized their imperial ancestral homeland, which should be held as "sacred," and "forbidden" to the world at large and to the Han civilians in particular (Taeuber 1945, 260; SMR Co. 1929, 1), because it was central to Manchu ethnic identity. They began a process of the territorialization of memory from which, as Anthony D. Smith puts it, an "historic ethnoscape" emerged, invested with collective significance (cited by Zhurzhenko 2011, 73). Consequently, its customs had to be preserved and its economic resources had to be reserved exclusively for the Manchu people, similarly to exclusionary efforts made by other ethnic migrant groups throughout history. Thus, Manchuria was turned into a kind of "crown land," which remained off-limits to Han, Mongol, and Korean peoples for most of the Qing era. It was legally re-made as "a vast vacuum area" (Ho 1967, 159), to serve as an ethnic and economic reservoir for the Manchus. Manchu territoriality was expressed in the region by external as well as internal physical re-borderings, in defense against external influences and incursions by alien immigrants.

Thus, at the end of the seventeenth century, demarcation lines separated traditional Manchu areas from China proper (the Great Wall) and from areas already settled by Han and Mongol people in Manchuria (the "Willow Palisade"). The Qing's exclusion policy attempted to prevent foreign immigration and explicitly prohibited Han civilians from crossing the Great Wall and settling in Manchuria, with mobility controlled by the system of issuing road passes (*lupiao*). It also forbade those already residing in the southern section of Manchuria to advance northwards beyond the Inner Willow Palisade, which was built during the late seventeenth century surrounding southeastern Manchuria to control illegal entrance, and followed approximately the layout of the old Ming Liaodong Wall. With the construction of these barriers, the Qing government tried to create territories and classified those who belonged and those who did not, excluding those who could threaten their home territory or the interests of their ethnic group.

However, despite these attempts of the Manchus to preserve Manchuria and to control the inflow of migrants through official bans and physical borders, they were quite ineffective; hence, the Chinese farmers' infiltration continued into the region of the lower Liao, attracted by the more favorable agricultural opportunities and Manchurian labor needs. Finally, seeking a way to check the Russian southern advance, the Qing government modified its official policy of closure and exclusion, and actively sponsored planned Han Chinese colonization northward into central and north Manchuria, opening the territory to massive Chinese settlement. As a result, a total emigration of over 25 million Chinese to Manchuria—mainly from the North China provinces of Hebei and Shandong—occurred between 1890 and 1942 (Gottschang and Lary 2000, 180). This was aided by the construction of railroads, which facilitated the transportation of migrants from North China, and the marketing of their produce (Lee 1970, 103). Although not all of them settled permanently, with this migration flow the Manchurian frontier territory was ultimately sinicized.

Carving out spheres of interests in Manchuria

Meanwhile, the Russian advances south of the Amur and Ussuri Rivers continued. There were also Japanese attempts to penetrate from the east, with both nations trying to expand their influence and to enforce control over Manchuria by pressing the government of China for exclusive economic and political rights in the region, like the British and other European powers had been doing earlier in China proper. Thus, after the Japanese victory in the Sino-Japanese War, Japan claimed control of Kwantung—the strategic tip of the Liaotung peninsula in southern Manchuria—among other territories and concessions, as well as a large war indemnity. These demands drove China to seek the intervention of Russia and other foreign powers for a rapid Japanese withdrawal from the Liaotung peninsula. For its successful intervention, Russia was compensated through the Li-Lobanov Treaty of 1896 with the use of Chinese ports and the right to construct the Trans-Siberian Railway across Manchuria, which became the Chinese Eastern Railway (CER). In addition, Russia acquired from China a 25-year lease of Kwantung, and also obtained the right to build a branch railway line connecting the CER in Harbin to the Kwantung leased territory, spreading its influence further over Manchuria. Soon afterward, the clash of Russian and Japanese interests over Korea and Manchuria led to the Russo-Japanese War (1904–1905). In the aftermath, by the terms of the Treaty of Portsmouth, Russia transferred to Japan the lease of Kwantung, and also ceded the branch line of the CER running from Changchun (Hsinking during Manchukuo era) to Kwantung (Kowner 2006), which was renamed the South Manchuria Railway (*Minami Manshû Tetsudô)*. By means of further conventions, both nations agreed to respect the integrity of each other's territory and to maintain the status quo. Thus, through these long-term leases of the colonial railways and their attached lands, Russia and Japan created two separate spheres of influence in what was usually called "North Manchuria" and "South Manchuria," respectively. Although there was no accepted line of demarcation between the two (Murakoshi and Trewartha 1930, 482), borders could be drawn along a meridian passing through Changchun, where the Japanese established a railway town which grew to became one of the largest Japanese settlements in Manchuria (Sewell 2002). North Manchuria came to mean the Russian sphere, served by the CER with Harbin as its center. South Manchuria was the Japanese sphere, run by the South Manchuria Railway (hereafter, SMR) owned by Japan and operated by the South Manchuria Railway Joint Stock Company (*Minami Manshû Tetsudô Kabushiki Kaisha*, abbreviated as *Mantetsu*). The SMR terminated in Kwantung and had Mukden (today Shenyang) as its hub

(Murakoshi and Trewartha 1930, 482). These Russian- and Japanese-controlled railways—which were originally built as parts of one system—would become a development axis of these spheres, and served as the main agents of colonialism until the establishment of Manchukuo, when the status quo would be altered in favor of Japan. Ultimately, Japan would control the whole territory and expand its settlement areas also to North Manchuria.

The agency of Japanese imperial expansionism

Thus, through these leaseholds and treaty concessions inherited from Russia in 1905, Japan was able to expand its hegemony in southern Manchuria. Initially, Japan's zone of control had been limited to the Kwantung leased territory, the SMR commercial right of way, and the railway's lands—the South Manchuria Railway Zone (hereafter, Railway Zone), or *tetsudô fuzokuchi* in Japanese—where Japan ruled directly through a formal colonial apparatus. From this foothold, Japan tried to exert influence indirectly over the rest of South Manchuria, "through the relationship with local Chinese rulers, through economic dominance of the market, and through the constant threat of force by its garrison army" (L. Young 1998, 3). The SMR Company operated the railways that became the main colonizing agent in South Manchuria (Myers 1989, 121). Additionally, it administered the Railway Zone, including the towns therein (see Figure 4.1).

The SMR Company was established in 1906 and began its operation in 1907. It was modeled to a great extent after the Russian-controlled CER, claiming the same jurisdictional rights in Manchuria, including the right to police the line or to establish municipal administration along its right of way (W. C. Young 1931, 100). The Japanese government owned one half of the shares of the capital stock, and the rest were widely distributed in Japan. The SMR Company was organized under Japanese law with private commercial, and public functions derived from the Japanese government. The functions of the company included not only the management of the railways, which constituted its main business, but also the operation of mines, harbors and wharves, water transportation systems, electrical enterprises, warehousing, real estate business in the railway lands, and the sell-on-commission of the principal goods carried by the railways. Moreover, it was given complete charge of the development and administration of public utilities within the railway lands, being responsible for engineering works, education, sanitation, and scientific experimentation in agriculture and industry in the municipal areas under its administration (W. C. Young 1931, 80–82; SMR Co. 1936b, 68; Avila-Tàpies 2013). It became the largest enterprise in Manchuria, ultimately dominating most of Manchuria's rail traffic and its general economic life in the first half of the twentieth century, and it was possibly the major concern in the Asia-Pacific region in terms of the breadth of its functions and the volume of its business (SMR Co. 1929, 65).

The SMR Company operated in the Railway Zone, which had also been transferred from Russia and was subject to extraterritorial rights until the formal abolition of extraterritoriality in Manchukuo on 1 December, 1937. The Railway

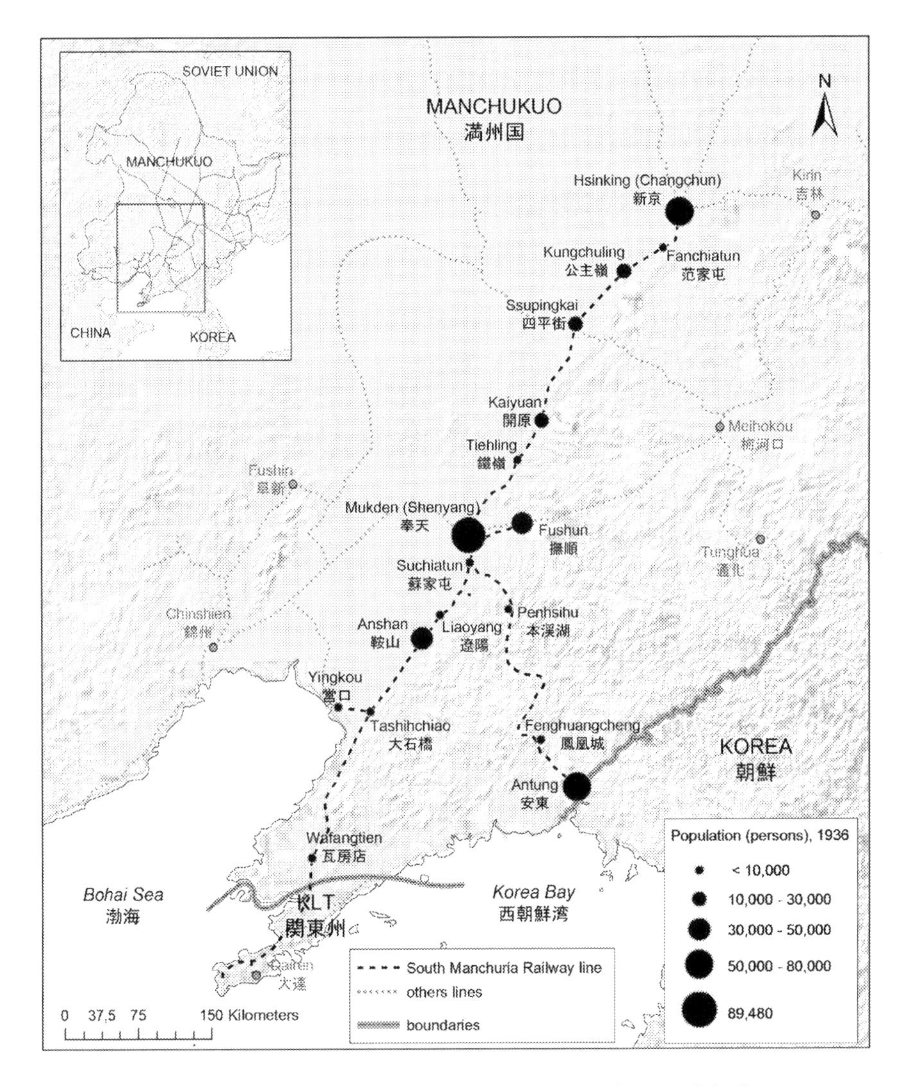

Figure 4.1 Map of the location and population of the SMR urban municipal areas on 31 December, 1936.

Source: Kwantung government (1936). Illustration by Tanausú Pérez García.

Zone was a narrow strip of land along the South Manchuria Railway line, enlarged around some station areas, which became an axis of Japanese activities and colonization in Manchuria (Matsusaka 2001, 70). Irregular in form, its extent was expanded over time—from 184 km² in 1908 to 524 km² in 1936 (SMR Co. 1939/1977, 33)—to include important municipalities and areas used for coal and iron mining. The Railway Zone was directly administered by the SMR Company from 1907 for a lease period of 99 years, although the lease was of shorter duration in some places. Once the lease commenced, the company started urbanizing the zone around major stations, erecting numerous planned cities following a rectangular block system to serve as residential and business activity centers for the new settlers (preferably) under the privilege of extraterritoriality. Hence, the SMR Company transformed its main railway stations into new urban hubs under Japanese jurisdiction and administration. It planned and built them following contemporary Japanese technological and building patterns, as examples of imperialist modernity (Matsusaka 2001; Yang 2010), on the model of technologically advanced, regulated urban spaces. The SMR Company also assumed responsibility for providing them with public services in education, health and sanitation, and urban construction to improve the quality of life for residents (Nishizawa 2000, 74). These towns became a showcase of the Japanese rule in Manchuria and a reflection of the colonizers' power and prestige, specifically designed to accentuate the differences with their environment (Avila-Tàpies 2013). They existed outside of the Chinese traditional order, and created a direct contrast between it and the modern colonial power.

Settlement and ethnicity in the SMR towns

The railway towns attracted most of the Japanese immigration, largely composed of employees of the SMR Company, its subsidiary and affiliated companies and other large private corporations, civil servants, military officers, manufacturers, builders, merchants, small business owners, restaurant operators, liberal professionals (teachers, doctors, and others), entertainers, and adventurers among others, and they arrived with their families. The railway towns remained the preferred residential district of the Japanese living in Manchuria even after their administration was transferred to the new state in 1937. These towns acted as Japanese enclaves that provided a security blanket for the Japanese to protect them from the insecurities of life in Manchuria; they were territories that helped build the feelings of confidence and security among the Japanese colonists. However, other nationalities and ethnic groups had also the right to reside in the towns—a right granted by China. As a result, the Chinese population[2] even slightly outnumbered the Japanese population. According to Kinnosuke Adachi, this is explained by the fact that Chinese merchants and citizens felt drawn to the prosperity and economic opportunities of towns that served as the business centers of Manchuria due to the transportation facilities offered by the railway running through. In addition, they were attracted to the towns' modern educational and sanitary facilities and public services, and the lower taxes, as well as their security under Japanese military protection. The latter can also explain the presence of Koreans, who sought safety, shelter, and sustenance in the SMR Zone (Adachi 1925, 130–131; Woodhead 1932, 58). On average, the majority of the Chinese living in the SMR Zone worked in the commercial sector, followed by the industrial and mining sectors in similar proportions, but with important variations depending on the town.

The towns presented differences regarding origin, function, area and population. Among them, there were important storage and distribution centers of agricultural products (Kaiyuan, Ssupingkai, Kungchuling, Tiehling, and Fenghuangcheng); others were producers of coal, iron ore, limestone, and other minerals (Anshan, Fushun, Penhsihu, Tashihchiao, and Wafangtien); and others were commercial centers located near older walled cities (Mukden, Changchun and Liaoyang) or trade ports (Yingkou and Antung). The largest municipalities in terms of area were Anshan (17.6 km^2), Mukden (10.1 km^2), and Kungchuling (8.2 km^2), while the smallest were Penshihu (0.9 km^2), Wafangtien (2.0 km^2), and Tashihchiao (2.7 km^2), according to the 1936 statistical annual of the SMR Company (SMR Co. 1936a). Over the 30 years that they spent under Japanese jurisdiction, new towns emerged and their population grew rapidly, particularly after the creation of Manchukuo. At the end of 1936—the last year for which statistics are available—the population of the 17 SMR urban municipal areas (*shigai*)[3] of the principal railway stations in the SMR Zone totaled some 415,635 persons, with the largest populations in Mukden (89,480), Antung (76,985), Changchun (64,025), Fushun (47,459), and Anshan (37,556) (see Figure 4.1).

The towns were mostly populated by Chinese (48.2 percent) and Japanese (45.0 percent), followed by Koreans (6.5 percent), who had recently immigrated to Manchuria. Other foreign nationals were almost non-existent, except in Mukden and Changchun; however, in both cases their proportion was less than 0.7 percent, being mostly exiled Russians ("White Russians") and Soviets. The ethnic composition of the railway towns varied greatly as a result of each town's migration history. Thus, some had a large majority of Japanese inhabitants, such as Mukden (73.3 percent), Suchiatun (68.8 percent), or Yingkou (62.6 percent), whereas others had a majority of Chinese, such as Fanchiatun (88.4 percent), Kaiyuan (81.3 percent), or Fenghuangcheng (64.4 percent). The massive influx of migrants had affected sex and age structures, turning them into predominantly masculine and young spaces. In particular, these towns presented on average an extraordinarily marked gender imbalance for the Chinese (sex ratio = 249.7), as a result of the specific labor demands for Chinese male workers in construction, factories, and mining; female migrants to the Manchurian frontier were a smaller proportion of about 15 percent (Gottschang and Lary 2000).

Due to their ethnic diversity, the railway towns constituted examples of multi-ethnic societies in Manchurian urban space. However, they presented a remarkable co-ethnic concentration in certain residential districts, and there were medium-to-high levels of residential segregation, demonstrated by the calculation of the Index of Dissimilarity; whose results and determinants of the causes of segregation (Avila-Tàpies 2003, 2004) are presented in the following section.

Measuring the residential segregation

The Index of Dissimilarity (hereafter I.D.) was calculated for all the cities for four years.[4] The I.D. measures the level of segregation or coexistence between pairs of ethnic groups across all the districts of a city, representing the part of the population that should change its place of residence to make pairs of ethnic groups evenly distributed across all districts (see Table 4.1). The I.D. ranges between 0 (no segregation) and 100 (total segregation). The study's interpretation follows Kantrowitz's classification of I.D. above 70 as "high," I.D. of 30 or less as "low," and variations in level of less than 5 points as unimportant, unless they are otherwise correlated (Kantrowitz 1969, 686–687). The data used were the statistics of the government of Kwantung for the Railway Zone which are available for the period until the date of December 31, 1936. Thus, the statistics examined here by year of recording include: 11 towns (1920), 14 towns (1926 and 1930), and 17 towns (1936).

General outline: average segregation values

The average I.D. values indicate that the separation of the Japanese in these towns with regard to other groups was between medium and medium-high levels. With the Chinese (labeled "Manchurian" in the statistics of 1936), these values reach 38.68 (1920), 43.09 (1926), 50.82 (1930), and 49.08 (1936); with the Koreans, they reach 56.53 (1930) and 57.52 (1936); and with the remaining foreigners, mostly exiled Russians and Soviets, they reach 59.79 (1926), 60.13 (1930), and 66.87 (1936).

In addition, a very high degree of inter-annual correlation is observed in the distribution of the I.D. values between towns, especially for the Japanese-Chinese pair. This means that spatial segregation trends did not change significantly over time, and most importantly it indicates that the levels of segregation in the railway towns did not change, either before or after the creation of Japanese-ruled Manchukuo.

According to these calculations, there was greater similarity in the distribution of population by districts between the Japanese and the Chinese than with any other pair of nationalities; hence, there was greater probability of contact and exchange between these two groups. Specifically, segregation between the Japanese and the Chinese for the years 1930 and 1936 obtained an approximate mean value of 50 (interpreted in the sense that 50 percent of the Japanese or the Chinese should move to another residential district to reach equidistribution).

It is extremely interesting to observe very similar values of segregation in the years 1930 and 1936, which means that the creation of Manchukuo and the resulting extension of Japanese control over all of Manchuria did not alter the distribution of ethnic groups by urban districts in the old Japanese enclaves. That is, the construction of a multi-ethnic and harmonious ideal state as one of the founding state principles of Manchukuo did not translate into a decrease of ethnic segregation in the urban space of the original Japanese enclaves (Avila-Tàpies forthcoming).

Segregation values of towns

The detailed analysis of each town shows a great inter-city variation in segregation levels. Focusing on the quantitatively most significant Japanese-Chinese pair for the year 1936, low levels of segregation (<33.3) were found in

Table 4.1 Indexes of Dissimilarity between Japanese and other ethnic groups, SMR urban municipal areas, years 1920, 1926, 1930 and 1936

Railway town	*Number of districts (1936)*	*Average size of district (1936)*	*Pairs*	*I.D. 1936*	*I.D. 1930*	*I.D. 1926*	*I.D. 1920*
Mukden	16	5,593	Japanese–Chinese	39.78	36.62	33.97	35.75
			Japanese–Korean	42.21	27.84		
			Japanese–foreigner	71.42	59.62	42.03	
Changchun	17	3,766	Japanese–Chinese	57.61	60.76	52.11	50.77
			Japanese–Korean	42.65	59.13		
			Japanese–foreigner	43.65	50.08	44.06	
Anshan	13	2,889	Japanese–Chinese	77.38	75.56	83.10	63.11
			Japanese–Korean	67.10	58.96		
			Japanese–foreigner	62.15			
Fushun	12	3,955	Japanese–Chinese	73.10	64.43	47.72	40.55
			Japanese–Korean	60.34	48.71		
			Japanese–foreigner	61.12	60.90	85.93	
Antung	17	4,277	Japanese–Chinese	66.79	66.71	49.10	37.83
			Japanese–Korean	57.24	51.58		
			Japanese–foreigner	65.35	42.01	68.93	
Ssupingkai	8	2,534	Japanese–Chinese	37.25	14.86	22.75	21.63
			Japanese–Korean	47.58	29.42		
			Japanese–foreigner	55.03	70.83	87.45	
Yingkou	5	1,285	Japanese–Chinese	30.49	30.38	6.47	
			Japanese–Korean	55.32	48.33		
			Japanese–foreigner	98.09	17.60		
Kungchuling	6	2,323	Japanese–Chinese	51.48	60.58	52.19	55.57
			Japanese–Korean	62.92	67.39		
			Japanese–foreigner			39.59	

Liaoyang	7	1,314	Japanese–Chinese	43.21	45.25	39.53	17.23
			Japanese–Korean	73.51	59.56		
			Japanese–foreigner	78.14			
Kaiyuan	7	2,252	Japanese–Chinese	52.91	47.92	34.46	21.87
			Japanese–Korean	56.93	63.99		
			Japanese–foreigner		58.43	53.58	
Wafangtien	3	2,350	Japanese–Chinese	65.25	75.90	67.11	64.08
			Japanese–Korean	82.95	84.72		
			Japanese–Foreigner		84.72		
Tashihchiao	4	1,776	Japanese–Chinese	70.04	74.44	68.54	
			Japanese–Korean	91.08	81.20		
			Japanese–foreigner		96.99		
Penhsihu	3	1,157	Japanese–Chinese	3.96	11.30	7.74	
			Japanese–Korean	26.40	27.94		
Tiehling	3	1,593	Japanese–Chinese	51.09	46.73	38.43	17.16
			Japanese–Korean	59.80	82.65		
Suchiatun	4	1,350	Japanese–Chinese	10.98			
			Japanese–Korean	61.17			
Fenghuangcheng	3	859	Japanese–Chinese	56.28			
			Japanese–Korean	65.33			
Fanchiatun	3	1,386	Japanese–Chinese	46.80			
			Japanese–Korean	25.37			

Source: Kwantung government (1920, 1926 and 1936), population data of 31 December for all years; Kwantung government (1930a), population data of 1 October, 1930.

Note: Blank space means that population data is insufficient or the ethnic group is not present. In the years 1920 and 1926, the Korean population is included under "Japanese."

Penhsihu, Suchiatun and Yingkou. In contrast, very high levels of segregation (>66.7) were found in Anshan, Fushun, Tashihchiao, and Antung. The rest of the 10 towns showed segregation values between 33.3, and 66.7. This situation followed the trend established in previous years, as the Pearson correlations indicated (Avila-Tàpies 2003).

Measuring the causes of segregation

The causes of variation of the I.D. values among these railway towns have been statistically analyzed. Particularly, the causes of residential segregation between the Japanese and the Chinese in the towns of the SMR Zone were analyzed, because these were the two largest ethnic groups residing in these railway cities, and therefore, their spatial separation or coexistence is the most significant for evaluating the general situation. To discover the predictors of segregation, multiple regression analysis was applied to two differentiated studies: initially, a cross-sectional data study was carried out with multivariate analysis of the levels of segregation between cities for the years 1930 and 1936, treated separately (see Table 4.2); and later, a longitudinal data study was carried out with multivariate analysis of the changes in levels of segregation between 1930 and 1936 (see Table 4.3). The independent variables considered were those that could affect the redistribution of the population in the cities. In the cross-sectional data analysis those variables included: sex ratio and the percentages of each ethnic group, workers in each sector of activity, and Mantetsu's employees and dependents in the city. In the longitudinal data analysis, the variables considered were: the percentages of each ethnic group, workers in each sector of activity, Mantetsu's employees and dependents, residential land increase between 1930 and 1936, and residential land rented by each ethnic group (the full results of the two studies are presented in Avila-Tàpies 2004).

The regression analyses revealed a negative and statistically significant relation between the levels of residential segregation and the percentage of domestic employees: that is, a higher presence of domestic employees (Japanese or Chinese) implied less segregation. In addition, there was a positive and significant relation between changes in the levels of segregation and changes in the percentage of Japanese population working in the transportation sector. The latter suggests that the increase of Japanese population in the transportation sector was related to an increase of segregation over time. The association between the presence of Japanese in the transportation sector (SMR Company employees) and residential segregation of the Japanese and Chinese residents was revealing. Therefore, the residential policy of the SMR Company—the entity which was considered to have the greatest responsibility for urban management—was studied. In spite of the regulations about the use of urban space, a specific residence zoning by races was not specified; in practice, the company reserved urban space for its Japanese employees, for whom it established reserved institutions and infrastructures, thus hindering both inter-ethnic contact and contact between the different socio-economic Japanese classes. The spatial segregation of the workers of the SMR Company fomented the creation and maintenance of stereotypes and social

Table 4.2 Results of the cross-sectional data study with multivariate analysis, years 1930 and 1936. Correlations between independent variables and I.D.

Independent variable	*Dependent variable (I.D.)*	
	1930	*1936*
Population (%)		
population–Japanese	–0.18	–0.43
population–Chinese	0.17	0.33
Sex Ratio		
sex ratio–Japanese	0.16	0.00
sex ratio–Chinese	–0.47	–0.23
Occupation (%)		
agriculture–Japanese	0.37	0.29
agriculture–Chinese	0.20	0.16
mining–Japanese	0.19	0.35
mining–Chinese	0.15	0.40
manufacturing–Japanese	0.27	0.35
manufacturing–Chinese	0.56	0.13
commerce–Japanese	–0.15	–0.06
commerce–Chinese	–0.19	–0.12
transport–Japanese	0.13	–0.38
transport–Chinese	0.31	–0.39
public ad. & professional serv.–Japanese	–0.23	–0.07
public ad. & professional serv.–Chinese	–0.31	–0.10
domestic service–Japanese	–0.66	–0.05
domestic service–Chinese	–0.45	–0.72
Mantetsu employees & dependents (%)		
Mantetsu–Japanese	0.45	–0.07
Mantetsu–Chinese	0.15	–0.05

Source: Kwantung government (1930b and 1936); S.M.R.Co. (1930 and 1936a). Statistical data of 31 December for all years.

Table 4.3 Results of the longitudinal data study with multivariate analysis, years 1930 and 1936. Correlations between independent variables and I.D. change (Japanese–Chinese)

Independent variable	*Dependent variable (I.D.)*
	1930–1936
Population (%)	
population–Japanese	0.24
population–Chinese	–0.23
Occupation (%)	
transport–Japanese	0.70
manufacturing–Chinese	0.62

(Continued)

Table 4.3 (Continued)

Independent variable	*Dependent variable (I.D.)*
	1930–1936
Mantetsu employees & dependents (%)	
Mantetsu–Japanese	0.22
Mantetsu–Chinese	–0.03
Residential land (%)	
residential land increase	0.34
residential land rented–Japanese	–0.38
residential land rented–Chinese	0.38

Source: Kwantung government (1930b and 1936); S.M.R.Co. (1930 and 1936a). Statistical data of 31 December for all years.

distance, maintaining inter-ethnic relations at an occupational level (functional integration), but rarely at a social level. The present study thus allows us to glimpse how the imperialist utopia of Manchukuo and the principle of inter-ethnic harmony did not affect the residential segregationist policy of the SMR Company, whose residential segregation policies were determined by ethnicity, as well as by social class (Mizuuchi 1985).

Conclusions

From the opening of the Manchurian frontier in the late Qing dynasty until the establishment of Manchukuo, Northeast China became an intensively contested border land, a crossroads of empires, where foreign capital investments, regional migration and inter-ethnic encounters transformed lives and landscapes definitively. In particular, the push of the Japanese empire in the Manchurian frontier entailed important industrial developments and major population movements, which initially started in the leased territory of Kwantung and the SMR Zone. The Japanese towns within the SMR Zone were delimited and separated from the rest of the Chinese territory, turning rapidly into modern settlements with ethnically heterogeneous populations. Along with Japanese colonizers, Chinese citizens also settled in the new bordered spaces. However, little was known until now about their degrees of spatial interaction. This study examined the level of physical contact or separation among different ethnic groups in these railway towns by measuring their residential segregation. Furthermore, the predictors of such segregation were also discovered both through cross-sectional data and longitudinal data studies with multivariate analysis. The statistical results revealed medium and high levels of segregation in the towns, and the crucial role played by the SMR Company in residential segregation, which was determined by ethnicity and status. Based upon these results, it can be said that the railway towns were spaces with barriers and socio-spatial separations where there were usually few opportunities for physical inter-ethnic contact. Moreover, the study showed how the establishment of Japanese-sponsored Manchukuo did not alter residential segregation levels in the Japanese railway towns, despite Manchukuo's strong defense of inter-ethnic coexistence and equality in its inclusive rhetoric of ethnic harmony, as presented in its proclamation of independence. This implies a persistent lack of commitment to equity by the SMR Company before and after the creation of Manchukuo, and gives evidence that the expansion of the Japanese empire in the Manchurian frontier was inseparable from the creation of territorial identities and the establishment of ethnic borders of inclusion and exclusion.

Acknowledgments

The author wishes to thank Tanausú Pérez García of the University of Las Palmas de Gran Canaria for his cartographic assistance.

Notes

1 Manchuria, now referred to as "Northeast China," is a term used herein as a historical convenience. The region had been usually called the "Three Eastern Provinces" (later also named "Three Northeastern Provinces")—Liaoning (Fengtien), Kirin and Heilungkiang—during the Qing dynasty. The author has primarily used the geographical names as Romanized in the coetaneous English documents of Japanese publishers.

2 In the statistics of the Kwantung bureau used for this analysis, the Chinese population was categorized as "Shina-jin" (years 1920, 1926 and 1930) or "Manshu-jin" (1936), without distinctions being made between Hans, Manchus, or other Chinese ethnic groups. However, by making use of the 1940 census of Manchukuo, it is possible to state that the Manchus were a very small proportion of the inhabitants of the cities, and that the Chinese population we refer to herein was overwhelmingly Han Chinese.

3 For the purpose of the analysis, only the "urban" area (*shigai*) in each case, as delimited in the 1935 Census of Kwantung, has been considered.

4 The Index of Dissimilarity for two groups, Japanese and Chinese, in a particular town:

$$\frac{1}{2}\sum_{i=1}^{N}\left|\frac{b_i}{B}-\frac{w_i}{W}\right|$$

Where: N = number of districts; b_i = number of Japanese in district i; B = total number of Japanese in the city; w_i = number of Chinese in district i; W = total number of Chinese in the city.

References

Adachi, Kinnosuke. 1925. *Manchuria: A Survey*. New York: McBride & Co.

Avila-Tàpies, Rosalia. 2003. "The Residential Segregation by Ethnic Groups in Colonial Manchuria: The Case of the Cities of the South Manchuria Railway (first part)." *Konan Daigaku Kiyô* (Bulletin of Konan University-Letters) 134: 1–17. In Japanese.

——. 2004. "The Residential Segregation by Ethnic Groups in Colonial Manchuria: The Case of the Cities of the South Manchuria Railway (second part)." *Konan Daigaku Kiyô* (Bulletin of Konan University-Letters) 139: 193–208. In Japanese.

——. 2013. "Imperial Expansion, Human Mobility and the Organization of Urban Space in Manchuria (Part I)." *Shakai Kagaku* (The Social Sciences) 43–3: 35–56.

——. [Forthcoming.] “Co-ethnic Spatial Concentrations and Japan’s 1930s Concord Project for Manchukuo.” *Geographical Review of Japan Series B* 88–2.

Gottschang Thomas R. and Diana Lary. 2000. *Swallows and Settlers: The Great Migration from North China to Manchuria*. Ann Arbor: University of Michigan Press.

Ho, Ping-ti. 1967. *Studies on the Population of China, 1368–1953*. Cambridge, MA: Harvard University Press.

Kantrowitz, Nathan. 1969. “Ethnic and Racial Segregation in the New York Metropolis, 1960.” *The American Journal of Sociology* 74–6: 685–95.

Kowner, Rotem. 2006. *Historical Dictionary of the Russo-Japanese War*. Plymouth: Scarecrow Press.

Kwantung government. 1920. *Kantô-chô Tôkei-sho. Dai 15 (Taishô 9 nen)*. In Japanese.

——. 1926. *Kantô-chô Tôkei-sho. Dai 21 (Shôwa 1 nen)*. In Japanese.

——. 1930a. *Kantô-chô Shôwa 5 nen Kokusei Chôsa Kekka-hyô*. In Japanese.

——. 1930b. *Kantô-chô Tôkei-sho. Dai 25 (Shôwa 5 nen)*. In Japanese.

——. 1936. *Kantô-kyoku Tôkei-sho. Dai 31 (Shôwa 11 nen)*. In Japanese.

Lattimore, Owen. 1931. *Manchuria, the Cradle of Conflict*. New York: Macmillan.

Lee, Robert H. G. 1970. *The Manchurian Frontier in Ch’ing History*. Cambridge, MA: Harvard University Press.

Matsusaka, Yoshihisa T. 2001. *The Making of Japanese Manchuria, 1904–1932*. Cambridge, MA: Harvard University Press.

Mizuuchi, Toshio. 1985. “Formation and Development of a Japanese Colonial City: Dalian from 1899 to 1945.” *Japanese Journal of Human Geography* 37–5: 50–67. In Japanese.

Murakoshi, Nobuo and Glenn T. Trewartha. 1930. “Land utilization maps of Manchuria.” *The Geographical Review* 20–3: 480–493.

Myers, Ramon H. 1989. “Japanese Imperialism in Manchuria: The South Manchuria Railway Company, 1906–1933.” In *The Japanese Informal Empire in China, 1895–1937*, edited by Peter Duus, Ramon H. Myers, and Mark Peattie, 101–132. Princeton, NJ: Princeton University Press.

Nishizawa, Yasuhiko. 2000. *Mantetsu: The Giant of “Manshû.”* Tokyo: Kawade Shobo Shinsha. In Japanese.

Paasi, Anssi. 2011. “A Border Theory: An Unattainable Dream or a Realistic Aim for Border Scholars?” In *The Ashgate Research Companion to Border Studies*, edited by Doris Wastl-Walter, 11–31. Farnham: Ashgate.

Reardon-Anderson, James. 2005. *Reluctant Pioneers: China’s Expansion Northward, 1644–1937*. Stanford: Stanford University Press.

Rhoads, Edward J. M. 2000. *Manchus and Han: Ethnic Relations and Political Power in Late Qing and Early Republican China, 1861–1928*. Seattle: University of Washington Press.

Sack, Robert D. 1983. “Human Territoriality: A Theory.” *Annals of the Association of American Geographers* 73–1: 55–74.

——. 1986. *Human Territoriality: Its Theory and History*. Cambridge: Cambridge University Press.

Sakatani, Yoshiro. 1932/1980. *Manchuria. A Survey of Its Economic Development*. New York: Garland.

Sewell, Bill. 2002. “Railway Outposts and Puppet Capital: Urban Expressions of Japanese Imperialism in Changchun, 1905–1945.” In *Colonialism and the Modern World. Selected Studies*, edited by Gregory Blue, Martin Bunton, and Ralph Croizer, 283–321. New York: Sharpe.

S.M.R.Co. 1929. *Report on Progress in Manchuria, 1907–1928*. Dairen: S.M.R.Co.

——. 1930. *Chihô Keiei Tôkei Nenpô. Shôwa 5 nendo.* In Japanese.
——. 1936a. *Chihô Keiei Tôkei Nenpô. Shôwa 11 nendo.* In Japanese.
——. 1936b. *Fifth Report on Progress in Manchuria to 1936.* Dairen: S.M.R.Co.
——. 1939/1977. *South Manchuria Railway Zone Management Total History.* Tokyo: Ryuukei-shosha. In Japanese.
Stewart, John R. 1936. *Manchuria since 1931.* New York: Institute of Pacific Relations.
Taeuber, Irene, B. 1945. "Manchuria as a Demographic Frontier." *Population Index* 11–4: 260–274.
Wastl-Walter, Doris. 2011. "Introduction." In *The Ashgate Research Companion to Border Studies*, edited by Doris Wastl-Walter, 1–8. Farnham: Ashgate.
Woodhead, Henry G. W. 1932. *A Visit to Manchukuo.* Shanghai: The Mercury Press.
Yamamuro, Shin'ichi. 2006. *Manchuria under Japanese Dominion.* Philadelphia: University of Pennsylvania Press.
Yang, Daqing. 2010. *Technology of Empire. Telecommunications and Japanese Expansion in Asia.* Cambridge, MA: Harvard University Press.
Young, Louise. 1998. *Japan's Total Empire: Manchuria and the Culture of Wartime Imperialism.* Berkeley: University of California Press.
Young, Walter C. 1931. *Japanese Jurisdiction in the South Manchuria Railway Areas.* Baltimore: The Johns Hopkins Press.
Zhurzhenko, Tatiana. 2011. "Borders and Memory." In *The Ashgate Research Companion to Border Studies*, edited by Doris Wastl-Walter, 63–84. Farnham: Ashgate.

Part III

States

5 Spatiality, jurisdiction, and sovereignty in early Latin American approaches to the Law of the Sea

Daniel S. Margolies

Jurisdictional claims over oceanic spaces in the immediate postwar period deserve close attention, since the oceanic spaces proximate to sovereign coastlines were a central focus of global competition for control of seabed resources and the ability to construct and regulate the overall jurisdictional order of the post-World War II world system. Latin American nations were at the center of this moment of political-economic competition and legal innovation, particularly Mexico, Argentina, Chile, and Peru. These nations individually and forcefully pursued expansive sovereignty and jurisdictional claims on the offshore continental shelf.

The purpose of this chapter is to explore the contingent involvement of the Latin American states in this project of reconceptualizing and restructuring global ocean spaces as new varieties of sovereign and jurisdictional space, or what can be conceptualized as new or freshly expanded *state space*. To do so, this chapter first presents an interdisciplinary exploration of interpretative and theoretical approaches to the questions of jurisdiction, sovereign territoriality, and governance. This discussion is followed by a focused narrative of legal and diplomatic assertions at the critical formative postwar moment. The results of this Latin American effort were newly imbricated claims to sovereignty and jurisdiction in new ocean spaces, which in turn triggered a half-century process of zonal reordering across the globe that remains ongoing, contentious, and differentially interpreted.

The jurisdictional reordering of Latin America sovereignty over oceanic spaces should be situated contextually with the emerging hegemonic power of the United States, but that can become a frame which is equally constraining as well as illuminating. The claims of individual Latin American states also should be considered within the broader questions of spatial governance across the hemisphere and across the Pacific Rim. This chapter deliberately decenters the story of a response to U.S. hegemony, and instead provides a consideration of key, regionally specific proclamations and jurisdictional innovations in Latin American after 1945. These sovereign jurisdictional assertions over a variety of oceanic spaces should be seen as constitutive to later agreements and zonal definitions rather than categorized as reactive or derivative of U.S. assertion and practice.

For both Latin American states and the United States in the period between 1945 and 1958, new technological innovations in strategic resource exploitation produced a desire for expanded jurisdiction over seabed resources and fisheries. These state ambitions combined with the newly fluid opportunities of the postwar period to invite legal innovations and bold assertions of jurisdiction over newly emergent continental shelf resources, led to new conceptions of oceanic spaces. At the time, the impact was so profound that, as Sir Hersch Lauterpacht observed in 1950, "the concept of the continental shelf had become virtually 'instant' customary international law" (quoted in Scharf 2013, 118). This was an unprecedented expansion of new sovereign claims around the globe.

From the start, most scholars of the Law of the Sea have attributed the foundation of the modern era of jurisdictional claims to the eye-opening United States assertion of sovereign rights articulated by President Harry Truman in September, 1945 in his "Proclamation with Respect to the Natural Resources of the Subsoil and Sea-bed of the Continental Shelf" (Proclamation 1946; Gidel 1951). Donald R. Rothwell and Tim Stephens call the Truman Proclamation "the first substantive claim by a coastal state to a distinctive offshore resources zone, which was completely separate from the territorial sea as it had been developed to that point" (Rothwell and Stephens 2010, 100–101). José Antonio de Yturriaga, a Law of the Sea arbiter for Spain, asserts that "it was the US . . . historically an opponent to fishing zones—which sowed the seeds of change" (Yturriaga 1997, 3) via the Truman Proclamation.

This traditional starting point of 1945 sometimes lends a United States-centric view, not just to the study of the genesis of the First United Nations Convention on the Law of the Sea (UNCLOS I) regime created in 1958, but also to the articulation of the major conceptual terrain of the negotiations. UNCLOS I was a global effort to freshly define and govern interstate sovereignty in oceanic spaces via "two distinct but not unrelated processes: *negotiation* and *dispute management*" (Johnston 1988, xiii). UNCLOS I turned out to be only the beginning of a process which stretched over the next four decades and ultimately instantiated many Latin American innovations in sovereignty questions into international legal practice. The first negotiation over continental shelf jurisdictional boundaries occurred in the region between Venezuela and Great Britain, which was colonial ruler of Trinidad and Tobago, in 1942 (Prescott 1985, 338).

Jurisdiction and Latin American sovereignty questions in the sea

Articulations of jurisdiction are central to the construction of space by the state. They are the essential beginning for understanding the Latin American approach to oceanic governance and sovereignty. Issues of sovereignty and space in policy terms can best be approached by focusing specifically on questions of territoriality and jurisdiction as they have been expressed or pursued by the state, or as they have been deployed in governance regimes. Sovereignty subsumes a shifting aggregation of territory, power, interest, and jurisdiction in diverse arenas, all operating in the context of longstanding, global constructions of legal order. States, regulatory systems, and institutions engage in transnational relations which very often occur in jurisdictional spaces. Sovereignties of the state are reflected in these spatial assemblages of power through what Foucault termed "governmentality." (Foucault 2009). Jurisdictional assertions are the means by which a state articulates power into new spaces and achieves the ordering of governmentality. In so doing, traditional territorial governance develops a newly creative cast reflecting the means and institutions by which a state makes manifest its control. The law, via jurisdictional assertion, becomes explicitly regulatory. Arnould Lagendijk, Bas Arts, and Henk van Houtum (2009, 5–6) called this governmentality "the 'art of governing' . . . an ensemble of power techniques that is used by the state to exercise 'government' . . . through institutions, bureaucracies, tactics, procedures, knowledge, technologies, etc." Their argument is that governmentality thus reveals the "actual meaning of state territoriality." A jurisdictional approach to governmentality as a technology of the state highlights the ways sovereignty needs to be viewed in analytically distinct ways when it is applied to the sea.

Assertions of spatial control in the oceans were, simply put, new. As Henri Lefebvre wrote in explaining the deliberately guided production of space, "the state has presided over this integration. Spaces that were once unoccupied—the mountains, the sea—enter into exchange, become commodities." Lefebvre provocatively argues that "space has a much stronger connection with the State than territory once had with the nation" (Lefebvre 2009, 212, 213–214). The new jurisdictional assertions in newly integrated ocean space can be seen as a form of enclosure in turn producing a new "structural regulatory role." Territorialization is a term of geographers rather than of international law, but the function of it can be readily traced in sovereign attempts to define and control space via jurisdictional assertion. As Alvaro Sevilla-Buitrago (2015, 1005) argues, "the notion of territory constitutes perhaps the most relevant conceptual bridge to enclosure's spatialities." Sovereigns assert jurisdiction over (enclose) a resource or zone, for example, without simultaneously establishing the exclusivity of sovereign territoriality.

Accordingly, understanding the embeddedness of sovereignty in oceanic governance and territoriality questions requires attention to the spatialities of jurisdiction. Historians have overlooked the significance of spatial orders in legal assertions despite the ongoing "spatial turn" (Massey 2005; Rosenberry 2002). However, the jurisdictional approach to the study of the UNCLOS regime as it emerged from Latin American practice is enriched by spatial analyses combining the perspectives of legal geography and social theory with legal doctrinal and historical approaches. Legal geographer David Delaney (2011, 15; 19) provocatively and abstractly calls for "three interrelated operations: (i) seeing space as imagined and discursively organized; (ii) seeing space as performed; [and] (iii) rethinking the materiality of space," particularly in the ways spaces interact with "legal imaginaries." This chapter argues that approaching oceanic spatiality in these relational and discursive terms highlighted by Delaney is especially useful for understanding how Latin American sovereignty and governance assertions in the context of UNCLOS has been constructed in a variety of historically contingent spaces. Oceanic spaces were reimagined (almost conjured) by postwar policymakers in ways which allowed the state to assert new sovereignty via specific extraterritorial assemblages and jurisdictional regimes. Placing the legal assertions within an interdisciplinary theoretical framework where governmentality can be understood in terms of performance, as Lagendijk, Arts, and van Houtum and Delaney variously do, assists in defining the spatial confines of oceanic governance in ways legible to historians concerned broadly with the politics of territoriality.

Writing law (the functional production of policy) manufactures jurisdiction. Mapping jurisdiction as was done in Law of the Sea spatializes power and also effectively manufactures it. Shaunnagh Dorsett and Shaun McVeigh (2013, 14) gracefully define jurisdiction as "a power to speak the law." This idea of adopting the power to assert the law and thus to conjure the power it bespeaks helps to frame the Latin American approach to jurisdictional claims over previously ungoverned or contested oceanic space. Critical emphasis in this chapter is thus placed on the jurisdictional assertion itself as well as on the doctrinal issues they evoke, which is the more common emphasis of legal scholars. Mariana Valverde (2009, 144–145) argues that "governing projects and the power knowledges that make them work are differentiated from one another and kept from overtly clashing by the workings of the machinery of 'jurisdiction,' which instantly sorts governance processes, knowledges, and powers into their proper slots."

This idea that jurisdictional claims could effectively manufacture authority necessarily underlies the approach to the meaning of Law of the Sea assertions in international practice and as an element of unilateral state actions. Latin American states making claims over territorial and continental-shelf space in the late 1940s and 1950s were engaging a loose tradition about territorial limits in ocean space. There was, in fact, no global consensus on the limits of the territorial sea. In 1924, the International Law Association "chose to rest the conception of territorial sea simply on 'a right of jurisdiction'" (Yang 2006, 118), which implied a different status than full sovereignty. The issue remained highly unsettled at the 1930 conference at The Hague for the Codification of International Law, which protected certain rights in the territorial area. The legacy was confusing, not least because the territorial sea itself can be seen in terms of the sovereign rights asserted over it and within it, and because the definition of the "breadth of the territorial sea was the most debatable issue regarding the law of the sea," as Yoshifumi Tanaka (2012, 20) has argued. As he observed, the territorial sea was considered to have stretched from the familiar three-mile limit, to the four miles favored by Scandinavian countries, or even to the "cannon-shot rule." Mexico announced in 1935 that since the 1930 Codification had "found that there was no uniform standard or practice amongst States with regard to the extent of the territorial waters," and since it had signed no treaties requiring otherwise, that its territorial sea was nine nautical miles (*Diario Oficial* 1935).

The most critical early Latin American jurisdictional claims over oceanic space came in the period between the end of the war in 1945 and 1952, continuing until UNCLOS I was settled in 1958 (Young, 1951). The literature on UNCLOS is staggeringly large, and the body of scholarship focused exclusively on Latin America is robust, though generally adopting the focus on the contemporary implementation and implications of the legal doctrines involved (Kirchner 2007; Szekely 1986; Zacklin 1974; Costa 2013). Alternatively, emphasizing the spatial and jurisdictional dimensions of the assertions will help to clarify the political-economic interests embedded in the state actions.

Legal spatiality and the postwar proclamations

The idea that there was a new way to extend sovereignty beyond territorial limits into the previously unencumbered sea via a "special contiguous zone for safeguarding law and property" (MacDougal and Feliciano 1994, 470) had been floated in limited form in the October, 1939 Declaration of Panama, which created a continental wide "zone of security" of up to 300 miles offshore. This wartime measure was seen at the time as possibly excessive and perhaps "even guilty of encroaching upon the 'freedom of the seas'" (Fenwick 1940, 116). Carl Schmitt argued in his postwar *The Nomos of the Earth* that this declaration "remains of extraordinary and fundamental significance for the spatial problem of international law" because it expanded a hemispheric interest from three miles to a hugely extensive limit. Schmitt called it "a new form of sea-appropriation" where the Western hemisphere became an abstraction wielding actual power in sea space. Citing Friedrich Ratzel of *lebensraum* fame, Schimitt argued that as a result of the declaration, "*space* had protruded into the expanse and evenness of the sea" (Schmitt 2003, 282, emphasis in original).

In addition to this delicate but important concept of "space" being applied to oceanic order, it is important to note that one of the lasting legacies of the 1939 declaration is one often overlooked in Law of the Sea scholarship but central to histories of neutrality law and kindred subjects. That is that the 1939 proclamation represented, as C. G. Fenwick (1940, 119) pointed out at the time (perhaps a bit wishfully), "the first application in inter-American relations of the procedure of consultation" as the fruit of the 1936 Inter-American Conference for the Maintenance of Peace in Buenos Aires in 1936 and the 1938 Eighth International Conference of American States in Lima. Though this consultation might have not been sustained, this moment contributed to the preparation for even broader claims of oceanic governance in the postwar period via the protrusion of space, if one accepts Schmittian terms.

The United States was not alone in supporting the idea of recreating sovereignty and jurisdiction over the continental shelf even though it was the first nation to do so. The idea was not a secret. Before the September 1945 Truman Proclamation, the US had in fact shared a draft of it with 11 nations, though Cuba and Mexico were the only ones in Latin America included in this group. The others were Canada, Denmark, France, Great Britain, Iceland, The Netherlands, Norway, Portugal, and the Soviet Union. No nation registered disapproval. (D'Amato 2011).

In the official proclamation, Truman said the reasons for it were straightforward. The US was "aware of the long range world-wide need for new sources of petroleum and other minerals" and understood that its own continental shelf had large reserves accessible. Truman said that "with modern technological progress their utilization is already practicable or will become so at an early date." Therefore, it was:

> the view of the Government of the United States that the exercise of jurisdiction over the natural resources of the subsoil and sea bed of the continental shelf by the contiguous nation is reasonable and just, since the effectiveness of measures to utilize or conserve these resources would be contingent upon cooperation and protection from the shore, since the continental shelf may be regarded as an extension of the land-mass of the coastal nation and thus naturally appurtenant to it, since these resources frequently form a seaward extension of a pool or deposit lying within the territory, and since self-protection compels the coastal nation to keep close watch over activities off its shores which are of the nature necessary for utilization of these resources; the Government of the United States regards the natural resources of the subsoil and sea bed of the continental shelf beneath the high seas but contiguous to the coasts of the United States as appertaining to the United States, subject to its jurisdiction and control . . . . The character as high seas of the waters above the continental shelf and the right to their free and unimpeded navigation are in no way thus affected.
>
> (Proclamation 1946, 45–48)

The linkages here are important. Truman fused sovereigntist claims over the resources with hoary territorial jurisdictional claims first invented to claim sovereignty over guano islands in the Pacific in the 1850s (the term "appurtenances") with national security assertions and strong support for "free and unimpeded navigation." Truman captured the complicated and contradictory assertions of American foreign policy with great precision.

Mexico had long maintained an expansive interpretation of its territorial sea to a non-standard limit of nine nautical miles. Just a month following Truman's Proclamation, Mexico released its own presidential declaration with respect to the continental shelf. Couched carefully in language of "the growing need for states to conserve these natural resources," this affirmed that the continental shelf, bound by the "isobaths," is an "integral part of the continental countries." It would not be "wise, prudent, or possible for Mexico to renounce jurisdiction and control over and utilization of that part of the shelf which adjoins its territory in both oceans." Like the United States, Mexico claimed jurisdiction over the resources of the shelf, which it enumerated clearly to include "liquid and gaseous minerals, phosphates, calcium, hydrocarbons, etc. of inestimable value whose legal incorporation into the national property is urgent and cannot be delayed." Mexico declared "control" over the resources, though it did not indicate that this meant sovereignty over the shelf territory, or merely control as an adjunct to jurisdiction. Mexico also asserted the right to govern and control the fisheries of the shelf, and all other resources "whether known or unknown." It did recognize "the lawful rights of third parties," but carefully added "based on reciprocity," which indicated a hedge against the likelihood of broad claims coming (*El Universal* 1945, 17).

Despite what might be seen as extensive jurisdictional claims, Mexico ultimately sought congruence with the international order and aligned its constitutional language to reflect the UNCLOS language regarding the continental shelf as it emerged. The precise nature of jurisdiction over the territory and the resources remained highly contested in Mexican law and state practice. Sovereignty, which Bernardo Sepúlveda Amor, a former Secretary of Foreign Affairs defined as "power to mandate," included jurisdiction both over access to the territorial sea and "control in an exclusive manner . . . in the territorial sea, its seabed and subsoil" (quoted in Vargas 2011, 7). This approach, which was not universally held in the Mexican state or among Mexican legal scholars, fuses sovereignty and jurisdiction in ways that other states sidestepped or elided. The situation was even more complex, as Mexican legal scholar Jorge A. Vargas points out, in the odd situation of Mexico's international stance on the continental shelf through UNCLOS I. Diplomatically Mexico followed the relatively limited position of separating what UNCLOS I called "sovereign rights for the purpose of exploring it and exploiting its natural resources," or jurisdiction over resources but not full sovereignty over territory, while internally amending its Political Constitution to claim "ownership of the waters of the seas that cover the continental shelf and the submarine shelves." However, this contradiction was not immediately resolved and the amendments were never fully promulgated despite passage (Vargas 2011, 26.).

Argentina grappled with the same issues with greater forcefulness and some creativity. It had in fact claimed rights over the resources of the continental shelf in January, 1944, though it restricted this claim to "temporary zones of mineral reserves." At this time, Argentina recognized "the zones at the international frontiers of the national territories," as well as "the zones of the oceans coasts," and "the zones of the epicontinental sea of Argentina" (*Boletín Oficial* 1944, 61). This jurisdictional claim was soon expanded when it was determined unilaterally that the nation owned the deposits to the limits of the continental shelf.

The concept of the "epicontinental sea," the linchpin of this assertion, was not entirely well defined outside of the Argentine declaration and was, in fact, novel in international law. It had evocative symbolic power wrapped in a sovereign declaration which should not be overlooked and which consequently should not be diminished as a mere "erroneous interpretation of the 1945 Truman Declaration," as has been claimed by some. (Zacklin 1974, 77). Rather than a derivation and a mistake, this decision signaled a broadened approach brought by the Latin American states. In fact the epicontinental sea idea would relatively soon evolve into the concept of a "patrimonial sea," which always included a pronounced unilateralism in its definition, and which signaled a connection between "maritime jurisdictional redefinitions" and "deep symbolic effect" (Breglia 2013, 290). Argentina rolled out the term to mean "the waters covering the submarine platform," which mirrored the continental shelf but also seemed to expand it (Novoa 1988, 27; 225–226). There was, after all, no "precise definition of the shelf" (Auguste 1960, 47).

While deliberately building on the new space provided by Truman and Mexican president Avila Camacho, the bold Argentine decree of September 11, 1946 seemingly invented a new spatial and legal category by claiming that the rights to the epicontinental sea were "implicitly accepted in modern international law" combined with "the realm of science." It pushed a new emphasis on national sovereignty on both the "epicontinental sea" and the continental shelf. This was a significantly more expansive statement than Truman had made, as the US had claimed jurisdiction but not sovereignty. Argentine indicated that "free navigation . . . remains unaffected" but the sea and shelf were "subject to the sovereign power of the nation" (*Concerning National Sovereignty* 1946).

In June 1947, Chile followed in the footsteps of Mexico and Argentina and made its own broad claims. Its proclamation is very instructive to read because it signals a very ambitious program, as well as the inaccurate assertion that "international consensus of opinion recognizes the right of every country to consider its national territory any adjacent extension of the epicontinental sea and the continental shelf." To begin, the Chilean proclamation rather sweepingly and inaccurately describes the U.S., Mexican, and Argentine positions as having "categorically proclaimed the sovereignty of their respective states over the land surface of [the] continental shelf adjacent to their coast, and over the adjacent seas within the limits necessary to preserve for the said states the natural resources belonging to them." In fact, as demonstrated here, there was quite a bit of variation in the sovereignty claims. Like the other states, Chile asserted its support for conservation of its fisheries. Most centrally it frankly noted in section 3:

> it is manifestly convenient . . . to issue a similar proclamation of sovereignty, not only by the fact of possessing and having already under exploitation natural riches essential to the life of the nation and contained in the continental shelf . . . but further because in view of its topography and the narrowness of its boundaries, the life of the country is linked to the sea and to all present and future natural riches contained within it, more so than in the case of any other country.
>
> (*El Mercurio* 1947, 27)

The Chilean proclamation made three major claims. The first was "national sovereignty over all the continental shelf adjacent to the continental and islands coasts of its national territory." The second claim was "national sovereignty over the seas adjacent to its coasts whatever may be their depths." All fishing and whaling activities were explicitly included in this national sovereignty. Finally, and most radically, Chile claimed "protection and control" of a zone stretching across the "perimeter formed by the coast and the mathematical parallel projected into the sea at a distance of 200 nautical miles." This included the 200-mile zone around all Chilean islands. Chile tacked on the same ending phrase as did Mexico, claiming that its "present declaration of sovereignty does not disregard the similar legitimate rights of other states on a basis of reciprocity, nor does it affect the rights of free navigation on the high seas" (*El Mercurio* 1947, 27).

This was a major bombshell of a claim, projecting a 200 nautical-mile enclosure around the sea and definitely extending space into the sea in novel ways. Where the Truman Proclamation had divided sovereignty and jurisdiction over the continental shelf into two discrete issues, and Mexico had mirrored this argument, Chile used and intensified the Argentine position to make entirely novel claims of full and expansive sovereignty. To complicate matters, the next year Chile passed a law that said that the "adjacent sea up to a distance of fifty kilometers, measured from the low-water mark, constitutes the territorial sea and belongs to the national domain, but the right of policing" in terms of customs and security "extends up to a distance of 100 kilometers measured in the same manner" (*Diario Oficial* 1948, 258). This further subdivided sovereignty and jurisdiction in another layer of national authority over this oceanic space.

Conclusion

This chapter examined sovereignty assertions over oceanic space pioneered by Mexico, Argentina, Chile, and Peru while also taking seriously the language employed in staking jurisdictional claims over the resources of the continental shelf. The chapter observed how these new sovereign state spaces were carved out from within a fluid international regime, and set the global legal scaffolding for unilateralist approaches to global oceanic order and resource exploitation. Accordingly it is possible to situate this legal story within the emerging literature on strategic raw materials (Ingulstad 2015), and to begin to uncoil the contingencies of the legal, the spatial, the economic, and the environmental in Latin American policymaking as they emerged in the early Law of the Sea regime.

All of these elements were apparent in August of 1952, when Chile, Ecuador, and Peru met in Santiago for the "First Conference on the Exploitations and Conservation of the Maritime Resources of the South Pacific." The subsequent Declaration became the basis for a worldwide revolution in sovereign and jurisdictional claims over the resources outside the traditional boundaries of the state. By enclosing the oceanic and subsoil resources of an enormous new area, these nations profoundly changed the terms of the discussion. The effort typifies the production of new state space discussed earlier, while also signaling new ways for the state to realize its sovereignty over newly articulated spaces. It is important to distinguish between the transformative implications of these assertions and the actual realization of the "complex, dualistic scheme of both national and supranational jurisdiction," which was incremental and heavily contested (Cottier 2015, 45).

Nevertheless, it is essential to recognize how dramatic the new assertions were in shifting the global legal discussion into new domains. Relying on "geological and biological factors" (and "having no continental shelf along their coasts" (Hjertonsson 1972, 22)), the states determined that "the former extent of the territorial sea and contiguous zone is insufficient" for conservation. The three states thus claimed "sole sovereignty and jurisdiction over the sea adjacent to the coast" to the 200 nautical-mile limit. They also claimed "sole sovereignty and jurisdiction" over the seafloor and the subsoil, and over the islands (Declaration of the Maritime Zone).

Possibly most interesting is that Chile, Ecuador, and Peru were not planning to wait for the international community to validate their claims. They considered this proclamation to be the initiation of a new international legal regime, and declared their intention to sign treaties to establish regulations to make it a reality. The Conference therefore also released conventions creating a conference on the conservation of marine resources, a Joint Declaration on Fishery Problems in the South Pacific, and Regulations Concerning Whaling in the Waters of the South Pacific. Two years later in Lima, the three reaffirmed their claims of sovereignty and agreed to consult on any issues arising. They also all pledged in article 4 "not to enter any agreements, arrangements, or conventions which imply a diminution of the sovereignty over the said zone" (Agreement Supplementary 1956, 730).

The United Nations' effort to codify the Law of the Sea was not a seamless line from the assertions discussed here to UNCLOS I. The International Law Commission worked from 1949 for seven years to produce a draft which only then became the basis of the final two years of global negotiations. The five parts of this Convention directly spoke to the approaches raised by the Latin American states in terms of defining the territorial sea, the high sea, and the continental shelf. Yoshifumi Tanaka (2012, 23) points out that UNCLOS I determined that the "continental shelf in the legal sense is part of the seabed and subsoil of the high seas." The fact that these definitions in terms raised new questions over the extent of the territorial sea while sidestepping or leaving open the most challenging aspect in terms of the growing consensus on the exclusive economic zone, which is traced in "genesis and development . . . back to Latin America" (Vargas 2011, 132), further underscores the reality that the major Latin American assertions laid the groundwork for a half-century of legal reordering in ways that permanently shifted the ideas of sovereignty and jurisdiction.

Nothing in the Law of the Sea assertions occurred in a discontinuous vacuum of interstate practice. The ordering of space via jurisdiction blossomed as a result not just of the new postwar diplomatic context and new technological developments, but also in a period of new approaches to governmentality. The novel legal assertions of Latin American governments grew from a long history of sovereign claims recapitulated and newly articulated in this era. The new territorial and extraterritorial order yielded radically transformative approaches to the definition of oceanic space for the rest of the century.

References

Agreement Supplementary to the Declaration of Sovereignty over the Maritime Zone of Two Hundred Miles, Second Conference on the Exploitation and Conservation of the Marine Resources of the South Pacific, Lima, December 1–4, 1954. Agreement

Supplementary to the Declaration of Sovereignty over the Maritime Zone of Two Hundred Miles, U.N. Legislative Series ST/LEG/SER.B/6, 1956, 729–730.

Auguste, Barry B. L. 1960. *The Continental Shelf: The Practice and Policy of the Latin American States with Special Reference to Chile, Ecuador, and Peru; A Study in International Relations.* Genève: Librairie E. Droz.

Boletín Oficial de la República Argentina. 17 March 1944, 52: 61

Breglia, Lisa. 2013. "Energy Politics on the 'Other' U.S.–Mexico Border," in Sarah Strauss, Stephanie Rupp, and Thomas Love, eds., *Cultures of Energy: Power, Practices, Technologies.* Walnut Creek, CA: Left Coast Press.

Concerning National Sovereignty over Epicontinental Sea and the Argentine Continental Shelf. 1951. U.N. ST/LEG/SER B/1, 4–5.

Convention Establishing an Organization of the Standing Committee of the Conference on the Use and Conservation of the Marine Resources of the South Pacific, First Conference on the Exploitation and Conservation of the Marine Resources of the South Pacific, Santiago, August 11–18, 1952. U.N. Legislative Series ST/LEG/SER.B/6, 1956, 724–726.

Costa, Eduardo Ferrero. 2013. "Latin America and the Law of the Sea," in Harry N. Scheiber and Jin-Hyun Paik, eds., *Regions, Institutions, and Law of the Sea: Studies in Ocean Governance.* Leiden: Martinus Nijhoff.

Cottier, Thomas. 2015. *Equitable Principles of Maritime Boundary Delimitation: The Quest for Distributive Justice in International Law.* Cambridge: Cambridge University Press.

D'Amato, Anthony, "What 'Counts' as Law?" (January 18, 2011) Northwestern Public Law Research Paper No. 11–02. Available at SSRN: http://ssrn.com/abstract=1743148, 104–105.

Declaration on the Maritime Zone. 1952. U.N. Legislative Series ST/LEG/SER.B./6. 1956, 723–724.

Delaney, David. 2011. *The Spatial, the Legal and the Pragmatics of World-Making: Nomospheric Investigations.* London: Routledge.

Diario Oficial. 71, (11 February 1948), p. 258, translated in U.N. ST/LEG/SER.B/1, 61.

——. vol. 1 (31 August 1935), pp. 28–29, translated in U.N. ST/LEG/SER.B/A, 86–87.

Dorsett, Shaunnagh and Shaun McVeigh. 2013. *Jurisdiction.* London: Routledge.

Fenwick, C.G. 1940. "The Declaration of Panama," *The American Journal of International Law* 34, No. 1 (Jan.): 116–119.

Foucault, Michel. 2009. *Security, Territory, Population: Lectures at the Collège de France 1977–1978.* New York: Picador.

Gidel, Gilbert. 1951. *La Plataforma Continental ante el Derecho Internacional.* Valladolid: Spain.

Hjertonsson, Karin. 1972. *The New Law of the Sea: Influence of the Latin American States on Recent Developments of the Law of the Sea.* Leiden: A.W. Sijthoff.

Ingulstad, Mats. 2015. "The Interdependent Hegemon: The United States and the Quest for Strategic Raw Materials during the Early Cold War." *International History Review* 37(1): 59–79.

Johnston, Douglas M. 1988. *Theory and History of Ocean Boundary-Making.* McGill-Queens University Press.

Joint Declaration on Fishery Problems in the South Pacific. 1952. First Conference on the Exploitation and Conservation of the Marine Resources of the South Pacific. U.N. Legislative Series ST/LEG/SER.B/6, 1956, 726–727.

Kirchner, Andree. 2007. "The Outer Continental Shelf: Background and Current Developments," in Tafsir Malick Ndiaye and Rüdiger Wolfrum, eds., *Law of the Sea, Environmental Law and Settlement of Disputes*. Leiden: Brill.

Lagendijk, Arnould, Bas Arts, and Henk van Houtum. 2009. "Shifts in Governmentality, Territoriality and Governance: An Introduction," in Arts, Bas, Lagendijk, and van Houtum, eds., *The Disoriented State: Shifts in Governmentality, Territoriality and Governance*. Berlin: Springer.

Lefebvre, Henri. 2009. *State, Space, World: Selected Essays*. Edited by Neil Brenner and Stuart Elden. Minneapolis: University of Minnesota Press.

MacDougal, Myres Smith and Florentino P. Feliciano. 1994. *The International Law of War: Transnational Coercion and World Public Order*. Dordrecht: Martinius Nijhoff.

Massey, Doreen. 2005. *For Space*. London: Sage.

El Mercurio, 29 June 1947, 27, translated in U.N. ST/LEG/SER.B/1, 6–7.

Novoa, Ezequiel Ramírez. 1988. *El nuevo derecho del mar y las 200 millas de mar territorial*. La Victoria: Amaru Editores.

Prescott, J. R. V. 1985. *The Maritime Political Boundaries of the World*. Methuen: London and New York.

"Proclamation by the President with Respect to the Natural Resources of the Subsoil and Sea Bed of the Continental Shelf," *The American Journal of International Law* 40: 1, Supplement: Official Documents (Jan. 1946): 45–48.

Rosenberry, William. 2002. "Understanding Capitalism—Historically, Structurally, Spatially," in David Nugent, ed., *Locating Capitalism in Time and Space: Global Restructurings, Politics, and Identity*. Stanford: Stanford University Press.

Rothwell, Donald R. and Tim Stephens. 2010. *The International Law of the Sea*. Oxford: Hart Publishing.

Scharf, Michael P. 2013. *Customary International Law in Times of Fundamental Change: Recognizing Grotian Moments*. Cambridge: Cambridge University Press.

Schmitt, Carl. 2003. *The Nomos of the Earth in the International Law of the Jus Publicum Europaeum*. New York: Telos Press.

Sevilla-Buitrago, Alvaro. 2015. "Capitalist Formations of Enclosure: Space and the Extinction of the Commons." *Antipode* 47(4): 999–1020.

Szekely, Alberto, ed. 1986. *Latin America and the Development of the Law of the Sea*, 2 vols, release 86–1. London: Oceana Publications.

Tanaka, Yoshifumi. 2012. *The International Law of the Sea*. Cambridge: Cambridge University Press.

United Nations. U.N. Legislative Series. *Regulations Concerning Whaling in the Waters of the South Pacific, First Conference on the Exploitation and Conservation of the Marine Resources of the South Pacific*, Santiago, August 11–18, 1952. ST/LEG/SER.B/6, 1956, 727–729.

El Universal, D.F., 116:10,541 (30 October 1945) 1, 17, translated in U.N. ST/LEG/SER.B/1, 13–15.

U.S. State Department Bulletin 1939 1: 331–333.

Valverde, Mariana. 2009. "Jurisdiction and Scale: Legal 'Technicalities' as Resources for Theory," *Social & Legal Studies* 18: 144–145.

Vargas, Jorge A. 2011. *Mexico and the Law of the Sea: Contributions and Compromises*. Leiden: Martinus Nijhoff.

Yang, Haijiang. 2006. *Jurisdiction of the Coastal State over Foreign Merchant Ships in Internal Waters and the Territorial Sea*. Berlin: Springer.

Young, R. 1951. "Legal Status of Submarine Areas Beneath the High Seas," *AJIL* 45: 225.
Yturriaga, José Antonio de, 1997. *The International Regime of Fisheries: From Unclos 1982 to the Presential Sea*. The Hague: Kluwer Law International.
Zacklin, Ralph, 1974. "Latin America and the Development of the Law of the Sea: An Overview," in Zacklin, ed., *The Changing Law of the Sea: Western Hemisphere Perspectives*. Leiden: Sijthoff.

6 State building and problematic geopolitical spaces in South Asia

The Himalayas and the extradition treaty of 1855

Alastair McClure

Introduction

As well as invigorating older models of study, the advent of transnational history has encouraged alternative focuses for scholarly work. Amongst these are the histories of geographical spaces, in particular border regions, which have fitted uncomfortably amongst orthodox histories of the formation of nation-states and the modern international order. In the midst of their respective state building drives of the nineteenth century, the Himalayan border between colonial India and Nepal (see Figure 6.1) remained a relatively unmanageable terrain, constantly threatening the quest of these states for sovereign power within a closed geographical territory. As these ruling powers began to take steps towards modern statehood, the protection of trade, control of movement, and concerns surrounding rebellious networks became increasingly important issues. Once it was clear that neat lines of sovereignty could not be drawn across the Himalayan border, the region became a site of power play between these emerging neighboring states and the individuals that transgressed the thresholds. These issues of security and power consistently spilled over ill-defined boundaries as criminals and men of "bad character" resided in or traveled through this borderland to escape punishment. The existence of such escapees put these states in a position where they had to balance the protection of their own carefully constructed perceptions of sovereign power through the retrieval of the absconders, with action that respected the sovereignty of their neighbors from whose territory they needed to be retrieved. Piecing together information from the correspondence between the Resident of Nepal and the British Government of India, this chapter seeks to offer insight into the "specific cultures of governance" possessed by these states, drawn out through an analysis of their behavior in this borderland (Michael 2012, 3). The chapter will seek to outline how the priorities of Nepal and colonial India converged and clashed through attempts to pacify this region, particularly in relation to the formation of laws of extradition and the policing and maintenance of their borders. In doing this, it is also important to highlight the emergence of an incipient language of modernity in relation to law and punishment on the international stage. Through this it will be shown how quickly this emerging language of modernity was manipulated as a tool to sway larger political battles.

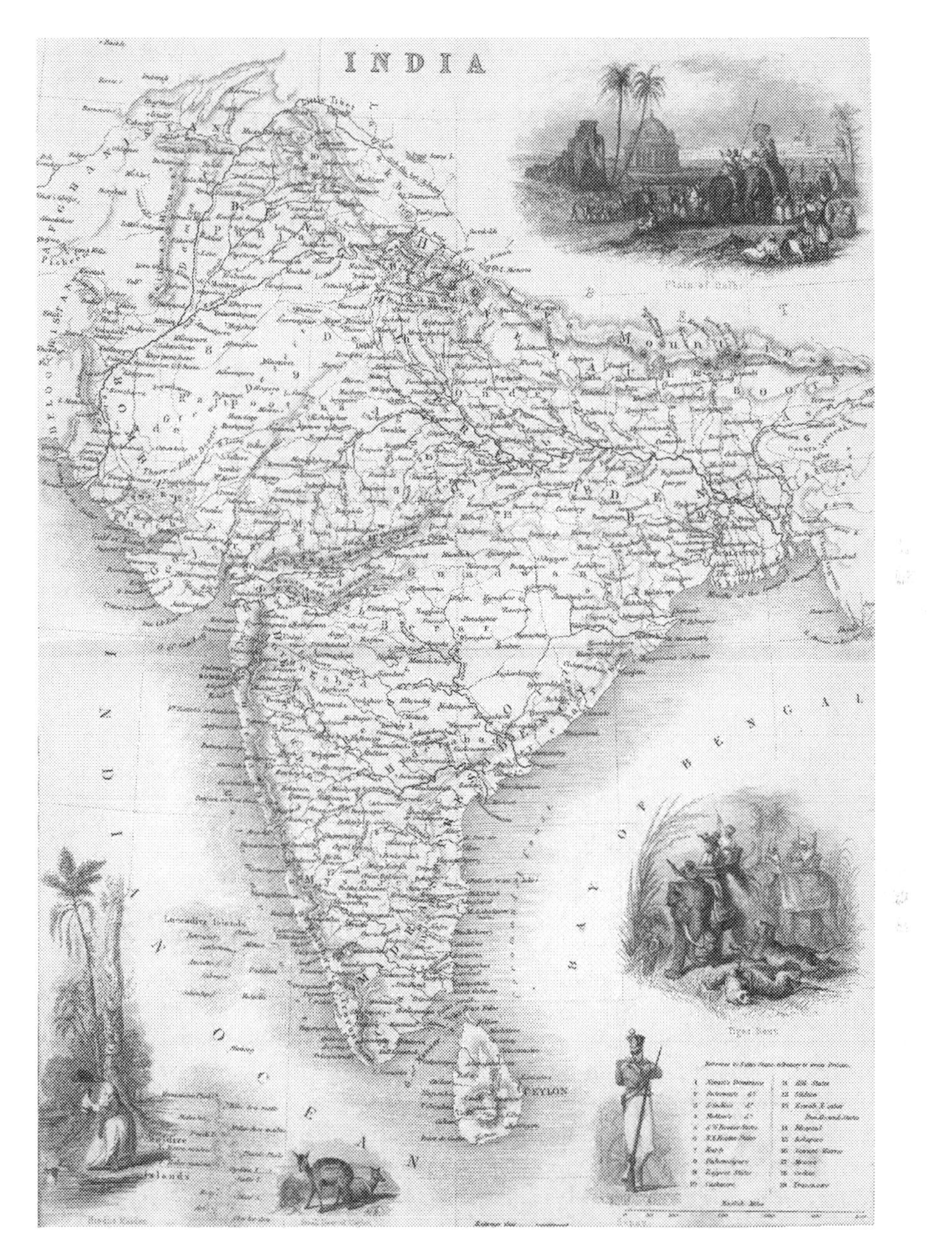

Figure 6.1 Map of the Himalayas. "India 1858," image, courtesy, Center for Study of the Life and Work of William Carey, D.D. (1761–1834), William Carey University, Hattiesburg, Mississippi, USA.

In the 1850s and 1860s, both India and Nepal saw significant changes to their structures of governance and rule as both began centralizing various institutions in response to violent episodes in their respective pasts. Throughout the eighteenth and nineteenth centuries the nature of the British presence in India had evolved from its humble roots as a purely commercial trading company into an organization progressively resembling that of a ruling government. The East India Company's informal empire had grown rapidly in the early nineteenth century, with military victories over regional powers in the subcontinent such as the Marathas and the Sikhs. To compound this military success, the Company's opportunistic Doctrine of Lapse policy had set out to aggressively engulf many princely states on any occasion in which an heir to the throne could not be shown. The effect of these policies was to bring large parts of the previously disparate subcontinent under one rule for the first time, but in so doing also destabilized many traditional structures of power. This rapid process of expansion was halted by the rebellion of 1857 in which large numbers of Indians revolted across North India, briefly reinstating a Mughal King to Delhi and rejecting the legitimacy of the British rule. Although it was eventually quelled, the British were deeply shaken by the hugely violent and seemingly spontaneous nature of this event. In response, the East India Company was dissolved and replaced by official British Government of India rule, with Queen Victoria now as its sovereign figurehead. Borders were consolidated rather than expanded, and the Doctrine of Lapse was ended as the authority of princely families was accepted. To manage this space which was now more static, a centralized state machine was built which included an Indian Penal Code, a police force, and a series of High Courts.

In Nepal similar changes were underway. Following the Anglo-Gorkha War in 1816 that ended the territorial expansion of the ruling Gorkhali class, the Nepalese political landscape underwent 30 years of considerable instability, with attempted coups, plots and assassinations. The situation was partially stabilized only after the Kot Massacre of 1846, instigated by Jung Bahadur who organized the murder of his key political rivals, which commenced a 104-year reign of uninterrupted Rana rule in Nepal (Pandey 1973; Vaidya and Manadhar 1985, 187). While, broadly speaking, borders had remained unchanged since 1816, his accession inaugurated a new stage of Nepalese governance for internal policy and foreign diplomacy. Once in power Jung Bahadur began his own state-building exercises. These included the consolidation of the judicial system through the introduction of a penal code in 1854 that reduced the use of capital punishment and discouraged the act of *sati* (widow immolation) (Adam 1950, 160–162; Adikhari 1976; Vaidya and Manadhar 1985, 187–283). State machinery was also expanded, and by the early 1860s Nepal had a civil bureaucracy which employed salaried government officials rather than private individuals to collect revenue (Regmi 1975, 106–107). In terms of foreign policy, Nepal had recognized a shift in power between their Chinese and British neighbors. As a landlocked buffer state wedged between these two much larger territories, Nepal had found its ability to play the two off against each other to be its biggest diplomatic strength. However, after British military victories over China in the Opium Wars and strengthened their position in the subcontinent, a conscious change was made to place emphasis on courting the British. In light of this, the Maharajah, somewhat of an Anglophile to begin with, made an official visit to England in 1854 in what was considered the first diplomatic mission of its sort by a Nepalese leader. He also acted as a consistent ally in the region, offering help to British India during the uprising of the Sikhs in 1848 and the rebellion of 1857 (Sever 1996, 89). His loyalty in the rebellion earned him a knighthood and also regained Nepal the lost and economically valuable Terai territory (Whelpton 1991).

In this period of state building, the number of those who had escaped across borders to avoid punishment had grown; this necessitated a point of conversation which during this period of increasingly cordial British–Nepalese relations was in the interests of both governments to engage in. Once officially enacted in 1855, the existence of a formal extradition law had immediate relevance for both parties. While many of the 1857 rebels had fled into the mountainous jungle terrain of Nepal to avoid punishment, Jung Bahadur had enemies who had escaped across the border and planned assassinations from the safety of northern Indian cities such as Benares. The fact that this border locality was only semi-governable represented an omnipresent threat, which could potentially expose the carefully constructed narratives of sovereign power within a closed geographical space as being artificial and weak. Both the British Raj and the Nepalese leadership underpinned their legitimacy of rule with the ability to project a perception of supreme power within their territories; hence, the individuals or groups who undermined this narrative would cause embarrassment by flouting laws and then disappearing unpunished across porous borders. The recovery of these individuals became a vital redemptive process of restoring the territorial supremacy of state power. To understand this process, I turn first to the formation of a hard extradition law and the emerging language of international relations encapsulated in this redemptive process.

Border politics and the construction of the extradition treaty of 1855

Nepal's relationship with British India had taken a decisive turn with the Treaty of Segouli in 1815 which saw Nepal give up the Terai region and accept a British Resident in Kathmandu. Both stipulations were held in high disdain by the Nepalese who felt they were colonial impositions on a strictly independent state (Thapliyal 1998, 19). Moreover, while the accession of Jung Bahadur had instigated a much warmer policy towards British rule, Nepal was still fiercely isolationist and its foreign diplomacy was primarily motivated by a desire to maintain its sovereignty as an independent nation. Nepal therefore still maintained many of the closed door policies of former governments. This meant a continued rejection of European trade and stringent rules regarding the access given to the British Resident of Nepal (Pradhan 1996, 123; Whelpton 2005, 47). As the two states became more closely engaged in forging treaties and laws together, an underlying tension arose between maintaining amicable relations in order to control movement across the border and the need to simultaneously ensure that their respective sovereignties were not undermined through the production of unequal international law. Through negotiation, enactment, and enforcement, the history of the extradition treaty in 1855 was one example that demonstrated international law's inability to provide the reciprocity and equality it claimed to offer.

The law of extradition takes its historical precedence from the Roman law of *noxae deditio* in which the father or master of a criminal could be held accountable for their criminal action if he or she was not brought to justice (Magnuson 2012, 847). Most historiography plots the history of extradition from a position of Eurocentrism, charting a noticeable swing in the focus of extradition from political to common crimes during the nineteenth century. This is explained due to the impact of the French and Industrial revolutions, the former acting to give legitimacy to political criminals who overthrew despotic governments,-while the latter had brought advances in faster modes of travel that made borders much more accessible for the common fleeing criminal (Magnuson 2012, 851). The politics of extradition between Nepal and British India in the mid-1850s offers a complicated reflection on both the internal and international political histories of the two countries but also an alternative to the Western-centric narrative in which, certainly for the British Indian Government, it was political criminals who remained the target.

Extradition across the Himalayas between Nepal and India had been based previously on informal agreement. However Nepal had remained dissatisfied at the state of this relationship. Colonel Ramsey, the Resident in Nepal, described the Bahadar's feelings concerning extradition practice in 1852 as follows:

> Much discontent prevails in Nepal in consequence of his invariably ordering the surrender to our local authorities of refugees of every description, when in all cases his own application for the surrender of fugitives from these territories has been unsuccessful. Many of the members of the council are incensed with the Minister for attending to these applications, unless he is assured by the resident that the fugitives applied for have been guilty of heinous crimes, in which cases they would invariably by surrendered without demur.[1]

In lieu of Nepal's concern over extradition in its informal status it was the Nepalese who were the driving force for the configuration of a formal law of extradition in 1855, seen as a solution to this inequality.[2] The treaty was a transition from an agreement in 1849 that extended only to British subjects or persons in the service of the East India Company, and then a subsequent treaty in 1854 which extended the discussion to all subjects.[3] The final treaty officially marked the transition from soft to hard law and was heralded as representing a new precedence of reciprocity and negotiation between two independent states.

However, even in the treaty negotiations the roots of the later occurring tensions were already visible. The first point of contention was defining how far the law would extend. From the start of negotiations British India was keen to minimize the number of applicable offenses to only heinous crimes, while Nepal sought to apply the law to all forms of crime. Though the earlier 1854 treaty had included the crimes of embezzlement and robbery, further negotiations saw Nepal's requests ignored as crimes were systematically removed to fit the wishes of the British Government.[4] After this setback the Nepalese Government then sought to prioritize cattle theft, which held significant religious importance and carried heavy punishments in a country ruled by Hindu royalty. After this was also refused, the Resident commented, "So serious was the Durbar that the crime should be included in the Treaty of Act 7 of 1854 I at one time feared that negotiations would be broken down in consequence."[5] This early political exchange which defined exactly what this "strictly reciprocal" hard law would resemble left Nepal with a heavily weakened treaty that accepted British wishes at the expense of the wishes of the Nepalese. While the treaty was formed on the notion of "reciprocity" between the two states as equals before the new law, the negotiations of what the treaty exactly meant foregrounded the dominant position which colonial India sought to maintain in the space of international politics in South Asia.

After the enactment of the treaty the behavior of the British in accordance with the law caused further problems. While criticism can be found in British quarters concerning the speed with which Nepal returned criminals, by 1858 the estimated 2,000 rebels that had passed through Nepal were said to have all been extradited, for which "great credit to Jung Bahadoor for his arrangement and exertions" was acknowledged. Yet the British attitude towards "reciprocity" had not changed. In 1863 Colonel Ramsey wrote:

> I think I am correct in saying that, for one reason or other, not one formal surrender has ever yet been made to Nepal since the Treaty came into force. There has either been an insufficiency of evidence, or as unfortunately happened more than once, the accused parties have escaped from custody.

Ramsey later commented the treaty was so disregarded within Nepal that it was seen as a "dead letter" in the eyes of the Nepalese Darbar.[6] The following year, when the British had requested the extradition of Mohun Singh who was accused of the theft of public money (a crime not even within the jurisdiction of the treaty), the Commissioner Ramsey commented:

> The Extradition Treaty is, as I have before reported, a very sore subject with the Goorkhas, and, to the best of my belief, not a single criminal has ever yet been given up, under its terms, to the Nipalese Government. The personal attendance of witnesses has always been exacted from them, and either the evidence tendered has been pronounced to be insufficient and unsatisfactory, or, in several instances, prisoners have made their escape from our Police.[7]

The British had spent significant effort reducing the treaty through political negotiations, enabling them to negotiate enough space in which they could comfortably reject extradition requests by stating reasons such as a lack of evidence. The law which had been at least rhetorically built on a foundation of equality had instead made the border a very visible site of disparity, such that the treaty had formalized the ability of British India to exert strength in relation to Nepal.

If skepticism over the evidence or nature of the crime was a common reason for the breakdown of extradition, the British also refused extradition because of perceived civilizational differences between Nepalese and British rule. In an earlier remark made in 1834, a magistrate had commented that, while a footing of reciprocity between independent governments was important, "With respect to Nipal the extreme severity of its criminal laws renders it desirable that such mutual surrender of fugitive should be subject to those restrictions which humanity would dictate."[8] This concern over Nepal as a barbaric and inhumane state remained a factor after the law had been formalized. While criticizing this position one British officer wrote in 1865:

> In its Extradition Treaties with the civilized nations of Europe, the British Government makes no reservation in favor of its own subjects yet it is right, no doubt, to make such a reservation in the case of a Treaty with a country in the condition of Nipal; it is a reservation which some European countries insist upon even in dealing with nations the most civilized. But I cannot, I confess, see the force of the argument that it is unjust on similar grounds i.e. because a country is not administered on comparatively enlightened, just, or humane principles of jurisprudence, to surrender to a Foreign State a criminal it *native born subject*.[9]

The notions of what "humanity would dictate" and international treaties attached to indefinable "reservations" represented the language of civilizational hierarchy creeping its way onto the stage of international lawmaking. The construction of an intangible moral compass that presumed British rule's natural connection to "humanity" saw the extension of their civilizing mission in the form of a narrative practiced internally and extended to relationships with independent states. For the British, Nepal would never represent anything other than a "semi barbarous" state.[10] This was a perception that remained unaffected even by Nepalese reforms that had purportedly consciously replicated Western values, an argument that Jung Bahadur pointed to in relation to the reforms of the judicial system.[11] The reality was that the acceptance of difference in forging a stage of equality was never the goal of these laws on the subcontinent. Instead, the treaty represented an opportunity for the consolidation of difference, allowing British rule to remain at the top of their self-created and self-enforced civilizational hierarchy.

While the border space could never be totally pacified from the dangers of rogue individuals, the negotiations held over extradition offered an opportunity for the space to be dominated politically. In light of this, British India's policy with the Nepalese border was created on a useful contradiction. In entering an agreement on the implicit acceptance of the sovereign character of both states, the British Government then retrospectively used the legal system of Nepal as a reason to refuse to cooperate. In doing so, an abstract field of equality had been created which immediately excluded Nepal from membership; as such, the ability of a nation to speak the language of civility in South Asia was only superficially attached to holding universal values, and was much more practically attached to power. Returning to the questions of the historiography of international law, this moralizing language struggles with some of the dominant narratives in the field. In extradition law this related to the rule of non-inquiry, the suggestion it was not the job of the government to impose their notions of morality and justice on other sovereign nations when considering extradition. The rule of non-inquiry has generally been seen as the status quo in extradition relationships until the impact of Nazi Germany and the influence of twentieth-century human rights movements forced change. William Magnuson, for instance, argued it was only after the Second World War that, "For the first time, the idea that international law should protect the rights of individuals started to gain traction in legal circles" (Magnuson 2012, 257). This episode, while only offering a small example, jars with Magnuson's argument and reminds us of the truly fragmented nature that anything claiming to be "international" must be, underlining the dangers of assuming a monolithic character, particularly in relation to laws acting across the world in countless and varied circumstances.

Policing borders, violence, and resistance

Beyond the high politics of international law, the Himalayan border being at the edge of both states also made forms of state violence and resistance to governmental control visible, which contributed to make resistance more manageable or containable in the metropolitan centers. While the treaty was supposed to regulate the correspondence between the states over the right to request criminals, the physical border was to be governed by mutually agreed rules to prevent either the police or army entering foreign soil illegitimately. These rules were described by one British official as being "neither clearly understood nor acted up to by some of our officers upon the frontier," and numerous complaints were made as British officers continually entered Nepalese land illegally, searching houses, making demands in villages for the surrender of rebels, and generally paying no respect to the Goorkha officers.[12] Another governmental report aired concern of "the mischievous effect of the disinclination which British Officers have to take up the complaints of Foreign States for fear of being driven into a complication."[13] At times these instances did not just represent the breaking of agreed border restrictions, but were acts of genuine criminality from the British border police. One interview records three police officers plundering a village and raping a villager.[14] Another incident in which five or six villagers were killed by police who had crossed the border in pursuit of a wanted dacoit was explained as "owing to the over zeal of the Sawars who had been sent in pursuit of Fuzul Alli who seem to have mistaken for Dakoits every person they fell in with."[15] These incidents frustrated and angered the Nepalese leader and his Darbar, and yet throughout the period showed no signs of stopping.

A lack of control in rural areas of the empire was not an uncommon thing, nor was it necessarily a hindrance to the colonial regimes stretched over large areas of geographically diverse land. Elizabeth Kolsky, for instance, has used the examples of tea plantations to engage with an idea of productive lawlessness. Rural and distant from metropolitan centers, Kolsky shows the plantations were located in regions in which a significant gulf existed between the rhetoric of just law and the reality of violence, simultaneous with the existence of ineffective legal structures. She explains these areas of contradiction as a necessary plank of colonialism in which unofficial violence and lenient criminal justice systems acted to support the enforcers of colonial power who existed on the fringes of empire, vulnerably placed away from the centers of its power (Kolsky 2010, 68). The argument can be stretched to these difficult geographical spaces on the borders. The border as a relatively unknown space was understood as a breeding ground for dacoits and criminals. Police reports were often found chastising these poorly governed areas, arguing they formed "a lurking place for bad characters, where they can plan their intended Dacoities, and also insuring them a safe retreat and harbor."[16] With the anxiety that the rebellion had produced, accentuating fears of the existence of rebellious criminal networks, the Himalayas could not be managed by the same methods practiced in other places where strong governmental structures were present. In response to this situation, the existence of less regulated violent policing continued to prosper as a response to fears of criminality in the region. The border thus became a physical embodiment of the contradictions present between rhetoric and practice in colonial rule, played out in through the difference and distance between metropolis and periphery.

Yet the border and the movement of people across it was not purely a space in which the British imposed their violent or political power. In the face of such domineering behavior from the British Government, pressure had been building within the Nepalese Durbar. Never completely free from potentially troublesome political groups within Nepal, elements of the Durbar had become increasingly vocal in their criticism of Jung Bahadur and his relationship with the British, exemplified by his trip in 1854. Aware of these murmurings and unhappy with the state of the law, Jung took a stand in 1863. Responding to an extradition request, the Maharajah expressed himself as being satisfied of the prisoner's guilt, but then observed:

> [A]s our Magistrates *always* refuse to receive documentary evidence and insist upon the personal attendance of the witnesses in *all* cases in which the surrender of fugitives from Nipal is applied for by the Goorkha Government, he will not dispense with their attendance in the Nipalese Courts when surrenders are demanded by the British Government, unless I will guarantee that for the future our Magistrates shall in like manner refrain from sending for them: this, of course, I have told him I cannot do, as our Officers upon the frontier are guided in such matters by the Regulations of their Government, from which they have no power to deviate.[17]

In response to this the British trod carefully, and the reply was written under orders to "be as conciliatory as possible in its tone."[18] Future requests were screened more heavily by the Resident who more vocally criticized the influx of criminal extradition requests that lay outside of the purview of the law, and by 1866 cattle theft was added to the treaty. Yet beyond the actual moment at which Nepal's patience snapped, the language of this rejection reflected the result of when an imbalance of power, in supposedly equal contracts, was constantly exploited. While British India had rejected numerous applications previously their dominance in the region meant they not only did not have to worry about the implications of these rejections, they could still expect Nepal's support with their own requests. In doing this there was never a need for the British to actively break the letter of the law they had signed. Through this practice, they could maintain their position as a "modern" state that implicitly supported modern and progressive laws. Alternatively, Nepal as the smaller and weaker state did not have the power to manipulate the treaty for its own gain or question the behavior of others when small liberties were frequently taken. The practicality of having such unequally weighted states in legal contracts led the less powerful Nepal into a more radical response in which it was forced to reject the "modern" language of law to regain a semblance of sovereignty.

Finally, while sources that offer significant information about the lives of the individuals who resided in these borderlands are hard to come by, a comment should be made concerning the Himalayan border as a space of relative agency for these non-state actors. While the extent of criminality in these areas was almost certainly exaggerated, and many criminals were caught, for some the Himalayas offered relative protection in which they could continue their lifestyles that would have been impossible in more strictly policed regions. Some criminals for instance made very conscious decisions to settle just over the border, which side depending on their crimes and the punishments they would receive. For Nepal in particular, many complaints were made that cattle thieves were committing crimes in both territories, but residing just over the British side because the proof of conviction was more stringent and the punishment lower.[19] With the Himalayan border proving itself unsuited to the methods of control favored by orthodox governance, the border became a space in which power relations were much more fluid than elsewhere, leading to a proliferation of violence and resistance from a variety of actors.

Conclusions

The unpacifiable nature of the mountainous terrain separating the two territories would remain a dark spot onto which the light of the power of the ruling states could never fully shine. In doing this the border as an analytical space can be a useful tool to uncover the inherent vulnerabilities of the emerging states, allowing scholars a deeper insight into their intentions, agendas, and character as they responded to this threat. As foreign rulers dependent on India's financial contribution to the metropole, British India relied heavily on two premises to ensure

British presence on the subcontinent, the immovable nature of their power and the benevolent and liberal nature of their laws. Through the politics of extradition negotiation and border control two things are particularly exposed from this period. First, the justificatory language of colonialism is starkly contrasted with the prevalence of illegal forms of violence by colonial officials in the territory. The colonial commitment to the rule of law is further undermined with the almost non-existent attempt to cooperate on a level of reciprocity with Nepal in respect to enforcing and managing the laws of extradition. Second, while exposing contradictions between British India's rhetoric and practice, the margins of the state also provided notable space for resistance. Non-state actors in particular were at times able to use these dark spaces to skirt the paradigms of nation-state politics and the relevant notions of crime and punishment, discernible with criminals making tactical decisions on which side of the border to stay depending upon what suited their criminal history the best.

Nepal alternatively was keen to assert itself as a friendly ally but also a fiercely independent sovereign state. With Nepal taking the lead in the push for formal law and then honoring British requests for extradition in the broad sense, the episode complicates assumptions that the more "modern" the state, the more emphasis it would place on border control. It was only after numerous examples in which Nepal's sovereignty was undermined by the marauding escapades of Indian border police, at times accused of serious criminal offenses, combined with the consistent refusal of British India to act on Nepalese requests, that Nepal finally rejected a British extradition request and actively rejected the premise of the treaty and the foundation of international law it was built upon. Through this experience, Nepal quickly began to understand extradition requests as the imposition of foreign power on domestic ground rather than reflecting the upholding of a shared international duty (Murchison 2007, 300). The incipient international law between Nepal and India in this period instead represented a theoretical language of equal exchange between two nations that once created, could barely conceal the complex and very unequal relations of power that were being reinforced from below. While in theory these treaties were built to provide mutual security for these states from dangerous individuals, in practice contrasting ideas of criminality and guilt and the internal politics of India and Nepal made the practice of the law in any genuinely reciprocal manner impossible.

Finally, this episode of Nepalese-British Indian history, often an outlier in South Asian scholarship let alone in the scholarship of international law and politics, offers an interesting layer of complexity to existing understandings of how the history of law and international relations constituted themselves. In the light of a tremendously violent twentieth century which saw a consistent breakdown of international order, the nineteenth century has often been broadly boiled down to a pre-modern era of classical international law between sovereigns. Here colonialism and imperialism set the ground for this future disintegration of international order epitomized by the two World Wars. In terms of legal theory, scholarship has focused on the incipient philosophical debates around statehood and sovereignty and has created a rather linear and Eurocentric narrative of modernity

(Kennedy 1999, 386). The shift from pre-First World War history of international law has thus often been analyzed as a separate and quite distinctive chapter in the genesis of international relations. In formulating the law to control this border, a commercial company under colonial orders was negotiating with a Nepalese Maharajah, both using the language of "international" to enact laws. The outcome of these negotiations could hardly fit comfortably with the overarching narratives of the period within international law, in particular with assumptions of a universal transition from political to common criminality, and the emphasis paid to non-inquiry in the historiography of extradition. The presence of these small episodes outside of the dominant focuses of international and world history provoke us to keep striving in a productive manner to fragment and complicate our existing understandings of terms such as "international" and "modern" that are too often easily taken for granted.

Notes

1 Extract from a despatch from the Court of Directors in the Political Department dated 18 February 1852, Foreign Department, Political A, 115–134 (Jan., 1853), *National Archives of India*. Henceforth *NAI*.
2 Letter from Captain G. Ramsey Officiating Resident at Nepal, dated 10 April 1852, Foreign Department, Political A, 115–134 (Jan., 1853), *NAI*.
3 Extract from a letter to the Hon'ble the Court of Directors in the Home Legislature Department, Foreign Department, Political A, 23–27 (July, 1854), *NAI*.
4 Letter from Colonel G. Ramsey, dated 23 February 1855, Foreign Department, Political A, 12–14, *NAI*.
5 Letter from Colonel G. Ramsey, dated, 5 July, No. 47 of 1855, *British Library*, IOR/R5/130, British Library.
6 Letter from Colonel G. Ramsey, dated 10 April 1863, Foreign Department, Political A, 113–115 (May, 1863), *NAI*.
7 Letter from Colonel G. Ramsey, Resident at Nepal, to Colonel H. M. Durand, C.B., Secretary to Government of India, dated 13 June 1864, Foreign Department, Political A, 62–63, (July, 1864), *NAI*.
8 Extract from a letter to the Hon'ble the Court of Directors in the Home Legislative Department dated 19 May 1854, Foreign Department, Political A, 23–27 (July, 1854), *NAI*.
9 Letter from J. D. Gordon to Secretary of Bengal, dated 11 April 1865, Foreign Department, Political A, 23–27 (August, 1865), 82–85, *NAI*.
10 Letter from J. D. Gordon to Secretary of Bengal, dated 18 July 1865, Foreign Department, Political A, 82–85 (August, 1865), *NAI*.
11 Letter from Colonel G Ramsey Officiating Resident at Nepal, dated 10 April 1852, Foreign Department, Political A, 115–134 (Jan., 1853), *NAI*.
12 Letter from Colonel G. Ramsey, Resident at Nepal, dated 20 July 1859, Foreign Department, Political A, 272–281 (August, 1859), *NAI*.
13 Letter from the Hon'ble A. Eden, Secretary to Government of Bengal, dated 18 July 1865, Political A, 82–85 (August, 1865), *NAI*.
14 Letter from Colonel G. Ramsey, Resident at Nepal, dated 20 July 1859, Foreign Department, Political A, 272–281 (August, 1859), *NAI*.
15 Letter from Colonel G. Ramsey dated 18 April, No. 29 of 1857, *Residency Records*, IOR/R/5/130, British Library.
16 Extract of Report of the Police Administration of Zillah Goruckpore for the year 1864, Judicial Proceedings B, 141 (March, 1865), *NAI*.

17 Letter from Colonel G. Ramsey, Resident at Nepal, dated 15 June 1863, Foreign Department, Political A, 111–112 (July, 1863), *NAI.*

18 Letter from Colonel Ramsey, Resident at Nepal, dated 15 June, Foreign Department, Political A, July 1863, No. 111–112, *NAI.*

19 Letter from Hon'ble A. Eden, Secretary to Government of Bengal to the Secretary to Government of India, 18 July 1865, Political A (August 1865), *NAI.*

References

Archives consulted

British Library, India Office Library and Records London, UK (*BL, IOR*).
National Archives of India, New Delhi, India (*NAI*).

Printed works

Adam, Leonhard. 1950. "Criminal Law and Procedure in Nepal a Century Ago: Notes Left by Brian H. Hodgson," *The Far Eastern Quarterly*, 9(2), 146–168.

Adikhari, Krishna Kant. 1976. "Criminal Cases and Their Punishments before and during the Period of Jang Bahadur," *Contributions to Nepalese Studies*, 3(1), 105–116.

Bahadur, Pudma Jung. 1909. *Life of Maharaja Sir Jung Bahadur, G.C.B., G.C.S.I., of Nepal*, Allahabad: Pioneer Press.

Kennedy, David. 1999. "International Law and the Nineteenth Century: History of Illusion," *Quinnipiac Law Journal*, 17, 385–402.

Kolsky, Elizabeth. 2010. *Colonial Justice in British India: White Violence and the Rule of Law*, Cambridge: Cambridge University Press.

Magnuson, William. 2012. "The Domestic Politics of International Extradition," *Virginia Journal of International Law*, 52(4), 839–901.

Michael, Bernado A. 2012. *Statemaking and Territory in South Asia: Lessons from the Anglo-Gorkha War (1814–1816)*, New York: Anthem Press.

Murchison, Matthew. 2007. "Extradition's Paradox: Duty, Discretion, and Rights in the World of Non-Inquiry," *Stanford Journal of International Law*, 43(295), 295–318.

Pandey, Madhav Raj. 1973. "How Jang Bahadur Established Rana Rule in Nepal," *Contributions to Nepalese Studies*, 1(1), 50–64.

Pradhan, Bishwa. 1996. *Behaviour of Nepalese Foreign Policy*, Kathmandu: D. D. Pradhan.

Regmi, Mahesh. 1975. "Preliminary Notes on the Nature of Rana Law and Government," *Contributions to Nepalese Studies*, 2(2), 103–115.

Sanwal, B.D. 1993. *Social and Political History of Nepal*, New Delhi: South Asia Books.

Sever, Adrian. 1996. *Aspects of Modern Nepalese History*, New Delhi: Vikas Publishing.

Thapliyal, Sangeeta. 1998. *Mutual Security: The Case of India Nepal*, New Delhi: Lacer Publishers.

Vaidya, Tulasi Ram and Tri Ratna Manandhar. 1985. *Crime and Punishment in Nepal: A Historical Perspective*, Kathmandu: Bini Vaidya and Purna Devi Manandhar.

Whelpton, John. 1991. *Kings, Soldiers and Priests: Nepalese Politics and the Rise of Jang Bahadur Rana, 1830–1857*, New Delhi: Manohar.

——. 2005. *A History of Nepal*, Cambridge: Cambridge University Press.

Part IV

Regionalisms and agency

7 Transnational communities in the Yunnan borderlands in the nineteenth and early twentieth centuries

Rethinking the Yunnan borderlands and frontier history writing in China

Diana Zhidan Duan

Introduction

This chapter provides a discussion of the transnational nature of the Yunnan borderlands where China's southwest merges with mainland Southeast Asia and South Asia (see Figure 7.1). I especially focus on the nineteenth and twentieth centuries, when border communities in Yunnan experienced the expansion and competition of Burma and Siam (Giersch 2006, 117–120), the European encroachment on China and mainland Southeast Asia, and the making and redefining of states and regional power-relations (Scott 2009, 10–13). These border communities were comprised not only of indigenous people but also of migrants including the Chinese, and existed across China's nominal, fuzzy boundary with mainland Southeast Asia and South Asia. As people "in-between," the loyalties and alliances of the residents of border communities shifted between multiple overlords, which included imperial China, the new European colonial powers, and the protectorates of these two, who likewise shifted between regional dominant powers (Lockhard 2009, 102; Winichakul 1994, 84–87).

The nature of the transnational border evolved during the nineteenth and twentieth centuries, as China and its peripheries underwent different stages of the state-making process. I define transnational as the nature of "sustained" (Wills et al. 2004, 1) cross-border interactions (Iriye 2013, 11) and ties of individuals or groups (Faist 2000, 10–11) that transcend the boundaries of regional political entities, such as the spheres of historical empires, colonial powers, and their protectorates, and which shape or cultivate the participants' collective identity as well as their socioeconomic inter-dependency, and cultural and religious affiliation with other regional polities. Akira Iriye further stresses that "transnational" particularly looks for "historical themes and concepts that are meaningful" (Iriye 2013, 8), and is "concerned with issues and phenomena that are of relevance to the whole of humanity, not just to a small number of countries or to the regions of the world" (Iriye 2013, 11).

In most scholarship, the term transnational suggests relatively contemporary interconnections and exchanges across national borders (Iriye 2013, 8, 11; 36–63). However, similar migrant communities and cross-border ties were long established

Figure 7.1 Map of Yunnan.

Source: Carto GIS, College of Asia and the Pacific, the Australian National University. URL: http://asiapacific.anu.edu.au/mapsonline/base-maps/china-yunnan-province-0. Accessed October 8, 2015.

in human history before their nature and legitimacy were redefined by modern nation-states and national boundaries. Thus, an extended definition of transnational serves to explore an enduring pattern of human relations. Today, when we discuss transnational we are not discovering a new phenomenon, but describing the continuity of a natural human mobility and its manifestations as well as restrictions in current terms.

To further understand the transnational nature of Yunnan itself and that of the border communities therein, I differentiate *functional transnational ties* from *intrinsic transnational ties*. I argue that the transnational nature of the Yunnan borderlands was not merely reinforced by its geographic advantages that facilitated cross-border trade and migration, but also by border communities that relied on various transnational ties for their existence and development.

Indigenous polities, migrant societies and "run-aways": decentering China and the diversity of border communities

Akira Iriye contends that "decentering" the West and "denationalizing" Western-centric history is essential to embrace a more global and transnational history (Iriye 2013, 31). For the study of the period prior to the mid-nineteenth century, when China was still the center of the Chinese tributary system in Asia (Stuart-Fox 2003, 2), decentering China and denationalizing the Sinocentric history is critical for recognition of the fluid nature of China's peripheries and the local populations' agency in shaping their own history.

Decentering China lies in at least two aspects. First, in the recognition that, before the establishment of the People's Republic of China (PRC) in 1949, "China itself has not always been unified and Southeast Asia is a wonderfully varied region that historically has comprised many more independent kingdoms and principalities" (Stuart-Fox 2003, 1). The central Chinese government failed to initiate consistent control over China's peripheries whose exact territory fluctuated throughout history. It largely relied on local Chinese officials, who cooperated with indigenous leaders, to establish indirect control. Second, it should be understood that the boundary between China and its mainland Southeast Asian tributaries was never clear-cut. Clearer boundaries between China and mainland Southeast Asia, including Vietnam, French Indochina and British Burma, were gradually drawn in the late nineteenth century, but also left segments disputed or unsolved until the 1980s, especially along the Sino-Burmese and Sino-Vietnamese border (Zhou 2002, 60–61; Feng 2008, 53–62). Thus, "border community" was actually a contingent concept, subject to the effectiveness of China's indirect rule and the fluctuation of states' boundaries. The national border was also a relatively new and contingent concept which local people would have found difficult to comprehend.

In the nineteenth and twentieth centuries, at least three main categories of border communities existed in the Yunnan borderlands: indigenous polities, migrant societies, and "run-aways" (Scott 2009, 24).

Indigenous polities varied in sizes and the coherence of their power structures. "Tribal chieftaincies" based on "men of prowess" formed the rudimentary societal structure commonly found in the early stages of East Asian and Southeast Asian peripheries (McRae 2009, 97–124; Junker 1999, 73). Meanwhile, the indigenous political structures were often established on the leaders' family and clan frameworks. Based on that, larger political units such as tribal confederations and states emerged. By the early twentieth century, the Yunnan borderlands hosted a variety of indigenous polities, including small family units led by the patriarch (Shi and Gong 1986, 1, 11; Song 1978, 158), family and clan groups led by collective village leadership (Honghe Minzu Zhi 1989, 193; Yuanjiang Xianzhi 1993, 105), tribal confederations (Shi and Gong 1986, 54–57, 78–94, 130), and small states. In the Kachin Hills area of northeastern Burma and western Yunnan prior to 1885, extremely small-sized and self-claimed independent units proliferated; at the same time, the Shan state of Hsenwi and Mengmao enclosed a large number of sub-states (Leach 1964, 6).

Indigenous populations' connection with central and local Chinese government usually was established through their leaders. Until the thirteenth century when the Mongols conquered Yunnan, the symbols of China's control were established (Barnett 1993, 489). The Yuan Empire granted indigenous leaders the title of *tusi,* which pertained to native officials (Hu 2005, 283), and initiated indirect rule over local populations. Both the Ming and the Qing rulers attempted to incorporate indigenous leadership into direct administration of the Chinese bureaucracy, namely *gaitu guiliu*. However, during the Ming period, both indigenous military and civil leaders still dominated most areas in the Yunnan borderlands where Chinese population was thin (Herman 2006, 136–137). In the Qing era, *gaitu guiliu* failed after much bloodshed (Yang 1997, 44; Giersch 2006, 45). Moreover, the Qing ruler's ambition to station standing armies in the borderlands, especially in southern Yunnan, became utterly unrealistic due to the prevalence of tropical diseases (Cao 1997, 165; Hu 2005, 285). By the mid-nineteenth century, the *tusi* system still dominated almost the entire Yunnan borderlands (Jindai Shi 1993, 6). The Chinese government officials assigned by the provincial government were usually stationed, and did not penetrate the prefecture level until the 1910s, especially in Xishuangbanna (Li 1933, 20).

In addition to the indigenous polities, transnational business networks and communities established by merchants and laborers flourished from the eighteenth century. Across western Yunnan and northern Burma, Chinese silver miners' societies were highly organized and militarized (Wu 1993, 178–179). The Qing rulers "tried to take advantage of the local hierarchy by appointing leading miners from each native places as overseers" (Giersch 2006, 133). Chinese jade merchants formed a vast mercantile network that connected the gem mines in northern Burma to the processing mills in western Yunnan and further with the markets and consumers in Hong Kong and Europe (Zhang 1986, 190–211). Chinese, Bai, Muslim, and Tibetan caravans as well as Akha traders carried commodities in and out of the Yunnan borderlands, and further distributed them to China proper, Tibet, Southeast Asia, and India (Duan 2015, 160–161; Toyota 2000, 208). Subject to Chinese authorities, indigenous political units, Western colonial forces, and the fluctuation of the international market, these temporary transnational communities were vulnerable and short-lived, especially in the twentieth century.

People who intended to run away from the "state-making process in the lowlands" (Scott 2009, 24) also found homes in the mountains of this area. James Scott contends that, as a part of the world-wide Zomia, an ungoverned periphery, Yunnan represented "a constant temptation, a constant alternative to life within the state" (Scott 2009, 6). "Hill populations of Zomia have actively resisted incorporation into the framework of the classic state, the colonial state, and the independent nation-state" (Scott 2009, 19). Scott's paradigm of Zomia is a crucial step to decenter

China and recognize "hybrid identities, movement, and the social fluidity that characterizes many frontier societies" (Scott 2009, 18). However, he goes too far in denationalizing history, and neglects the active role of the indigenous communities in shaping complicated local dynamics, hierarchies, and interstate relations.

Many local historical records suggest that the concept of Zomia is a generalization yet to be carefully applied. First, the ambiguous concept of hill societies seems too narrow to address the diversity of the indigenous communities, and too vague to specify their geographic locations. Scott's employment of hill societies not only includes mountainous communities on which his concept of hill people intended to focus, but also valley-based polities that he hesitated to include, such as the Nanzhao kingdom (Scott 2009, 19). Given the ambiguity of this term, hill communities could be "flexible and more egalitarian than in the hierarchical, codified valley societies" as suggested by Scott (2009, 18), but also could be highly organized with strict hierarchies and social codes as suggested by Leach's account of the political system of the Kachin communities in highland Burma (Leach 1964).

Second, Scott idealizes the runaway life style and simplifies the social and political structure of the hill societies. For family unit-oriented swidden practitioners such as the Kucong people, drifting around for suitable farming land was nowhere near a "temptation." Most Kucong families suffered food shortage and hunger an average of three to four months each year (Shi and Gong 1986, 5–6; Song 1978, 152–154). In fact, instead of locating settlements with the goal of avoiding states' power, the choice of geographic settlement largely hinged on the settlers' adaptability to tropical climate. Except the Dai people, Chinese migrants and almost all the other indigenous populations whose ancestors migrated to Yunnan as early as the seventh century BC, chose to settle in the mountains to avoid malaria (Cang 2004, 182; Eickstedt 1944, 115–130, cited by Wiens 1954, 73). Seasonal migrants shifted between their homes in the highlands and lowlands due to climate change (Cang 2004, 98). In Luxi, southwestern Yunnan, the Chinese settlers stayed in the mountains above 1600 meters' elevation to avoid the threat of malaria lower down (Luxi Xian Zhi 1993, 395). They sustained the Chinese educational system in the valleys and participated in the civil service examination. By the late Qing period, each private school in these Chinese settlements provided education for 40 to 50 families on average (Luxi Xian Zhi 1993, 56–57).

Overall, the Yunnan borderlands was not "knitted together" by a "political unity" (Scott 2009, 19), but it was more than merely a cluster of anarchist communities. The "comparable patterns of diverse hill agricultural, disposal and mobility" (Scott 2009, 19) emerged because of the social, cultural and political diversity. The communities formed a dynamic local order and hierarchy that was based on competition, cooperation, and connections with each other and with larger regional powers. Thus, the concept of "statelessness" (Scott 2009, 3) accurately describes the indigenes' relation with external larger powers. However, in terms of the indigenous internal social structure, "statelessness" neglects the existence of well-organized indigenous communities, which could be qualified as smaller states by Scott's definition that emphasizes the individuals' obligations and duties to the state (Scott 2009, 7).

Indigenous polities: autonomous communities at the margin of larger regional powers

Despite the central Chinese government's loosely exercised control over peripheries, the indigenous and the Chinese domestic political system belonged to two different and relatively independent institutions for over 700 years after the Yuan rulers established indirect control in this area. The power and authority of the indigenous communities in the Yunnan borderlands was deeply rooted in their local economic, religious and cultural soil. The indigenous authorities, such as village heads, tribal leaders, or native officials, did not rely on the approval of external powers to legitimize their rule.

The capability to obtain and monopolize natural resources and laborers contributed to the charisma or merit (Winichakul 1994, 82) of indigenous leaders and determined the extent of their social control. Thus, trading and raiding became common in addition to regular farming and hunting (McRae 2009, 97–124; Scott 2009, 88–89). Family and clan ties further bonded individuals into coherent social networks and basic collective identities and obligations. The combination of secular and religious authorities established both private and public discourses regarding the legitimacy of the religious and secular leaders, the protocol of the religious and secular relationship, and the obligations and expectations of the commoners. Leaders of various animist communities were also the high priests and judges. They not only organized religious rituals and ceremonies but also enforced the customary and moral rules (Honghe Minzu Zhi 1989, 109; Gong and Shi 1986, 10). The Buddhist authority and the secular authority of the Dai polities were sustained as a joint power structure through advanced religious, social, and cultural institutions, such as the mutual sanction and sponsorship of the two authorities, moral laws, donation, and monastic education. The Buddhist monastic education produced a sustainable pool of religious and social elites who were loyal to the secular and religious authorities (Li et al. 2005, 249; Lishi Yanjiusuo 1979, 11–12, 15–16) and responsible for cultivating commoners' loyalty toward both authorities through rituals, social connections and informal education.

In the 1930s and 1940s, the Kuomintang (KMT) cooperated with the indigenous military forces in Yunnan to eliminate the Chinese Communist Party during the Long March and later to fight against the Japanese. At this time, the upward circulation of the indigenous elites reached its highest point within the indigenous society, and only occasionally intermingled with Chinese domestic politics. Layers of indigenous authorities (Winichakul 1994, 81) cooperated with or submitted to each other before they eventually reached the bottom of the Chinese imperial hierarchy, usually at the prefecture level where the head native official ruled (Honghe Minzu Zhi 1989, 109; Renda Minwei 1957, 2, 15). Chinese officials at the local level overlapped with the same ranking indigenous leaders, forming a shared-power zone (see Figure 7.2). This shared-power zone had at least two distinct functions. First, it cushioned the influence of the Chinese central government and the provincial authority. Second, it sealed the channel for the indigenous elites to directly participate in domestic Chinese politics.

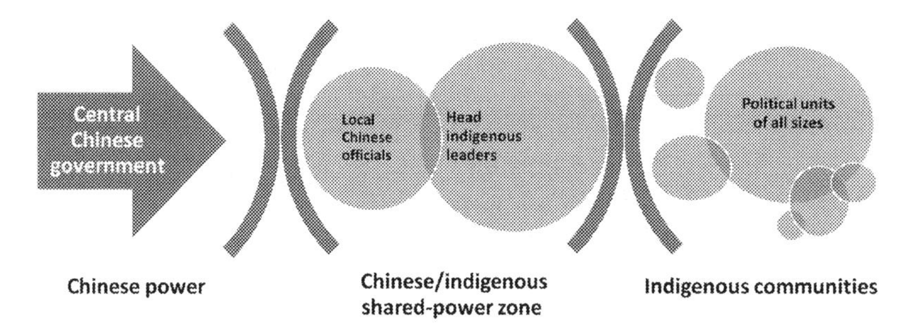

Figure 7.2 Simplified structure of central–local power relations.
Source: Figure by the author.

However, the Chinese officials themselves were not immune from internal factional conflicts. Thus, the influence from the central government was filtered, sifted, and significantly undermined not only by the indigenous authorities, but by the Chinese bureaucracies as well.

The selection of indigenous elites at the local level was also different and independent from the Chinese system and influence, until Chinese educational institutions took root in the eighteenth century, bringing opportunities for small numbers of the indigenous population (Pan and Pan 1997, 37–43; Yuan 1993, 36–39). Limited numbers of indigenous students, especially in northeastern Yunnan, took the civil service examination and improved their social status (Pan and Pan 1997, 37–43). Indigenous officials' residences hosted private Chinese language classes to train bilingual local elites and facilitate the communication between the Chinese authority and indigenous leadership (Longchuanxian Jiaowei Bian Zhi Weiyuanhui 2000, 42–43). Nevertheless, by the early twentieth century, the Chinese-Dai bilingual population remained very limited, and communications between Chinese and Dai officials still largely relied upon interpreters (Dao et al. 1979, 33).

On the eve of the Qing dynasty's fall, civil wars in Xishuangbanna that involved indigenous rivals, merchants, and Chinese officials (Dao et al. 1979, 41–50; Sun 1979, 51–55) enabled the Qing provincial authority to pacify the riots and establish direct administration at the county level (Dao et al. 1979, 33). After the 1911 Revolution, the newly established Yunnan provincial government continued such efforts and expected to subtly initiate institutional reforms modeled after those of China proper. New institutions such as the Chinese Pacification Committees and Periphery Colonization Bureau oversaw local court, schools, trade, and police, leaving other responsibilities to the indigenous leaders (Sheng Zhi Ban 1985, Vol. 1, 19). However, these institutions largely failed during the 1910s and 1920s, because of strong local dissatisfaction. Meanwhile, Yunnan provincial officials failed to actively initiate their colonizing plans because of their focus on factional conflicts and military expansion in southern and southwestern China.

The indigenes' acceptance toward a more coherent Chinese national identification did not appear until the late 1920s when Chinese public schools began to replace private schools. In 1930, indigenous leaders in Xishuangbanna openly supported the Chinese educational institutions and warmly advocated for the Periphery Education Program conducted by the provincial government. Chinese language education became available for young apprentices at Buddhist temples (Guan and Wang 1998, 13, 46). Citizen education that was in accordance with the KMT's stress on party principle, patriotism and national unity also received more attention from local educators (Guan and Wang 1998, 13, 36, 104; Yuan 1993, 87–88.)

However, indigenous leaders' warmth toward the concept of a Chinese nation did not represent the indigenous communities' willingness to give up their autonomy. From the late nineteenth century to the early part of the twentieth, as the provincial authority of Yunnan pushed its colonization into the borderlands, it placed more financial obligations and interfered more in local and border affairs, prompting indigenous communities to adopt various forms of resistance.

Indigenous leaders protested or rebelled against local Chinese officials to express their antagonism toward the aggressive expansion of the Yunnan provincial authority. In the spring of 1912, Chinese armies marched into the indigenous settlements in Nujiang and caused tremendous terror among the local population. Immediately, these new colonizers were attacked by the Lisu and Nu people, who either fought spontaneously or under the command of their indigenous leaders (Ma 2012, 54–58, 89). In Dehong, the Pacification Committee was dissolved in the spring of 1912 and replaced by the Administration Committee of Five Districts (ACFD), ushering in a period of miniscule local influence; in the following years, the ACFD committee chair was frequently replaced. Each who assumed the chair position then frequently requested the provincial government to grant more control over local politics and economy (Yang 1997, 43–45). In the 1930s and 1940s, the provincial government's policies to tighten direct administration over the borderlands triggered more violence. The native official in Mengmao asserted that indigenous autonomy was necessary in order to handle local affairs in the borderlands (Yang 1997, 45). Unsurprisingly, the Chinese colonizing institutions remained largely nominal in the early twentieth century (Gong and Shi 1986, 140; Yang 1997, 47). In some cases, indigenous leaders looked for better alternatives to cooperating with the Chinese. In Nujiang prefecture near northern Burma, the Denggeng Tusi, who was supported by the Chinese, exacted heavy taxes, which caused the lower-ranked indigenous leaders in Pianma to defect to the British in 1910. This incident flamed future border disputes between China and Burma (Liu 1946, 43).

Intrinsic transnational ties: shifting loyalty and cross-border alliances

Prior to the late nineteenth century, before the Qing influence over mainland Southeast Asia was challenged by the Europeans, the indigenous communities, along with the Southeast Asian kingdoms and China, formed a triangular hierarchy that involved two types of tributary systems. Being the people "in-between," the indigenous communities in Yunnan's borderlands, and the small mainland Southeast Asian states vowed allegiance to multiple larger powers of the region (Winichakul 1994, 81–84, 96–97). For indigenous polities, submission to larger powers secured political alliances, which were essential to survival and to successful competition with local rivals. Specific strategies of alliance formation included intermarriage, gift exchange (Winichakul 1994, 87), and tribute. I define intrinsic transnational ties as the social, political, and cultural interconnections and ties that indigenous communities established with their cross-border kin and allies. Intrinsic transnational ties provide cultural, political, and social capital (Faist 2000, 13–15) to bond similar border communities together, thus forming a security sphere to sustain their survival and even autonomy. Regardless of variations in tribal identities, affiliations to larger powers, religious differences, and what would be understood as "racial, ethnic and national differences" in twentieth-century terms, "there was always ample room for pragmatism and cooperation between different groups" (Gonzalez-Quiroga 2010, 52).

The tributary system, in which indigenous communities formed important intrinsic transnational ties, was perceived by the Chinese and the Southeast Asian kingdoms in different terms. For the Chinese, tribute symbolized the "recognition and reinforcement of China's superior status in its own Sinocentric world order" (Stuart-Fox 2003, 33), instead of wealth accumulation. It was a formality which was owed because of China's hegemony in the regional political hierarchy, which required the submission of the tribute payers (Stuart-Fox 2003, 33). For the Southeast Asian kings (except in the case of Vietnam), tribute brought "mutually beneficial trade" under the condition of politely recognizing China's superior status; thus tribute symbolized "a formality that went with mutually beneficial trade" (Stuart-Fox 2003, 33–34).

Within the realm of mainland Southeast Asia, the kingdoms and small polities (including border communities in Yunnan), maintained their relations by a "hierarchical world order in which the supreme overlord . . . cast his influence over the inferior kingships" (Winichakul 1994, 82). Although the tributary system here enabled the overlord to "enforce his demand or intervene in the affairs of the inferior kingdom," such as approving or conferring the appointment of tributary ruler, it did not guarantee a consistent relationship between the overlord and tributary. Tributaries shifted between various competing overlords to secure their best interests in regional power-struggles. Seeking and securing protections from a more powerful overlord against the threat posed by another was a pragmatic strategy for communities "in-between" to maintain relative autonomy at the expense of sacrificing limited independence and resources such as manpower, goods and money (Winichakul 1994, 82–84, 88).

Subjugated to imperial China and other regional powers, these in-between indigenous polities demonstrated exceptional diplomacy to manipulate and benefit from multiple overlords. In the south, indigenous communities swore equal loyalty to Siam and Burma after the thirteenth century (Li 1933, 20). During the eighteenth century, intersectional areas "between Lanna and the Burmese kingdoms, between Luang Phrangbang, Yunnan, Tonkin, and between Vientiane and Tonkin or Annam" hosted chiefdoms of "the Shan, Lu, Karen, Lao, Phuan, Phutain, Chinese and several other ethnic peoples" who were subjected to multiple overlords (Winichakul 1994, 96). To northern Lanna, Kengtung was the most powerful domain among all the indigenous polities in Xishuangbanna, and was also a tributary of the Burmese, the Chinese, and (at times) the Siamese overlords (Winichakul 1994, 96–97).

In the eighteenth and nineteenth centuries, the Burmese and Siamese influences rose in southern Yunnan when the Qing dynasty's frontier activism declined (Giersch 2006, 124). The Qing court hesitated to involve itself in the local affairs of its tributaries and protectorates. By the early nineteenth century, the indigenous leaders in Xishuangbanna were sucked into the Siamese sphere of influence (Giersch 2006, 117–118). Simultaneously, both the Qing and the Burmese were involved in the Dai leadership's factional conflicts. Ironically, the Qing dynasty withheld military intervention because its influence was weak in Xihuangbanna. In 1805, to rescue the Burmese who encountered the Siamese expansion, the indigenous official in Menglian (Muong Laem) violated the Qing ruler's prohibition to leave his domain. Siamese troops ambushed and killed him in Burma, and then pursued his forces back to Menglian. The Qing court "merely notified Burma and Siam to keep their allies under control and out of imperial territory," because the Siamese troops did not linger in Menglian. Yunnan officials firmly recommended the court to "stay out of Southeast Asian affairs" because the indigenes seemed to "pretend to be loyal to the Qing and other powers at the same time" (Giersch 2006, 118–119). As the Qing continued to avoid local confrontations, acting as the "overlord" who favored neither Burma nor Siam (Giersch 2006, 120), the indigenous communities' preference of alliances kept shifting (Giersch 2006, 120–122). In the 1860s, Louis de Carne observed that the coexistence of "the Burman, Laotian and Chinese mandarins" in the same area was not a rare phenomenon. Meanwhile, "as tributaries to two rival states," the indigenous political units in Xishuangbanna enjoyed a considerable degree of self-government (Carne 1995, 203). Around the same time, even the Muslim regime that emerged during the mid-nineteenth century sought British and French aid to escape Chinese control (Jindai Shi 1993, 55–58).

Cross-border alliances were also established between indigenous communities of equal caliber. The Dai tribes' mutual military assistance agreements specified the liabilities of the indigenous polities, as well as matters of religious leadership; these agreements remained effective until the mid-twentieth century (Xibian Gongwei 1979, 6). Inter-marriages eroded the valley–hill distinctions (Li 1947, 13) as well as states' boundaries. Usually, cross-border alliances between indigenous communities remained a local issue. Laborers, resources, and land properties flowed across the boundary between China and Southeast Asian states without the interference of state authorities. In the nineteenth century, the Guangnan ruler's daughter married the ruler of Baole whose jurisdiction fell within the boundary of Vietnam. As part of a dowry, the Guangnan ruler granted his daughter the rights of taxation in the Sanpeng area, which caused border disputes between China and France in the late 1880s when the line between Yunnan and Tonkin was drawn. The French believed that Sanpeng was part of Tonkin, whereas the Chinese insisted on their sovereignty over Sanpeng. After a 40-day quarrel, the French claimed 90 percent of Sanpeng whereas China occupied five villages (Jindai Shi 1993, 85).

In addition to the intrinsic transnational ties that accumulated the political and social capital for survival and autonomy of indigenous communities, cultural and religious connections sustained indigenous communities' cultural capital and strengthened their kinship with people across the border. Various forms of long-term integration, including migration, intermarriage, trade, and confrontation, reorganized the indigenous political units into culturally and religiously homogeneous[1] or heterogeneous communities. Political, cultural, and religious boundaries were overcome by various connections among indigenous communities and redefined. For example, the diversity of religious and cultural zones in this area resembled that of mainland Southeast Asia. The existence of the pattern of Confucian Southeast Asia (Vietnam) and Theravada Buddhist Southeast Asia (Burma, Thailand, Cambodia, and Laos) transcended states' boundaries (Legge 1992, 13). In fact, various forms of intrinsic transnational ties, regardless of their political, social, cultural, and religious functions, were inseparable, playing a part in shaping the transnational features of indigenous communities and maintaining the regional political hierarchy. For example, by exerting influence on indigenous polities, a Bangkok overlord also fulfilled his mission to protect religion and prevent weaker powers from falling into the "domain of evil" (Winichakul 1994, 83–84).

The sphere of transnational communities

C. Patterson Giersch argues that the social boundaries and cultural practices in the Yunnan borderlands, the Crescent (Giersch 2006, 4), were in flux because of the region's function as a "meeting place" or "middle ground." The Yunnan borderlands belonged to the "persistent frontiers," including China's northern and northwestern peripheries, where the first contact between the Chinese and the indigenes was uncertain, and the date for the decisive closing of the frontier remained undefined (Giersch 2006, 7–9). Similar to Giersch, Guo Xiaolin considers the Yunnan frontier a "marketplace" for the exchange of all forms of commodities, where "amalgamation," instead of elimination, dominated cultural change, providing for social and cultural variation (Guo 2008, 68).

Both scholars emphasized the transnational function of the Yunnan borderlands in terms of its geographic nature as a traffic intersection or trans-regional/transnational corridor. Various cross-border ties, including large-scale and long-distance trans-regional trade and migration of non-border people, relied on the geography of the Yunnan borderlands. Unlike the intrinsic transnational ties that focus on the interconnections between border communities, functional transnational ties underscore the Yunnan borderlands' outward geographic function and the active role of the "outsiders" who brought transnational economic and social flows into this area. Most migrant societies of merchants and laborers involved in trans-regional and transnational trading networks were largely sustained by functional transnational ties.

The indigenous communities' transnational nature was partially but not entirely determined by the geographic location of the Yunnan borderlands. In other words, the transnational nature of this area was not merely an "add-up" feature oriented by the "outsiders" and external forces, but the very essence of the indigenous communities who existed amidst the larger regional powers. Thus, the functional transnational ties of the indigenous communities, such as their economic ties with other border communities or their material obligations and exchanges in the tributary system, were closely associated with and even incorporated into their intrinsic transnational ties. However, the indigenes' intrinsic transnational ties did not just include idealistic harmonious associations, but also confrontational relationships: cultural, social, and religious antagonisms as well as political competitions. A good example can be found in Giersch's account on the conflicts in the Mengmeng and Menglian (Muong Laem) hills, involving the Mahayana and Theravada Buddhists, as well as the Dai, Chinese, Lahu, and Wa people (Giersch 2006, 113–117).

The intensive Chinese nation-building era beginning in the 1950s eliminated the "nonstate spaces," as has been suggested by James Scott (Scott 2009, 11). Transnational communities of various sizes and forms in this area were endangered. The Yunnan borderlands was quickly occupied by state farms and relocated farmers and workers, cultivating cash crops and foodstuffs for the Chinese domestic market and war supply effort. The government closely directed and monitored transnational interactions, only opening to China's communist allies and serving China's national interest. The nation-building process of the PRC redirected the indigenous communities' political and cultural affiliations and economic dependencies toward the new socialist country. Border communities' intrinsic transnational ties were essentially undermined, and their functional transnational ties were considerably restricted.

After the 1980s, China opened and engaged Yunnan in frequent economic cooperation within the Greater Mekong Sub-region, or GMS (Chen 2005, 10–12), allowing the indigenous people to restore transnational ties with their kin across the border. The functional transnational ties between Yunnan and neighboring countries was increasingly discussed (Chen 2005, 201–202), which further underscores the Yunnan's geographic and economic function of connecting the GMS with the Chinese domestic market as well as with the adjacent Great Southeast China Sub-region (GSCS) (Chen 2005, 11).

However, the intrinsic transnational ties are not encouraged, for they create a centrifugal force that pulls the border communities from the core of the Chinese national identity. As part of the restored intrinsic transnational ties, intrinsic confrontations, regarding cultural, political and religious identities as well as their relationship with the state and other groups, have become common in the past two decades. Some scholars suggest that the conflict between identity groups—"religious, ethnic, cultural and others"—has become a new pattern that defines the core of confrontations (Demmers 2005, 11). And yet these "new" patterns of conflict between identity groups are continuations or mutations of former disputes that were interrupted or replaced by national and international conflicts centering on their new states' identities formed especially during the twentieth century. These conflicts seem rare in today's Yunnan borderlands, because the intrinsic ties are simplified and replaced by the discourse of "a bottom-up force in economic cooperation" (Chen 2005, 201–202). In other words, intrinsic transnational ties of the Yunnan borderlands combine with its functional transnational feature, a neutral concept that serves more for the economic interests of the state and local population. In contrast, in China's borderlands with Central Asia, the intrinsic transnational ties and/or conflicts of the local communities have persisted despite the government's efforts to eliminate them for decades.

Conclusion

The existence of the transnational borderlands of Yunnan implies two aspects which require understanding. First, the functional transnational ties were sustained by the geographic advantage and the collaboration of outsiders, a situation which illustrates the transnational nature of the "place" itself as a platform for external economic, social, and cultural exchanges. Second, intrinsic transnational ties, which are the focus of this chapter, prompt us to emphasize the transnational nature of the people who act as an internal force to shape cross-border interconnections.

Fundamentally, transnationalizing the Yunnan borderlands relies upon decentering China and identifying the autonomous indigenous communities in China's peripheries. This further distinguishes the influence of the Chinese migrants and the impact of the Chinese state's power in the peripheries prior to the 1950s. However, decentering China can be a longer process than decentering the West (Iriye 2013, 34). The indigenous communities in China's peripheries have already participated in a constructed concept of a grand multi-ethnic nation of China for more than six decades. In a long-term view, the dichotomy of China-centrism adopted by "mainstream" Chinese scholars and decentering China will produce two types of distinct approaches to access China's peripheries.

Note

1 Giersch points out that homogeneous villages existed in this area by the early eighteenth century and formed a cultural patchwork of humanity, providing homes for "diverse arrays of people" and groups who "did not occupy geographically discrete territories" (Giersch 2006, 21). This pattern continued as regional trade and wars in the nineteenth century further integrated the indigenous communities in the Yunnan borderlands with each other and with those in Burma and Siam. In the early twentieth century, Kachins (Animist believers) and Shans (Buddhist practitioners) in the western Yunnan and northeastern Burma were "almost everywhere close neighbors and in the ordinary affairs of life they were (are) much mixed up together" (Leach 1964, 2). A member of the Kachin community living in Mong Hko simultaneously identified himself as a Shan and a Buddhist, and showed his loyalty to the Shan state of Muong Mao (Leach 1964, 2).

References

Barnett, A. Doak. 1993. *China's West: Four Decades of Change.* Boulder, San Francisco and Oxford: Westview Press.

Cang Ming. 2004. *Yunnan Biandi Yimin Shi* (The History of the Migration in the Yunnan Periphery). Beijing: Minzu chubanshe.

Cao Shuji. 1997. *Zhongguo Yimin Shi, Di 6 Juan* (The History of the Migration in China Vol. 6). Fuzhou: Fujian renmin chubanshe.

Carne, Louis de. 1995. *Travels on the Mekong, Cambodia, Laos and Yunnan: The Political and Trade Report of the Mekong Exploration Commission (June 1866–June 1868).* Bangkok and Cheney: White Lotus Co.

Chen Xiangming. 2005. *As Borders Bend: Transnational Spaces on the Pacific Rim.* Lanham, Boulder, New York, Toronto and Oxford: Rowman & Littlefield.

Demmers, Jolle. 2005. "Nationalism from Without: Theorizing the Role of Diasporas in Contemporary Conflict." *Central Asia and the Caucasus: Transnationalism and Diaspora*, edited by Touraj Atabaki and Sanjyot Mehendale. Longdon and New York: Routledge.

Duan Zhidan. 2015. *At the Edge of Mandalas: The Transformation of the China's Yunnan Borderlands in the 19th and 20th Century.* Ph.D. dissertation. Tempe: Arizona State University.

Faist, Thomas. 2000. *The Volume and Dynamics of International Migration and Transnational Social Spaces.* Oxford: Oxford University Press.

Feng Yue. 2008. "Zhong-Mian Bianjie Wenti Yanjiu Shulve" (A Historiography on the Sino-Burmese Border Studies). *Dangdai Zhongguo Shi Yanjiu*, 2008, 03, Di 15 Juan Di 2 Qi (Contemporary China History Studies, Vol. 15–2): 53–62.

Giersch, C. Patterson. 2006. *Asian Borderlands: The Transformation of Qing China's Yunnan Frontier*. Cambridge, MA and London: Harvard University Press.

Gong Peihua and Shi Jizhong. 1986. "Jingpo Zu de Shanguan Zhidu" (The Jingpo Mountain Official System). *Zhongguo Nanfang Shaoshu Minzu Shehui Xingtai Yanjiu* (The Study on the Social Structure of the Ethnic Communities of China's South), edited by Guizhou Minzu Xueyuan Yanjiu Suo. Guiyang: Guizhou Renmin chubanshe, 33–148.

Gonzalez-Quiroga, Miguel Angel. 2010. "Conflict and Cooperation in the Making of Texas–Mexico Border Society, 1840–1880." *Bridging National Borders in North America: Transnational and Comparative Histories,* edited by Benjamin H. Johnson and Andrew R. Graybill. Durham and London: Duke University Press, 33–58.

Guan Kairong and Wang Jianjun. 1998. *Xishuangbanna Dai Zu Zizhi Zhou Jiaoyu Zhi* (The Gazetteer of the Education of Xishuangbanna Dai People's Ethnic Autonomous Prefecture). Kunming: Yunnan Minzu Chubanshe.

Guo, Xiaoling. 2008. *State and Ethnicity in China's Southwest.* Leiden: Brill.

Herman, John E. 2006. "The Cant of Conquest: Tusi Officer and China's Political Incorporation of the Southwest Frontier." *Empire at the Margins: Culture, Ethnicity, and Frontier in Early Modern China,* edited by Pamela Kyle Crossley. London, Berkley and Los Angeles: University of California Press, 135–170.

Honghe Hani Zu Yi Zu Zizhizhou Minzu Zhi Bianxie Bangongshi (Honghe Minzu Zhi). 1989. *Yunnan Sheng Honghe Hani Zu Yi Zu Zizhi Zhou Minzu Zhi* (The Gazetteer of the Ethnic Minorities of the Honghe Hani and Yi People's Ethnic Autonomous Prefecture of Yunnan). Kunming: Yunnan daxue chubanshe.

Hu Shaohua. 2005. *Zhongguo Nanfang Minzu Lishi Wenhua Tansuo* (A Study on the History and Culture of the People in Southern China). Beijing: Minzu chubanshe.

Iriye, Akira. 2013. *Global and Transnational History: The Past, Present and Future*. Kindle version.

Junker, Laura Lee. 1999. *Raiding, Trading, and Feasting: The Political Economy of Philippine Chiefdoms*. Honolulu: University of Hawaii Press.

Leach, E.B. 1964. *Political Systems of Highland Burma: A Study of Kachin Social Structure*. Norwtch: Fletcher and Son Ltd.

Legge, J.D. 1992. "The Writing of Southeast Asian History." *The Cambridge History of Southeast Asia Volume 1: From Early Times to c.1800*, edited by Nicholas Tarling. Cambridge: Cambridge University Press, 1–43.

Li Fuyi. 1933. *Cheli*. Shanghai: Shangwu yinshuguan.

Li Wenhai, Xia Mingfang and Huang Xingtao. 2005. *Minguo Shiqi Shehui Diaocha Congbian Shaoshu Minzu Juan* (Social Surveys of the Republican China Era, the Volumes on Ethnic Minorities). Fuzhou: Fujian jiaoyu chubanshe.

Li Zhengcai. 1979. "Tantan Situo Tusi" (The Situo Native Official). *Yunnan Sheng Wenshi Ziliao Xuanji, Di 11 Ji* (The Historical Documents of the Yunnan Province, Vol. 11), edited by Zhongguo Renmin Zhengzhi Xieshang Huiyi Yunnan Sheng Wenshi Ziliao Weiyuanhui. Kunming: Yunnan renmin chubanshe, 66–83.

Liu Bokui. 1946. *Zhongmian Jiewu Wenti* (The Sino-Burmese Border Issues). Shanghai: Zhengzhong shuju.

Lockhard, Craig. 2009. *Southeast Asia in World History*. New York: Oxford University Press.

Longchuan Xian Jiaowei Bian Zhi Weiyuan Hui. Longchuan Xian Jiaoyu Zhi (The Gazetteer of the Education of the Longchuan County). 2000. Kunming: Yunnan jiaoyu chubanshe, 42–43.

Ma Ya-hui. 2012. "Hereditary Chieftaincy Reform in Yunnan during the Republic of China." *Journal of Wenshan University*, Vol. 25, No. 2: 54–58 and 89.

McRae, John R. 2009. "Comparing East Asian and Southeast Asian Buddhism." *Chung-Hwa Buddhist Journal*, Vol. 22: 97–124.

Pan Xianlin and Pan Xianyin. 1997. "Gaituguiiu yilai Dian Chuan Qian Jiaojie Diqu Yi Zu Shehui de Fazhan Bianhua" (The Transformation of the Yi Communities in the Intersectional Area of Yunnan, Sichuan and Guizhou since the *gaitu guiliu*). *Yunnan minzu xueyuan xuebao (zhexue shehui kexue ban)* (Journal of Yunnan Nationalities University, Social Sciences Edition), Vol. 4: 37–43.

Quanguo Remin Daibiao Dahui Minzu Weiyuanhui Bangongshi (Renda Minwei). 1957. *Yunnan Sheng Dehong Dai Zu Jingpo Zu Zizhi Zhou Shehui Gaikuang: Jingpo Zu Diaocha Cailiao Zhi 2* (Introduction to the Dai and Jingpo Communities in the Dehong Ethnic Autonomous Prefecture of Yunnan: The 2nd Investigation Report of the Jingpo People). Beijing: Quanguo Remin Daibiao Dahui Minzu Weiyuanhui Bangongshi.

Song Enchang. 1978. *Yunnan Shaoshu Minzu Shehui yu Jiating Zhidu Yanjiu* (Studies on the Ethnic Social and Family System in Yunnan). Kunming: Yunnan daxue lishi yanjiusuo minzu zu.

Scott, James C. 2009. *The Art of Not Being Governed: An Anarchist History of Upland Southeast Asia*. New Haven and London: Yale University Press.

Shi Jizhong and Gong Peihua. 1986. "Jinping 'Lahu Xi' de 'Ka': Lun Kucong Ren Fuxi Jiating Gongshe de Xingzhi he Tezheng" (The Ka of the Lahu Xi in Jinping: The Nature and Characteristics of the Patriarch Communes of the Kucong People). *Zhongguo Nanfang Shaoshu Minzu Shehui Xingtai Yanjiu* (The Study on the Social Structure of the Ethnic Communities of China's South), edited by Guizhou Minzu Xueyuan Yanjiu Suo. Guiyang: Guizhou Renmin chubanshe, 1–18.

Stuart-Fox, Martin. 2003. *A Short History of China and Southeast Asia: Tribute, Trade and Influence.* Crows Nest, NSW: Allen & Unwin.

Sun Tianlin. 1979. "Ke Shuxun Zhili Pu-Si-Bian Shaoshu Minzu Diqu Shimo" (Ke Shuxun's Governance in the Pu-Si Ethnic Periphery). *Yunnan Sheng Wenshi Ziliao Xuanji, Di 11 Ji* (The Historical Documents of the Yunnan Province, Vol. 11), edited by Zhongguo Renmin Zhengzhi Xieshang Huiyi Yunnan Sheng Wenshi Ziliao Weiyuanhui. Kunming: Yunnan Renmin chubanshe, 41–50.

Toyota, Mika. 2000. "Cross-Border Mobility and Social Networks: Akha Caravan Traders." *Where China Meets Southeast Asia: Social and Cultural Change in the Border Regions*, edited by Grant Evans, Christopher Hutton, and Kuah Khun Eng. New York: St. Martin's Press; Singapore: Institute of Southeast Asian Studies, 204–221.

Wiens, Herold J. 1954. *China's March toward the Tropics*. Hamden: The Shoe String Press.

Willis, Kate, Yeoh, Brenda S. A., and Fakhri, S. M. Abdul Khader. 2004. "Transnationalism as a Challenge to the Nation." *State/Nation/Transnation: Perspectives on Transnationalism in the Asia Pacific*, edited by Katie Willis and Brenda S. A. Yeoh. London: Routledge, 1–15.

Winichakul, Thongchai. 1994. *Siam Mapped: A History of the Geo-Body of A Nation.* Honolulu: University of Hawaii Press.

Wu Fengbin. 1993. *Dongnan Ya Huaqiao Tongshi* (The History of the Chinese Diaspora in Southeast Asia). Fuzhou: Fujian renmin chubanshe.

Xibian Gongwei Diachaozu (Xibian Gongwei). 1979. "Xishangbanna Dai Zu Fengjian Lingzhu de Falv" (The Dai Feudal Laws in Xishuangbanna). *Yunnan Sheng Wenshi Ziliao Xuanji, Di 11 Ji* (The Historical Documents of the Yunnan Province, Vol. 11), edited by Zhongguo Renmin Zhengzhi Xieshang Huiyi Yunnan Sheng Wenshi Ziliao Weiyuanhui. Kunming: Yunnan renmin chubanshe, 1–18.

Yang Changsuo. 1997. "Cong Mengmao Guozhanbi dao Mengmao Anfusi: Mengmao Dai Zu Tusi Shilve" (From the Guozhanbi of Mengmao to the Mengmao Pacification Commissioner: A History of the Native Officials of the Dai People), in *Dehong Zhou Wenshi Ziliao Xuanji, Di 10 Ji* (Selections on the Historical Documents of the Dehong Prefecture, Vol. 10), edited by Zhang Guolong. Jinghong: Dehong minzu chubanshe, 1–48.

Yuan Qingguang. 1993. *Simao Diqu Jiaoyu Zhi* (The Gazetteer of the Education of Simao District). Kunming: Yunnan Minzu Chubanshe.

Yunnan Jindai Shi Bianxie Zu (Jindai Shi). 1993. *Yunnan Jindai Shi* (The Modern History of Yunnan). Kunming: Yunnan renmin chubanshe.

Yunnan Sheng Lishi Yanjiusuo (Lishi Yanjiusuo). 1979. *Xishuangbanna Dai Zu Xiaocheng Fojiao ji Yuanshi Zongjiao de Diaocha Cailiao* (The Investigation Reports on the Theravada Buddhism and Primitive Religions of the Dai People in Xishuangbanna). Kunming: Yunnan lishi yanjiusuo.

Yunnan Sheng Luxi Xian Zhi Bianzuan Weiyuanhui (Luxi Xian Zhi). 1993. *Luxi Xianzhi* (The Gazetteer of the Luxi County). Kunming: Yunnan jiaoyu chubanshe.

Yunnan Sheng Yuanjiang Haini Zu Yi Zu Dai Zu Zizhi Xianzhi Bianzuan Weiyuanhui (Yuanjiang Xianzhi). 1993. *Yuanjiang Hani Zu Yi Zu Dai Zu Zizhi Xianzhi* (The Gazetteer of the Yuanjiang Hani, Yi and Dai People's Ethnic Autonomous Prefecture). Beijing: Zhonghua shuju.

Yunnan Sheng Zhi Bianzuan Weiyuanhui Bangongshi (Sheng Zhi Ban). 1985. *Xu Yunnan Tongzhi Changbian Shang Ce* (The Yunnan Chronicles Sequels, Vol. 1). Kunming: Yunnan Shengzhi Bianzuan Weiyuanhui Bangongshi.

Zhang Zhubang. 1986. "Dianxi Bianjing de Zhongyao Maoyi—Yushi Ye" (The Jade Trade: An Important Business in the Border of the Western Yunnan). *Yunnan Sheng Wenshi Ziliao Xuanji, Di 28 Ji* (The Historical Documents of the Yunnan Province, Vol. 28), edited by Zhongguo Renmin Zhengzhi Xieshang Huiyi Yunnan Sheng Wenshi Ziliao Weiyuanhui. Kunming: Yunnan renmin chubanshe, 190–211.

Zhou Jianxin. *Zhong Yue Zhong Lao Kua Guo Minzu jiqi Zuqun Guanxi Yanjiu*. Beijing: Minzu chubanshe, 2002.

8 The other sides of the frontier

Indigenous agency in the construction of borders in southwest Amazonia

Louise Cardoso de Mello

Introduction

Much like Asia's, Amazonian borders are still ambivalent. Encompassing a territory that spans nine nations, they go beyond the political boundaries. The upper Madeira region, which is formed by its four main river courses, the Madeira, Guaporé, Mamoré and Beni rivers, is one of Amazonia's transnational border areas, between modern-day Brazil and northeastern Bolivia (see Figure 8.1). Until fairly recently the southwest of Amazonia was an undefined and unknown region, located initially in the confines of the Iberian domains in South America, and remaining in the distant boundaries of the two American nations.

In recent decades, much has been written about the frontier expansion in southwest Amazonia in terms of land settlement and agricultural ecology, inspired by migration waves in the twentieth century. However, little is known about the indigenous agency in this Amazonian appropriation and diaspora during the course of the 1800s. Also, the roles attributed to and assumed by Amerindians in the process of constructing borders remains as yet understudied, and furthermore, misconceived.

From this perspective, this chapter presents a study of the indigenous agency in the process of constructing borders in southwest Amazonia in the nineteenth century. Its primary objective is to analyze the strategies developed by Amerindians in the context of colonial territorial agreements and political reforms, as well as during the construction of the Brazilian and Bolivian nation-states. Through this approach, this chapter further addresses the emergence and reinvention of indigenous identities in the process of construction of both the Brazilian and the Bolivian national identities.

Even though this Amazonian transnational border may be seen as the byproduct of governmental relations and agreements of the last three centuries, this chapter proposes to look at it as the result of interethnic and power relations, which developed on the fringes of both the colonial and independence processes. Therefore, it aims at showcasing the large gap that existed between what the treaties stated and what they actually brought about. Second, the argument is presented that it was within the multiple inner frontiers, as opposed to the rather porous outer border, where strategies and asymmetries were negotiated. This negotiation of power relations is what, in agreement with Van Valen (2013, 2), may be defined as agency.

According to Lévi-Strauss (1948), the Guaporé River was a "natural" frontier that divided the Amazonian culture from the Moxeño one in the northeastern Bolivian lowlands, extending all the way to the foothills of the Andes to the west. On the one hand, the concepts of Moxeño and Amazonian cultures homogenize a much more diverse and complex reality, while on the other hand, they reflect a far more recent scenario perceived by some as rather fictitious and even nationalist notions of fixed identities and rigid frontiers. In contrast, this chapter additionally intends to demonstrate that interethnic relations in southwest Amazonia were not only fluid, but also overflowed on both sides of said political border.

Utilizing a methodological and theoretical framework that is transdisciplinary and informed by the viewpoints of ethnohistory and anthropology, this study contributes towards bringing to light the indigenous participation in the construction of Latin American nation-states, in a political history otherwise devoid of actors or action. Also, by focusing on the interethnic and power relations, this work is intended to outline the creative and adaptive capacities of the indigenous groups, granting them not only voice but also agency.

Occupation and initial claims of fictitious frontiers

The process of constructing borders in southwest Amazonia takes us back to at least the middle of the eighteenth century. The colonization process of the Llanos de Moxos region in northeastern Bolivia may be traced back to the sixteenth century, with the establishment of the first Jesuit missions at the end of the 1600s under Spanish jurisdiction (see Figure 8.1). However, according to ethnohistorical sources, the Portuguese colonial penetration in the backlands (*sertões*) of the Madeira River started significantly later, in the second quarter of the eighteenth century, with rather unstable settlements. Contrary to other frontier areas between the Iberian crowns in America, the Treaty of Tordesillas (see Table 8.1), which had been signed in 1494, still dictated the demarcation of their eastern/western borders. This meant that by exploring Amazonia, the Portuguese were actually entering an area under Spanish jurisdiction.[1]

The incipient colonization of this region was led mainly by Jesuits from the few missions founded in the confluence of the Madeira and the Amazon rivers, and by explorers often referred to as *sertanistas* (or "backland dwellers"), who followed either the same routes of the Amazon and Madeira, or else came up through the south from the Captaincy of São Paulo; which is why to this latter route can be attributed the reason why the members of such expeditions or *bandeiras* were also known as *paulistas* or *bandeirantes*. However, both missionary and private fronts were seeking the same thing: the resources of the land, be it the gold to be extracted, the famous *drogas do sertão* (autochthonous goods[2]) or the infamous exploitation of indigenous peoples as workforce.

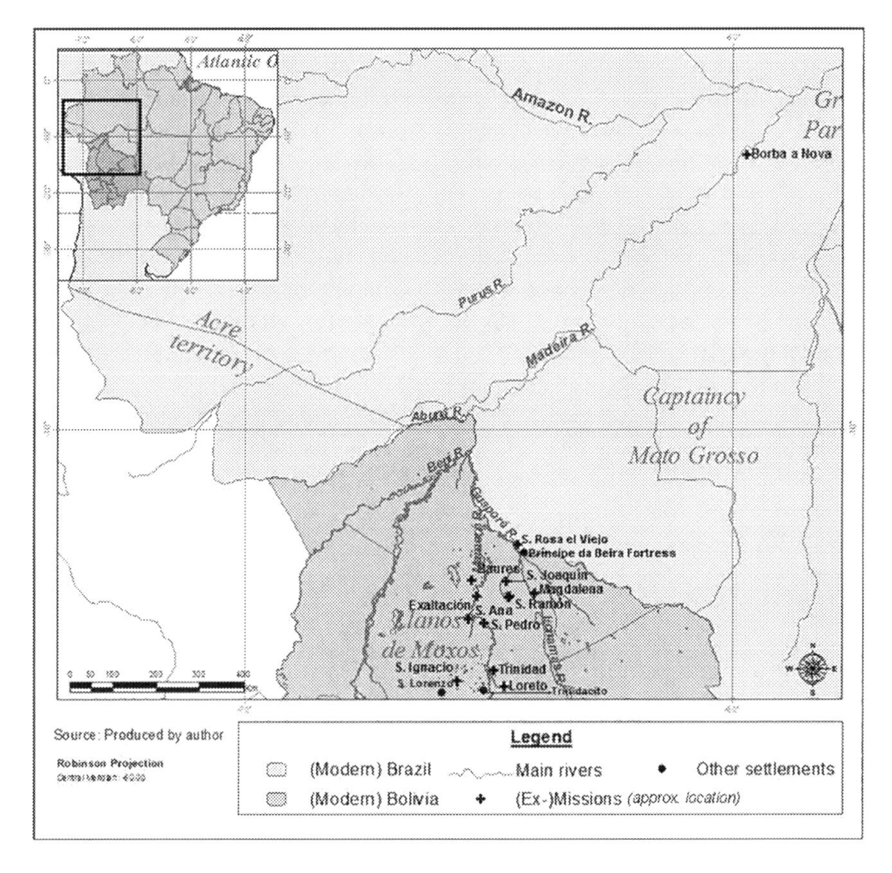

Figure 8.1 Map of southwest Amazonia showing cited locations.

Notwithstanding, it was only in the second half of the eighteenth century when the Portuguese Government decided to join in the undertaking in Amazonia, resulting in its more systematic exploitation. The historian Domingues (2000, 207) draws attention to the colonial perception of Amazonia's potential as an endless source of resources, much like that of the backlands and its gold-bearing reserves. Therefore, the signing of the Treaty of Madrid by the Iberian crowns in 1750 served to legally regulate this vast no-man's-land, whose territory was already being occupied, its jurisdiction being claimed, and more importantly, its resources exploited.

The Treaty of Madrid annulled all previous treaties and was based on the Roman legal principle of *uti possidetis iure*, so that "each party should remain with what it currently possesses."[3] However, one cannot help but question the Spanish decision to agree to cede a territory of such large extent (see Figure 8.2). Stemming from textual evidence, it was based on yet another principle, that of "equivalency" applied at the Treaty of Utrecht (1715), by means of which Spain

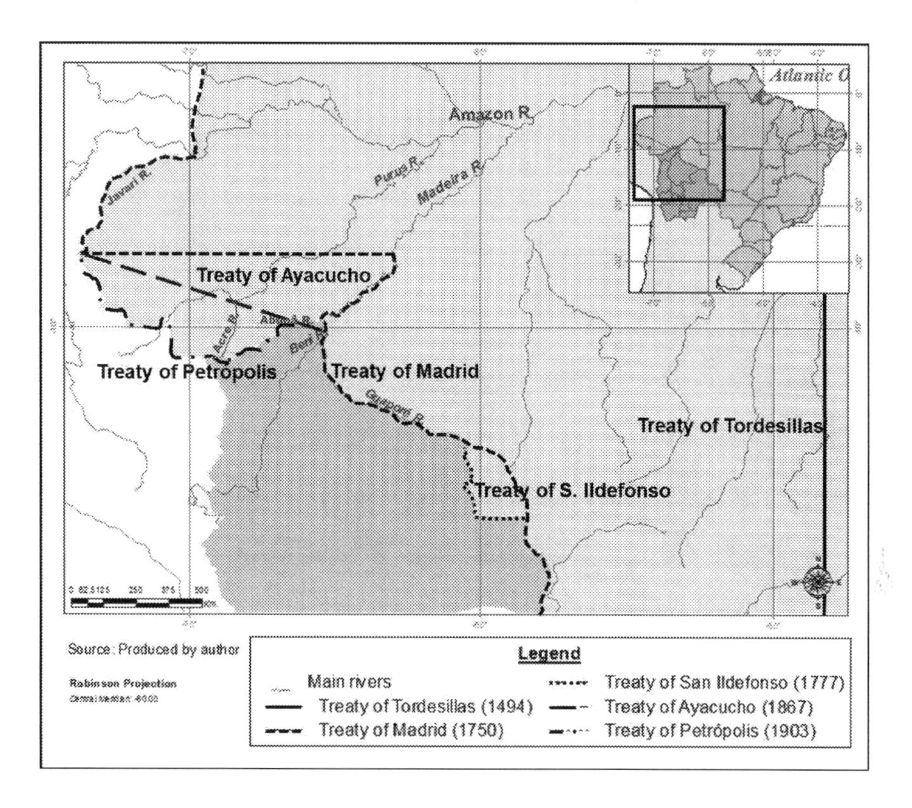

Figure 8.2 Map showing rough outline of the evolution of borders in southwest Amazonia.

had either to cede to Portugal its colony of Sacramento in the southern region of Río de la Plata, or "give equivalence."[4] The Portuguese author of the Treaty of Madrid, Alexandre Gusmão, would abide by the principle as long as it proved advantageous. Hence, trading the colony of Sacramento for half of the state of Grão-Pará, the area of Mato Grosso and Cuiabá, and part of the Captaincy of Goiás seemed like a great deal.[5]

Nevertheless, the signing of the Treaty of Madrid on paper was not nearly as effective in *praxis*. As a matter of fact, the border demarcation was hindered not only by a general lack of knowledge of the area, and thus by the inaccuracies of geographic locations, but also by the difficulties in accessing them; in many cases, this would only happen one century later, such as the initial assessment of the Javari headwaters and the course of the Beni River. In addition to this, the non-fulfilment of more than one clause of the treaty accelerated the increase of hostilities as well as its revocation.[6] In truth, over a decade later, Charles III of Spain annulled the Treaty of Madrid, which was officially substituted in 1761 by the Treaty of El Pardo in a far-fetched attempt to revert to the situation prior to 1750.

Table 8.1 Border treaties signed in southwest Amazonia

Treaty	*Year*	*Signees*	*Demarcation*	*Implications*
Treaty of Tordesillas	1494	Portugal & Spain	Meridian 370 leagues west of the Cape Verde Islands dividing Portuguese domains to the east and the Spanish to the west.	All land west of the mouth of the Amazon River belonged to Spain and that to the east to Portugal.
Treaty of Madrid	1750	Portugal & Spain	Border line to follow the course of the Guaporé River and the Madeira River (starting in the junction of the Guaporé and Mamoré rivers) until a point equidistant to the Amazon River and the mouth of the Mamoré River, from where a straight E–W line was drawn until it met the Javari River, following its course downstream to the Amazon River (art. 7 and 8).	Portugal's domain in Amazonia was extended towards the west, gaining jurisdiction over most of its territory.
Treaty of El Pardo	1761	Portugal & Spain	Annulled the Treaty of Madrid.	Borders were restored to situation prior to 1750.
Preliminary Treaty of San Ildefonso	1777	Portugal & Spain	Articles 10 and 11 ratified articles 7 and 8 of the Treaty of Madrid with the addition that the border could be delimited by other geographical features if considered more appropriate by the demarcation commissions.	It showcases not only the preliminary feature of the treaty but also its inaccuracy as well as the lack of knowledge of the region.
Treaty of Ayacucho	1867	Brazil & Bolivia	Moved the E–W borderline towards the southwest upstream the Madeira River, from the confluence of the Beni and Guaporé rivers up to the headwaters of the Javari River (art. 2).	Converted the left bank of the upper Madeira River into Brazilian jurisdiction and the Acre territory remained Bolivian.
Treaty of Petrópolis	1903	Brazil & Bolivia	Extended Brazil's western border towards the south, from the confluence of the Madeira and Abunã rivers through the headwaters of the Acre River until it reached the headwaters of the Javari River, hence ceding the territory of Acre to Brazil (art. 1)	Last territorial agreement signed for the border between the two countries in southwest Amazonia.

Indigenous roles in the securing of borders: from intrepid guardians to resourceful negotiators

In addition to the hostilities in the international relations, especially regarding the dispute over the mission of Santa Rosa el Viejo in the Guaporé River and the mines in the region of Mato Grosso (see Figure 8.1), there were further consequences that outlived the otherwise brief Treaty of Madrid. These included the intensification of trans-border smuggling, especially of basic goods, and the desertion of reduced Indians in the process of relocation of settlements and their population.[7] Therefore, even though the movements of both people and goods on the border was officially forbidden by the Iberian crowns, these strategies grew to be essential for the region's self-maintenance and subsistence, as a consequence of the difficulties that were faced in relation to access and supply. This further serves as evidence of the large gap between what the treaties dictated and what they actually brought into effect.

In point of fact, migrations and desertions increased as a result of the recruitment of armed contingents on both sides of the disputed frontier. According to primary textual sources, Spain's Moxeño troops consisted mainly of resettled Moxo, Movima, Cayuvava, and Baure Indians,[8] together with mestizos, black slaves, and Indians introduced from other parts of the Upper Peru, such as the missions of Chiquitos.[9] As for the Portuguese Mato Grosso troops, they were formed primarily by *sertanistas* and black slaves.[10] Although it does not exclude the likely participation of Indians (both resettled and independent) on the Portuguese side, it depicts a different demographic and colonization scenario on the eastern side of the border, with a population that was approximately 75 percent black, mestizo, and mulatto in the Portuguese Captaincy of Mato Grosso,[11] and a fairly small number of missions (see Figure 8.1) and reduced Indians. On the other hand, textual sources from one century later still observe that the majority of the population in the Llanos de Moxos region was indigenous, although they also note that the process of miscegenation was underway (Mathews 1879, 151–155).

Thus, the 16-year period between the annulment of the Treaty of Madrid and the signing of the Treaty of San Ildefonso in 1777 by the Iberian Empires (see Figure 8.2) is seen by the literature as one of "armed peace" (Domingues 2000, 17). Whereas the border settlements were encouraged to be fortified and the Indians armed, reliance on diplomatic tools for resolving the disputes prevailed.[12] As a matter of fact, although the Preliminary Treaty of San Ildefonso did not mark a significant reform, given that it retained many of the deficiencies and inaccuracies of the Treaty of Madrid (see Figure 8.2), it did manage to placate hostilities and substitute the "armed peace" with a *status quo* of sorts (Beerman 1996, 9). However, the Treaty of San Ildefonso, to which most of the current national border of Brazil owes its current form, left unresolved matters such as the contestable construction of the Portuguese fortress of Príncipe da Beira on the right shore of the Guaporé River (see Figure 8.1), one of the "natural" borders between the Iberian domains.

Regardless of its clear military functionality, the Príncipe da Beira Fortress also played the important role of attracting population in its surrounding area. It was a strategy intended to secure borders not so much by force, but by populating them. Taking into account the scarcity of Portuguese settlements in the upper Madeira region, the fortress became a focus of population and trade (or smuggling), and developed into a new space for interaction where mestizos, whites, Amerindians (military or non-military), and black slaves coexisted. The Amerindians that inhabited the vicinity of the fortress in the middle course of the Guaporé River and its hinterland, and have been identified as members of various tribes: the Pareci, Cautarios, Pakáas Novos, Txapakura, Baure, Itonama, and Cayuvava, among many others. Notwithstanding, the scrutiny of ethnohistorical evidence has allowed for the identification of more than 50 different ethnic groups in the upper Madeira region in the eighteenth century, the more recurrent being the Karipuna, Pama, Torá, and Mura Indians (Cardoso de Mello 2014).

These groups were not only in contact, but they also actively participated in the regional economy and trade networks of the colonial society, as well as interacting with groups of independent Indians and maroon communities (*quilombos),* which were multiethnic and multicultural groups whose origins stemmed from runaway slaves. Additionally, within this context of territorial disputes and agreements, an institutionalized competition arose between the Iberian crowns to attract deserters on both sides of the border, more often than not under false promises (Domingues 2000, 217). Upon their arrival in the fortress or bordering mission towns in search of better life conditions, many Indians were given a line of credit by the locals in order to settle, thus becoming debtors, and in addition, a valuable workforce (Maldi Meireles 1989, 173–175); it was a very similar dynamic to the one encountered almost a century later during the rubber boom.

Illustrious responses to illustrated reforms

As previously stated, it was not until the mandate of the Portuguese Secretary of State, the Marquis of Pombal (1750–1777), that the colonization of Amazonia started to appear in Portugal's political agenda (Chambouleyron 2006, 16). Similarly to Bourbon Spain and driven by an enlightened reformism, this was the period of efforts to increase central control and, when possible, to suppress the intermediaries that competed with the state by means of a series of reforms that comprehended the demarcation of the state's jurisdiction. Other important measures implemented by the Portuguese included monopolizing the navigation of the Madeira River, with the creation of the Grão Pará and Maranhão General Trading Company in 1755, the expulsion of the Jesuits (by the Portuguese in 1757 and by the Spanish in 1767), and the replacement of their regime by the Law of Directory of Indians (1755–1798).

Nevertheless, the scenario at the turn of the nineteenth century shows that not only did these reforms fail to achieve the goals they aimed for, but they also left the region in a situation of increased violence and governmental neglect (Cardoso de Mello 2014). For instance, the expulsion of the Jesuit order on the one hand, and the forbiddance of marriage between whites and members of the chieftains' families on the other, were measures taken against missionaries and *sertanistas* which would disrupt the somewhat stable balance of interethnic relations by breaking the alliances with and amongst indigenous groups. Moreover, the secularization of the missions with the introduction of priests and directors (or *corregidores* in Spanish domains) is cited in the primary textual sources as a cause for a significant rise in complaints from the indigenous population of all sorts of abuses, from economic to physical ones. Additionally, the demise of the Trading Company around 1775—its aim being to obliterate riverine competition and the intensive smuggling—resulted in a crisis in the supply of goods and in the flow of people; all of which contributed towards the decay and depopulation of the upper Madeira region.

Nevertheless, the aforementioned consequences of these reforms illustrate just one side of the picture. As an expression of the agency of indigenous peoples, new alliances were formed among indigenous groups, the colonial society, and other intermediaries, whose role was not eliminated but refashioned and assumed by different actors, such as black slaves, fluvial traders (*regatões*), indigenous and mestizo deserters, and others. New strategies and response mechanisms were developed as a result of increased governmental control, which, apart from smuggling and desertions, included marital alliances in exchange for privileges, the establishment of new settlements, legal appeals, and other tactics. New spaces of resistance emerged, where runaway slaves, Indians, mestizos, and whites found refuge from either compulsory work or military service, as well as from violent and hard life conditions. Some examples were the maroon communities and independent indigenous groups who inhabited the hinterlands of the upper Madeira region; or the Mura Indians, who offered an alternative to the civilizing colonial model, exerting an opposite force of attraction, insofar as they would incorporate deserters into their groups (Amoroso 1998, 309).

Therefore, new ethnic and social identities were constructed and reinvented not only within these maroon communities and independent indigenous groups, but also in the space of the ex-mission villages and other settlements such as military outposts like the Príncipe da Beira Fortress, where these groups interacted and coexisted (Cardoso de Mello 2015, 387). This process, also known as ethnogenesis, further showcases the creative strategies and agency not only of Amerindians but also of other groups often referred to as "subaltern." However, this has been a problematic and categorical division between the Indians and the whites; the situation has been more successfully tackled by means of the notion of "middle ground," coined by White (1991) to draw attention to the communication and creation of a common culture between Indians and Europeans (Boccara 2005). Along the same lines, Moreland (2006) further develops the concept of middle ground to address the social space "between those with power and those subjected to power, where resistance manifests itself in the appropriation, transformation, and negation of accepted codes."

Evidence of said means of resistance developed in this middle ground may be found in the form of legal appeals, which were lodged by the Cayuvava Indians in the ex-mission of Santa Ana in the Llanos de Moxos[13] and by community associations of both indigenous and non-indigenous individuals in the ex-mission of Borba a Nova (see Figure 8.1), in the lower Madeira River (Domingues 2000, 254). These legal appeals have been interpreted in the literature as evidence of a rise in the abuses carried out by both the clerical and public administrations against the indigenous peoples. Alternatively, but without discarding the latter interpretation, it may be argued that these documental sources might actually indicate an enlarged indigenous agency, insofar as they identified the benefits that having access to the colonial legal system could offer them in return.

By the end of the eighteenth century, despite the policy of assimilation and miscegenation, or in other words of "caboclization"[14] stimulated by the Directory of Indians, although many were mestizos, they were still Indians, for being an Indian was the colonial ethnic category that granted the means for negotiating better life conditions, such as the right to communal ownership of land and its yields (Almeida 2008). Therefore, it may be concluded that this empowerment process culminated with the transformation of both the mestizo and the Indian from an ethnic category to a social-political one. As a result of that process, Amerindians would often try to maintain or recraft their cultural differences in relation to the rest of colonial society (Almeida 2008). In this negotiation of identities insofar as social groups, the space of the ex-mission villages was essential for their territorial adscription and legitimation. This reverse process, referred to as indigenization (or "uncaboclization") would perpetuate throughout Imperial Brazil, and increased in liberal Bolivia in the nineteenth century, as will be shown below.

Indigenous participation in the construction of Brazilian and Bolivian nation-states

During the first two decades of Brazilian independence, the Regent Government was busy with a rebellion in Amazonia that grew to an enormous scale, considering the amount of casualties and the ethnocide of entire indigenous groups. The *Cabanagem* (1835–1840) consisted of a revolt against the central power by an alliance among Indians, mestizos, black slaves (*cabanos*), and the local elite due to miserable life conditions and the loss of political power of the historically autonomous northern province of Grão-Pará. Among the many Indians who fought in this rebellion, the Mura from the Madeira River are remembered for their active role in the conflict, and also for losing their lives *en masse*.

Similarly to the upper Madeira in Brazil, the Llanos de Moxos basically remained on the margins of the wars of independence (1807–1825). During this fairly long period, the liberation armies were manned by black slaves under the promise of emancipation which was not always honored (Sena 2013). However, it is important to highlight that the slave workforce in Bolivia was much less significant than the indigenous one. Although the effective abolition of slavery would not take place until 1851 (Sena 2013), Bolivia forged its emerging identity as "free soil" in contrast with its giant pro-slavery neighbor, Brazil. Moreover, this political strategy was perceived by the Brazilian Government not only as an affront, but also as a way to attract deserters.

As point of fact, on his journey to South America around 1830, D'Orbigny gave an account of the Llanos de Moxos region that was not very optimistic; he highlighted how he had "never seen more slavery and despotism under a free government."[15] Further textual sources from the nineteenth century also reported the increase of rebellions, showcasing on the other side of the coin, the counter-strategy applied by Amerindians. The uprisings were not only against the government authorities, such as the one led by the Cayuvava in 1830 and another in San Ignacio in 1855 (see Figure 8.1), but they also took place within different indigenous groups, who actively engaged in these conflicts while taking sides with either parties, whether indigenous or not.

Intertribal conflicts may be traced back to historic—and prehistoric—rivalries between indigenous groups, such as the Canichana from the ex-mission of San Pedro who, in the beginning of the nineteenth century, allied with the Moxo Indians from San Ignacio against the Moxo from Trinidad, who in turn had an alliance with the Moxo from Loreto (Van Valen 2013, 19). This demonstrates that such alliances went beyond ethnic boundaries and adopted a more tribal and thus powerful territorial adscription. Therefore, instead of looking at the conflicts in the manner most common in the historical literature, which generally assumes that colonial society took advantage of and incited indigenous rivalries, these rebellions may be analyzed as interethnic and intertribal alliances which were manipulated by both indigenous and non-indigenous groups in accordance with their own interests. Additionally, there are references to rebellions in the towns of Exaltación and Trinidad (see Figure 8.1) led by Indians against their own chieftains for non-payment of paddlers (Keller 1875). This observation further supports the importance of the role of intermediaries, a role that was also played by the chieftains in the relations and negotiations between Indians and creoles in independent Bolivia.

The acknowledgment of labor rights such as regulation and payment, especially for the paddling activity which was one of the most important in the region, came as part of a body of liberal decrees that were implemented from 1842 onward. Among these measures it is important to outline the granting of citizenship for the inhabitants of the newly created Bolivian Department of Beni (hitherto Province of Moxos), as well as the granting of access to property and commerce (Guiteras Mombiola 2012). The liberal reforms were put in place nationwide in order to favor the termination of communal property, as well as to incentivize white migration to the Llanos de Moxos area by assigning them parcels of the communal land (Van Valen 2013, 43, 99).

Nonetheless, there exists a debate over the actual effectiveness and benefits of the reforms for the indigenous population. Based on Guiteras Mombiola (2012), it may be sustained that in a manner similar to the Portuguese case previously analyzed, the acknowledgment of the rights of citizenship was appropriated by a relevant number of the indigenous population, who relied on the rights of citizenship in order to negotiate their effective access to land, their way upward in social hierarchy, and their way out of abusive situations. On the other hand, it is important to highlight that, from the last quarter of the nineteenth century, the frenzy of rubber extraction would bring about an important shift in the southwest Amazonian livelihood, in detriment of indigenous privileges, rights, and life conditions; as will be addressed next.

Reconstructing indigenous identities in rubber-boom Amazonia

The second half of the nineteenth century is perceived as a period of economic growth in southwest Amazonia. The rubber boom period (approx. 1860–1910) relied a great deal on the indigenous workforce. Nevertheless, despite its sudden flourishing, it drastically affected indigenous populations in terms of violence and exploitation of both their workforce and land. On the one hand, workforce recruitment often happened through physical coercion, capture, and traffic of Amerindians, who in order to perform the required labors were forced to engage in a debt system; this system tied the debtor to the rubber field owner (*seringalista*) by a patron–client bond. The obtainment of credit enabled them to furnish themselves with tools and goods, which had to be acquired at the rubber field owner's storehouse (*barracão*). The payment of such credit, or rather debt, was often made in rubber, whose price, as well as that of the provisions, was fixed by the owner to his own advantage. On the other hand, the search for new rubber fields (*seringais*) often resulted in land expropriations leading to violent incursions against indigenous groups. These incursions frequently involved the capture of Indian infants, who were then raised by the families of rubber tappers (*seringueiros*) or plantation owners.

By the 1860s, the rubber owners, who were mainly Bolivian, and their workforce, who were mainly Indians from the Llanos de Moxos area, occupied practically the whole course of the Madeira River. In fact, in the treacherous rapids section of the upper Madeira, settlements and rubber fields belonged almost exclusively to Bolivians (Teixeira and Fonseca 2003, 106). Many of the shacks where the indigenous workers lived during the rubber-gathering activity were named after ex-missions. However, even though some scholars see this as evidence of a mono-ethnic housing configuration (Van Valen 2013, 87), it is important to emphasize not only the diversity of the recruitment process, but also the multiethnic character of the ex-mission towns, masked by the homogenizing labels of Trinitarios, Loretanos, or even Moxos. For instance, the 1861 list of absent taxpayers (*ausentes*) from Exaltación portrayed a highly diverse population comprised of Cayuvava, Canichana, and Moxo Indians, as well as Cruceños and at least two Brazilians and one Mura Indian from Brazil (Van Valen 2013, 99).

Furthermore, nineteenth-century sources report many independent indigenous groups in the upper Madeira region that had established different types of alliances with both the creole and the Christianized indigenous society. Mainly inhabiting the rapids stretch were the Karipuna, Pama, Sanabó, Jacaria, Arara, Parintintin,[16] Pakawara, Acanga Piranga (Uru-Eu-Wau-Wau), Moré, or Itenéz (near the confluence of the Guaporé and Mamoré rivers), and many more. Whereas the Moxo from Trinidad, for example, were associated with the Sirionó, the Cayuvava had relations with the Arara, Karipuna, Pakawara, Chácobo, and Moré (Van Valen 2013, 100–101). Therefore, the forms of interethnic relations and alliances carried out were as varied as the groups themselves. Additionally, they relied on different types of adaptive strategies by means of which these groups of independent Indians would hire out their services, trade for manufactured goods, or else make use of violence in the same way as the non-indigenous would, when the chances were good.

The southwest of Amazonia was known to have one of the best and largest reserves of *hevea* trees in the whole region, and therefore attracted a number of migrants from different parts of Bolivia and Brazil (especially from the northeastern region of Brazil) to explore the hinterlands and interfluves of the Madeira and Purus rivers. In the midst of the rubber boom, the 1867 Treaty of Ayacucho (see Figure 8.2) turned the left bank of the Madeira River and its hinterland into Brazilian territory by means of a straight line that connected the confluence between the Beni and Guaporé rivers to the allegedly known headwaters of the Javari River. Regardless of the principle of *uti possidetis* on which previous treaties were based—since the region was mostly occupied by Bolivians—the Treaty of Ayacucho was signed between the two neighboring countries in exchange for free navigation and commerce on the rivers that connected the republican country to the Atlantic Ocean. This underlines the importance of an outlet for the transport of the products of northeastern Bolivia, as well as the benefits that access to the Atlantic could bring to its economy.

On the other hand, the case of the territory of Acre was slightly different, if not the opposite. Due to its hydrography, Brazilians had easier access to this area mainly through the Purus—one of the Amazon's main tributaries—and the Acre River. Consequently, during the rubber boom, the region was occupied predominantly by Brazilians. Nonetheless, it was only 40 years after the Treaty of Ayacucho, upon the long-awaited establishment of the location of the Javari headwaters that governments realized that the current territory of Acre actually belonged to Bolivia (Teixeira and Fonseca 2003, 102). Once again, in exchange for the construction of a railroad contouring the rapids of the upper Madeira River to expedite the discharge of Bolivian products, the Treaty of Petrópolis was agreed upon in 1903 (see Figure 8.2). The Madeira–Mamoré Railroad (EFMM), however, turned out to be a project doomed to failure, being abandoned at least three times, mostly on account of the harshness of the surroundings and the prevalence of diseases.

In his expedition to Amazonia in the 1860s, British explorer Henry Bates (1863, 160) was astonished to find that practically every Indian or mestizo was indebted, either in terms of money or labor. Nonetheless, Van Valen (2013, 86) draws attention to the fact that the Amazonian economy during that period had no cash flow, but rather it was based on the exchange of goods. Other authors go as far as supporting the claim that the shortage of cash flow was the reason why rubber field owners kept their workers in debt. The American historian also suggests that Amerindians may have perceived the labor in the rubber fields as an alternative that was preferable to the high work and tribute demands of ex-mission towns (Van Valen 2013, 87). In any case, as debts increased, the Amerindians became more tied to the storehouses, to such an extent that they were forced to abandon their own farming activities. Consequently, the rubber boom led not only to the depopulation of many towns in northeastern Bolivia, but also to a crisis of self-sufficiency in southwest Amazonia.

In view of these developments, the Amerindians carried out and "recycled" different strategies such as founding new settlements, of which Trinidacito (see Figure 8.1) is a well-known example, and deserting into independent Indian groups. For instance, primary sources document that Indians from Exaltación had deserted to live among the Chácobo; that Baure Indians from San Joaquin, and the Itonama from San Ramón and Magdalena had joined with the Moré Indians; and also that the Sirionó had taken in whites that had been kidnapped as children as well as descendants of runaway slaves (Cardús 1886, 290–291). In addition, there is one specific account which makes reference to a group of Pakawara Indians in the rapids stretch of the Madeira River, whose chieftain was a Cayuvava who had deserted from Exaltación (Mathews 1879, 58). This showcases not only how fluid interethnic relations were, but also the extent to which outsiders could be incorporated and assimilated within a different ethnic group. It was a scenario in many ways similar to that experienced among the Mura Indians and maroon communities at the end of the previous century.

Finally, in this ongoing process of reconfiguration of ethnic identities or ethnogenesis, it may be observed that elements of Christendom would be used and recrafted by the otherwise acculturated Indians as cultural traits in order to redefine and reunite themselves as a group while distinguishing themselves from others. In fact, the experience of Christianity is a distinctive feature observed in the ethnohistorical discourse amongst ex-mission Indians and non-indigenous groups. It is within that logic that the millenarian movement that took place in 1887 in the town of San Lorenzo should be understood. Like other messianic phenomena in Amazonia, the Guayochería may be analyzed as a revivalist and nativist manifestation rooted in the indigenous mythological background (Porro 1996, 71–72) as a resistance to the disruptive acculturation process that was being imposed by the rubber boom and the resulting penurious state of life. Therefore, being Christian Indians may well have been an alternative developed in the "middle ground" between either being savages, or not being Indians at all.

Conclusions

By revisiting some of the main ethnohistorical sources available for southwest Amazonia, it is possible to outline the strategies and responses developed by the Amerindians in the process of construction of national borders, of which desertions, smuggling, rebellions, interethnic alliances, legal appeals, and refuge in maroon communities and independent indigenous groups are just a few. Second, as a consequence of this ongoing negotiation of both interethnic and power relations, new spaces of interaction emerged where processes of ethnogenesis flourished. This hybridity resulted in the construction of new identities based on the remodeling of old ones. Third, this chapter may have succeeded in demonstrating the impracticability of most of the attempted treaties and agreements in southwest Amazonia as well as the porosity of its external borders, which were constantly defied by a far deeper one: its internal colonial frontier.

Finally, by analyzing the dialectic of power between the state and its intermediaries, this study has refuted the passive and secondary role attributed to the indigenous groups not only by the government but also by the national society. Quite on the contrary, it may be concluded that the indigenous peoples actively partook of the construction of said borders, not so much as "guardians of the frontiers," as they are often thought of by historiography, but rather as negotiators of ethno-social relations and identities in the homogenizing process of the rise of nationalism, by exerting pressure towards a much more plural cultural reality where multiple frontiers overlapped and intertwined. A role they still play to this day.

Acknowledgments

I wish to thank the editor of this book, Dr. Jaime Moreno Tejada, for his comments on this chapter and for the opportunity of sharing these pages with such outstanding researchers. I would also like to extend my gratitude to Dr. Elizabeth DeMarrais, not only for her suggestions but especially because I still have not had the opportunity to properly thank her for the supervision, conversations, and insights during my recent academic stay in Cambridge, which meant enormous progress for my research. Thanks are also due to Dr. Juan Marchena, director of my research at the Universidad Pablo de Olavide, in Seville, who is also a great appreciator of these maps, and lastly, to Christian Biggie, my dedicated reviser and, most importantly, avid reader.

Notes

1 For other border areas between Portugal and Spain in South America, such as the northern and southern ones, preliminary and provisional territorial agreements had already been subscribed to: respectively, the Treaty of Lisbon (1681), and the Provisional and Suspensive Treaties of Lisbon (1700) and the Treaty of Utrecht (1713–1715).
2 Such as cocoa, clove, cinnamon, sarsaparilla, Brazil nut, etc.
3 Relación de los acontecimientos que se produjeron a raíz del establecimiento de los límites en América del Sur, realizada por un oficial por encargo de S. M., 1750, Archivo General de Indias (AGI), Seville, Buenos Aires, leg. 535, doc. 2 (translated by author).
4 (*"dar equivalencia"*). Relación de los acontecimientos, AGI, Buenos Aires, leg. 535, doc. 2 (translated by author).
5 Relación de los acontecimientos, AGI, Buenos Aires, leg. 535, doc. 2.
6 Julián de Arriaga al gobernador de Buenos Aires y Marqués de Valdelirios, 12 de junio de 1760, AGI, Buenos Aires, leg. 536, doc. 2.
7 Instrucciones de Pedro de Ceballos al Marqués de Valdelirios, 19 de septiembre de 1760, AGI, Buenos Aires, leg. 536, doc. 33.
8 Juan de Pestaña a Pedro de Ceballos, 19 de octubre de 1766, AGI, Charcas, leg. 433, doc. 57.

9 Juan de Pestaña a Manuel Amat, virrey del Perú, 11 de abril de 1766, AGI, Charcas, leg. 433, doc. 1.
10 La Audiencia de La Plata al Virrey del Perú, 6 de diciembre de 1760, AGI, Charcas, leg. 433, doc. 5.
11 Luis de Albuquerque de Mello Pereira e Cáceres, Relação de toda a povoação da capitania de Mato grosso e Cuyabá, Instituto Histórico e Geográfico Brasileiro, Rio de Janeiro, ARQ. 1.2.10, p. 410.
12 Juan de Albarelos a Alonso Berdugo, 28 de marzo de 1761, AGI, Buenos Aires, leg. 537, doc. 14.
13 Antonio Aymerich y Villajuana a Fray Cayetano Tudela, 30 de noviembre de 1769, AGI, Charcas, leg. 623, doc. 49.
14 This concept refers to the process of "*caboclização*" or miscegenation of the "*caboclo*," which is a Portuguese term for an individual of mixed white and Indian ancestry.
15 D'Orbigny, Alcide. 1945. *Viaje a la América Meridional: Brasil, República del Uruguay, República Argentina, La Patagonia, República de Chile, República de Bolivia, República del Perú: realizado de 1826 a 1833*. Vol. 4. Buenos Aires: Futuro, 1312 (translated by the author).
16 Descrição das diversas nações de índios que residem em diversos lugares da província de Mato Grosso, de Fr. José Maria de Macerata, Cuyabá, 5 de dezembro de 1843, Instituto Histórico e Geográfico Brasileiro, Rio de Janeiro, Lata 763, Pasta 19.

References

Primary sources

Manuscripts

Antonio Aymerich y Villajuana a Fray Cayetano Tudela, 30 de noviembre de 1769, Archivo General de Indias, Charcas, leg. 623, doc. 49.
Descrição das diversas nações de índios que residem em diversos lugares da província de Mato Grosso, de Fr. José Maria de Macerata, Cuyabá, 5 de dezembro de 1843, Instituto Histórico e Geográfico Brasileiro, Rio de Janeiro, Lata 763, Pasta 19.
Instrucciones de Pedro de Ceballos al Marqués de Valdelirios, 19 de septiembre de 1760, Archivo General de Indias, Buenos Aires, leg. 536, doc. 33.
Juan de Albarelos a Alonso Berdugo, 28 de marzo de 1761, Archivo General de Indias, Buenos Aires, leg. 537, doc. 14.
Julián de Arriaga al gobernador de Buenos Aires y Marqués de Valdelirios, 12 de junio de 1760, Archivo General de Indias, Buenos Aires, leg. 536, doc. 2.
Juan de Pestaña a Manuel Amat, Virrey del Perú, 11 de abril de 1766, Archivo General de Indias, Charcas, leg. 433, doc. 1.
Juan de Pestaña a Pedro de Ceballos, 19 de octubre de 1766, Archivo General de Indias, Charcas, leg. 433, doc. 57.
La Audiencia de La Plata al Virrey del Perú, 6 de diciembre de 1760, Archivo General de Indias, Charcas, leg. 433, doc. 5.
Luis de Albuquerque de Mello Pereira e Cáceres, Relação de toda a povoação da capitania de Mato grosso e Cuyabá, Instituto Histórico e Geográfico Brasileiro, Rio de Janeiro, ARQ. 1.2.10, p. 410.
Relación de los acontecimientos que se produjeron a raíz del establecimiento de los límites en América del Sur, realizada por un oficial por encargo de S. M., 1750, Archivo General de Indias, Buenos Aires, leg. 535, doc. 2.

Printed works

Bates, Henry W. 1979 [1863]. *O naturalista no rio Amazonas*. Belo Horizonte: Itatiaia.

Cardús, José. 1886. *Las misiones franciscanas entre los infieles de Bolivia*. Barcelona: Librería de la Inmaculada Concepción.

D'Orbigny, Alcide. 1945. *Viaje a la América Meridional: Brasil, República del Uruguay, República Argentina, La Patagonia, República de Chile, República de Bolivia, República del Perú: realizado de 1826 a 1833*. Vol. 4. Buenos Aires: Futuro.

Keller, Franz. 1875. *The Amazon and Madeira Rivers*. Philadelphia: J. B. Lippincott and Co.

Mathews, Edward D. 1879. *Up the Amazon and the Madeira River through Bolivia and Peru*. London: S. Low, Marston, Searle & Rivington.

Secondary sources

Almeida, Maria Regina C. de. 2008. "Índios e mestiços no Rio de Janeiro: significados plurais e cambiantes (séculos XVIII–XIX)." *Memoria Americana*, 16(1): 19–40.

Amoroso, Marta Rosa. 1998. "Corsários no caminho fluvial." In *História dos Índios no Brasil*, edited by Manuela Carneiro da Cunha. São Paulo: Cia. das Letras, pp. 297–310.

Beerman, Eric. 1996. *Francisco Requena: la expedición de límites, Amazonía 1779–1795*. Madrid: Compañía Literaria.

Boccara, Guillaume. 2005. "Mundos nuevos en las fronteras del Nuevo Mundo." *Nuevo Mundo Mundos Nuevos*. Debates. Accessed: December 30, 2013. http://nuevomundo.revues.org/426.

Cardoso de Mello, Louise. 2014. "Os outros lados da fronteira: La historia del alto Madeira en el siglo XVIII desde el estudio de sus relaciones interétnicas." MA diss., Universidad Pablo de Olavide, Seville.

——. 2015. "De Cayari a Madeira: Procesos históricos de etnogénesis en el suroeste amazónico durante el período colonial." *Memorias*, 11(26): 360–391. Accessed: August 20, 2015. http://rcientificas.uninorte.edu.co/index.php/memorias/article/view/7353/7235.

Chambouleyron, Rafael. 2006. "Conquista y colonización de la Amazonía Brasileña (s. XVII)." In *La Amazonía Brasileña en perspectiva histórica*, edited by José Manuel Santos Pérez and Pere Petit. Salamanca: Ediciones Universidad de Salamanca, pp. 11–22.

Domingues, Ângela. 2000. *Quando os índios eram vassalos. Colonização e relações de poder no Norte do Brasil na segunda metade do século XVIII*. Lisboa: Comissão Nacional para as Comemorações dos Descobrimentos Portugueses.

Guiteras Mombiola, Anna. 2012. "'Los naturales son ciudadanos de la gran familia boliviana.' La participación indígena en la construcción del departamento del Beni, siglo XIX." *Anuario de Estudios Americanos*, 69(2). Sevilla: 451–475.

Lévi-Strauss, Claude. 1948. "La vie familiale et sociale des indiens nambikwara." *Journal de la Société des Américanistes*, 37(1): 1–132.

Maldi Meireles, Denise. 1989. *Guardiães da Fronteira Rio Guaporé, século XVIII*. Petrópolis: Editora Vozes.

Moreland, J. 2006. "Archaeology and Texts: Subservience or Enlightenment." *Annual Review of Anthropology* 35: 135–151. Accessed: March 7, 2015. http://www.annualreviews.org/doi/pdf/10.1146/annurev.anthro.35.081705.123132.

Porro, Antônio. 1996. *O Povo das Águas*. Petrópolis: Editora Vozes.

Sena, Ernesto C. de. 2013. "Fugas e reescravizações em região fronteiriça—Bolívia e Brasil nas primeiras décadas dos Estados nacionais." *Estudos Ibero-Americanos*, PUCRS, 39(1): 262–284.

Teixeira, Marco Antônio D. and Dante Ribeiro da Fonseca. 2003. *História Regional (Rondônia).* Porto Velho: Rondoniana.

Van Valen, Gary. 2013. *Indigenous Agency in the Amazon: The Mojos in Liberal and Rubber-boom Bolivia 1842–1932.* Tucson: University of Arizona Press.

White, Richard. 1991. *The Middle Ground. Indians, Empires, and Republics in the Great Lakes Region, 1650–1815*. Cambridge: Cambridge University Press.

Part V

Representations

9 Walking with the Gods

The Himalayas as (dis)enchanted landscape

Christopher A. Howard

> Modernity seems to signal the final triumph of theoretical consciousness, yet we humans remain inexorably mimetic and mythic creatures.
>
> Bellah (2006, 11)

Introduction

Travel destinations, be they ancient pilgrimage or tourist centres, do not present themselves; rather, they are always represented. This chapter explores how transnational representations of the Himalayas have been produced and reproduced throughout the history of travel in the region. Briefly surveying the religious significance and pilgrimage tradition of the Himalayas, the focus will primarily be on Western encounters in the modern period. Beginning with colonial explorers and early mountaineers, imagery of the Himalayas a mysterious frontier and "lost horizon" circulated back to Europe and beyond, attracting increasing numbers of international visitors by the late twentieth and early twenty-first centuries. The values and ideals that continue to bring Western travelers to the Himalayas, especially those of wild nature, adventure and spirituality are shown to be consistent with media representations (e.g. films, guidebooks, travel literature) of the region, as part of the more general discourse of modernity and a "mystical East."

Yet rather than being viewed as a set of purely Western, orientalist projections, from a perspective of transnational and cultural mobilities, the chapter considers how such imagery and forms of travel overlap with those produced in the regional context. Long before Western encounters, the Himalayas were represented as a sacred landscape, a remote frontier where pilgrims and ascetics were said to "walk with the Gods." Such representations emanate from the physical and symbolic geography of the region; namely, its peripherality to urban centres and the fact that it houses the highest mountains on Earth. This chapter will thus give special attention to how, across cultures, mountains and remote wilderness areas are often imbued with mythological qualities, especially timelessness.

Viewed in the *longue durée*, the images and meanings of the Himalayas have been produced through transcultural flows and relationships between what Appadurai (2010, 7) calls the "forms of circulation and the circulation of forms."

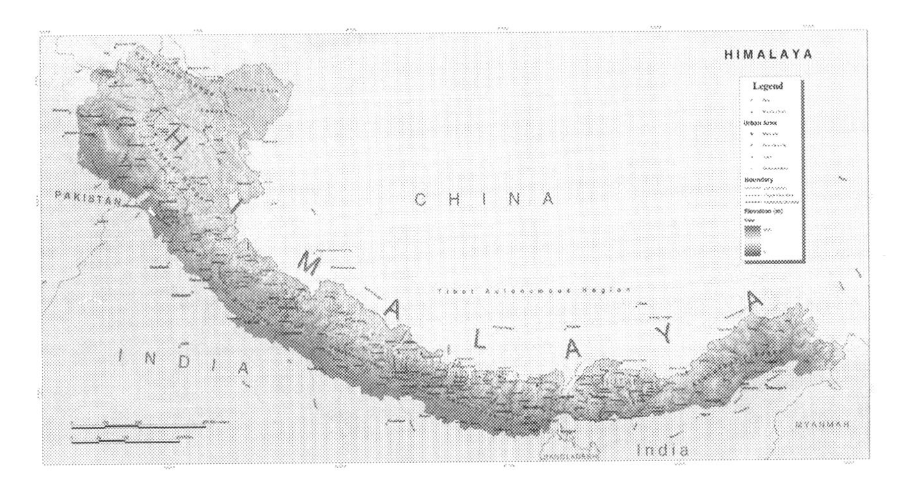

Figure 9.1 Map of the Himalayas.

Source: Digital Himalaya Project.

In contemporary times, the image of the Himalayas as an enchanted landscape standing outside of modernity circulates across global mediascapes, facilitating forms of "virtual tourism." While contemporary Himalayan journeys are often inspired and oriented by a search for authenticity, spirituality and difference (see Howard 2012), such ideals are shown to be paradoxically contested by the homogenizing and commodifying forces of globalization. The steady growth of transnational tourism in the region will be shown to have consequences not only for the (dis)enchanted imaginary, but for the local environment and cultural identities of Himalayan peoples.

Tracing early Himalayan representations

Encompassing northern India, Nepal, Tibet and Bhutan, the Himalayan region has one of the longest pilgrimage traditions in the world and, apart from Mecca, continues to see the largest number of pilgrims of any place on Earth (Singh 2004). Amidst the Earth's youngest and highest mountains, there are peaks which have been sacred to millions of people over the past two millennia. In the last 30 to 40 years, the region has become a significant destination for international tourism, marketed for its awe-inspiring natural beauty and opportunities for adventure, spirituality, cultural exoticism, and relative affordability (Kolås 2004; Lim 2008). But centuries before Western mountaineers, hippies, and today's adventure tourists, the first Himalayan travelers were nomadic tribes and religious pilgrims.

From at least the eleventh century BCE, wandering Hindu ascetics told of holy sanctuaries amidst the secluded valleys, glacial rivers, and hot springs of the world's highest mountain range. Bhardwaj (1973) notes how "the exquisite locations of the Himalayan-scapes were consecrated as shrines for Hindu pilgrimages where nature overwhelmed the human psyche into prayerful submission and existential humility." Over time, specific mountains became designated pilgrimage sites, often times for more than one religion. Mount Kailas in Tibet, for example, is one of the holiest sites in Buddhism, but historically has also been important for adherents to Bön (the indigenous religion of Tibet), Hinduism, and Jainism (Zafren 2007, 279). In recent times, the sacred mountain has attracted a small stream of international travelers whose journeys blend cultural and adventure tourism with aspects of pilgrimage. The *Lonely Planet Tibet* guidebook outlines a journey to Mount Kailas in a section entitled, "the trek at a glance":

> The circuit, or *kora*, of Mount Kailash (6714m) is one of the most important pilgrimages in Asia. A religious sanctuary since pre-Buddhist times, a trek here wonderfully integrates the spiritual, cultural and physical dimensions of any trip to Tibet, which explains its growing attraction. Being able to meet pilgrims from across Tibet and other countries is one of the many allures of this walk.
>
> (Mayhew 1999)

The longstanding "spiritual magnetism" (Preston 1992) of sacred sites such as Kailas appears to spill over into a broader context of global tourism and arguably, the contemporary discourse of "wellness." Other sacred mountains such as Mount Everest (*Chomolungma* in Tibetan and *Sagarmata* in Nepali, both meaning "Mother goddess of the world") and Himalayan pilgrimage sites such as Gangatori (the source of the Ganges River) in late modernity appear to be undergoing a process of incorporation into a global network of "power places"; that is, places perceived as having extraordinary spiritual energy (Arellano 2004). Often integrating indigenous traditions with forms of new-age or neo-pagan spirituality, other global "power places" are Machu Picchu, Delphi and Stonehenge (Rountree 2002). Remote wilderness frontiers such as the Australian Outback and Antarctica are other travel destinations that symbolize power and function as "therapeutic landscapes" in late modernity (see Frost 2010; Hoyez 2007; Picard and Giovine 2014; Picard and Zuev 2014).

The earliest representations of the Himalayas as a spiritual utopia come from a blending of pilgrim accounts and religious mythology, from which the Himalayas came to be represented as the abode of the gods and a sacred intermediary space connecting heaven and earth (Singh 1992). As the spiritual renown of the region grew, Himalayan pilgrimage became a Hindu institution, marked by the construction of temples and shrines, formal employment of priests, and the establishment of site specific rituals, prescribed routes, ceremonies, and a donation system (Jacobsen 2013). The institutionalization of Himalayan pilgrimage occurred in conjunction with the spiritual fervor that caught India from the eleventh to first centuries BCE with the spread of the great Hindu epic, the *Mahabharata*. With the popularization of the *Mahabharata* and the *Puranas* (Hindu scriptures on myths), pilgrimage became an integral aspect of Hindu life (Singh 2005, 219), with the

Himalayas being particularly revered by those on the plains. *Arjuna*, one of the four mythic *Pandavas* (heroes) of the *Mahabharata*, at one point addresses a Himalayan summit:

> Mountain, thou art always the refuge of the good who practice the law of righteousness, the hermits of holy deeds, who seek out the road that leads to heaven. It is by thy grace, Mountain, that priests, warriors, and commoners attain to heaven and, devoid of pain, walk with the gods.
>
> (in Cooper 1997, 128)

As Bhardwaj notes, the Hindu pilgrimage system, which criss-crosses the entire Indian subcontinent, creates and maintains a "pan-Indian holy space" that ultimately serves a socially-unifying function (1973, 95). Over the centuries, through expanded trade and communication networks and especially through the transmission of Buddhism, the Himalayas came to be known throughout Asia. Only much later, did this isolated, mountain frontier find its way into the Western imagination.

Western representations of the Himalayas

A survey of popular literature, films, art, and other cultural productions reveals a consistent imagery of the Himalayas, and the Far East more generally, as the antithesis of the modern West (Bishop 2000; Clarke 1997). This is evident in novels such as James Hilton's *Lost Horizon* (1933, followed by Frank Capra's film version in 1937), E. M. Forster's *A Passage to India* (1924), and Herman Hesse's *Siddhartha* (1922) and *Journey to the East* (1932). More recently, Elizabeth Gilbert's international best-selling memoir cum Hollywood block-buster, *Eat, Pray, Love* (2006/2010), and films such as *Seven Years in Tibet* (1997, based on the 1953 autobiographical account by Heinrich Harrer), *Kundun* (1997), and *The Darjeeling Limited* (2007) continue to reproduce the image of a spiritually superior, mystical East. Since the 1970s a profusion of New Age and self-help books inspired by Eastern spiritual philosophies, as well as travel literature, guidebooks, blogs, and other media, have also reproduced this dominant representation. For instance, on the Lonely Planet website for Nepal, we find the following description:

> Wedged between the high wall of the Himalaya and the steamy jungles of the Indian plains, Nepal is a land of snow peaks and Sherpas, yaks and yetis, monasteries and mantras. Ever since Nepal first opened its borders to outsiders in the 1950s, this tiny mountain nation has had an irresistible mystical allure for travellers. Today, legions of trekkers are drawn to the Himalaya's most iconic and accessible hiking, some of the world's best, with rugged trails to Everest, the Annapurnas and beyond. Other travelers prefer to see Nepal at a more gentle pace, admiring the peaks over a gin and tonic from a Himalayan viewpoint, strolling through the temple-lined medieval city

> squares of Kathmandu, Patan and Bhaktapur, and joining Buddhist pilgrims on a spiritual stroll around the centuries-old stupas and temples that lie scattered across the Kathmandu Valley. There are few countries in the world that are as well set up for independent travel as Nepal. Wandering the trekking shops, bakeries and pizzerias of Thamel and Pokhara, it's easy to feel that you have somehow landed in a kind of backpacker Disneyland. Out in the countryside lies a quite different Nepal, where traditional mountain life continues stoically and at a slower pace, and a million potential adventures glimmer on the mountain horizons. The biggest problem faced by visitors to Nepal is how to fit everything in.[1]

This snapshot from the world's premier travel guide publisher reproduces a "tourism imaginary" (Graburn and Salazar 2014) of an exotic and mystical Himalayan nation where travelers are spoiled for choice. Such a description demonstrates how linguistic imagery operates by re-presenting what is removed in space and time before our eyes. As an ordered deployment of meanings, representations make certain things visible, while simultaneously keep other possible meanings hidden from view (Rancière 2007, 114). Thus, through both image and text the Lonely Planet website makes visible a selected version of Nepal that is meant to appeal to potential travelers with particular tastes. At the same time, this representation hides other, less attractive realities, such as the fact that Nepal is a politically unstable nation and the fourth poorest in Asia.

If representations shape our imaginings and expectations of how places will be, then we are "virtual tourists" before we actually set off on physical journeys (Berger 2004, 17; Mercille 2005; Urry and Larsen 2012). As Bohm (1996, 55) points out, "representation is not only present in thought or in the imagination, but *it fuses with the actual perception or experience* . . . so that what is 'presented' (as perception) is already in large part a re-presentation" (emphasis original). In this sense, reality cannot exist outside re-presentations, which are always prefigured by power relations. Vidal Clarmonte (2005, 260) observes how "all construction of meaning implies acceptance of the power generating it, that is, acceptance of whoever chose the pieces and classified the materials for us." The dominant representations of places such as the Himalayas have thus been generated selectively and transmitted, though these go back further than Lonely Planet guidebooks or mass tourism.

Early encounters with the Himalayas

The first Western encounters with the Himalayan region began in medieval times with journeys by legendary explorers such as Marco Polo, Friar Odoric, and Sven Hedin, who were followed by Jesuit and Capuchin missionaries. However, as Bishop (1989: 25) observes, the British encounter with the region, beginning in the late seventeenth century, marks the beginning of the creation of the Himalayas as an "imaginal landscape" for Western societies. Through travel reports and geographic surveys by British colonialists, the Himalayas entered the Western lexicon, accompanied by a discourse of mystery and enchantment. It is significant to recall that it was during this same period that the Grand Tour had come into fashion for aristocratic, secularizing Europe, marking the beginning of both modern tourism and the genre "travel literature." Exemplified by Goethe's (1892) *Travels in Italy*, which took place between 1816 and 1817, it was important that travel accounts were not only highly descriptive and informative, but expressive and aesthetic. Central to the Grand Tour was a discourse of education and self-cultivation, in which sites of the Renaissance such as Florence and Venice became key destinations for young aristocrats (Turner and Ash 1975; Wang 2000, 176).

While travel through continental Europe was becoming more commonplace, colonial powers also explored the boundaries of their acquired territories. Bishop (1989, 33) notes that implicit in early British accounts of the Himalayas was "a late residue of medieval European fantasies of Asia and the East" which connected back to "a totally different conception of European identity." For Europeans, the other side of the Tartars represented not only uncharted territory, but also a historical discontinuity. This lack of historical and cultural continuity led early European travel accounts—often blending geographic description, rudimentary ethnography and lyrical expressivity—to exemplify the orientalist tendency of placing non-Western cultures outside of time (Fabian 2002; Said 1978).

Mountains as symbolic landscapes

As one of the last blank spaces on the map of Western exploration and conquest, the geographic isolation of the Himalayas has had much to do with the allure the region has exercised on the Western imagination (Schell 2000, 16). The Himalayas are isolated in terms of being a landlocked region surrounded by vast plains and high desert on three sides and rainforest on the fourth, while also being the highest and most climatically severe mountain range on Earth. Throughout history, mountains have been places of pilgrimage for many cultures around the world. Frequently lying in remote areas and requiring long journeys from villages or cities, mountains have often been represented in mythology as the sacred domains of gods and spirits. Yi-Fu Tuan (2001, 122) notes how peripheral geographic locations, especially mountains, have frequently symbolized anti-structure and timelessness in different cultural traditions: "In Taoist lore, timeless paradises are located myriad miles from any known human settlement . . . [t]he European mind also envisions atemporal Isles of the Blest, Edens, and Utopias in remote and inaccessible places."

In his analysis of Thomas Mann's *The Magic Mountain*, Ricouer (1985, 112–115) observes how the narrative is marked by a sustained contrast between "those up here" and "those down below." The spatial opposition between "up" and "down" also reinforces a temporal opposition between those on the mountain who are acclimatized and have lost their sense of time, and those down on the flatlands whose mundane occupations and social roles force them to follow clock time. Playing on the symbolic geography of the mountain, the novel is, according to Ricouer, fundamentally an exploration of the conflicting relations between internal, chronological and monumental time, a global theme running through mountain mythology (Eliade 1991).

In the context of post-Enlightenment, modernizing Europe, feelings towards mountains in Europe had changed. Previously understood as places to be avoided, the spirit of scientific discovery, followed by the Romantic sacralization of nature produced new representations of mountain landscapes as beautiful and sublime. As Mathieu (2009, 348) points out, "during the eighteenth and nineteenth centuries a vocabulary took root that came from the border of aesthetic and religious experience . . . [p]eople spoke of a feeling of 'delightful horror,' the idea of 'the sublime' was central, mountains were designated as the 'cathedrals of the world.'" As symbolic landscapes, mountains were ascribed morally and spiritually uplifting qualities. In 1857, John Ruskin, for instance, described mountains with a timely blend of scientific analogy and romantic reverence: "Mountains are to the rest of the body of the earth, what violent muscular action is to the body of man. The muscles and tendons of its anatomy are, in the mountain, brought out with force and convulsive energy, full of expression, passion, and strength" (Ruskin 2010, 38).

Aldous Huxley, whose father was an early mountaineering enthusiast, once remarked, "[m]y father considered a walk among the mountains as the equivalent of churchgoing" (in Armstrong 2006, 86). By literally and symbolically taking climbers up above the profane concerns of everyday life in the town or city—climbing mountains became a ritualized act. In secularizing Europe, however, this practice came not to be referred to as pilgrimage, but mountaineering.

The birth of modern mountaineering

The history of mountaineering is important for understanding how the Himalayas evolved from a remote frontier to a transnational travel destination. Desires for conquering mountains and escaping modern urban life emerged during a period of rapid and dramatic social change in late eighteenth- and nineteenth-century Europe. Mathieu (2009, 349) observes how while the Enlightenment period is usually portrayed as a secular, if not anti-religious movement, an apparent counter reaction led to the sacralization of mountains and other natural phenomena by the early Romantics. Analyzing representations from scientific and travel literature, he demonstrates how, from the fifteenth to seventeenth centuries, mountains in Europe were viewed as considerably less sacred than in the eighteenth to twentieth centuries.

This sacralization—occurring during a period of intense modernization, industrialization and nation-state building—contradicts the Western assumption that only "simple," traditional societies are religious, while modern "complex" societies undergo secularization. Thus, while a "religious instinct" does not necessarily disappear in the modern West, there is a displacement of sacredness from the *supernatural* to the *natural*, reflecting what Taylor (2007) has called a shift from a transcendent to an "imminent frame." As Beedie and Hudson

(2003, 635) observe, "mountain mythology is embedded with romantic notions of exploration, journey, and searching . . . ideas which become attractive in the modern social world where fragmentation and complexity are the norm." Since searching implies finding, the authors suggest that mountaineering and other forms of frontier travel were in part quests for stability in an increasingly destabilized world.

Colonial explorers and mountaineers were often the first Westerners to come into contact with many areas and peoples of the Himalayan region. Their travel accounts and the publicity their expeditions received in Europe did much to transmit representations of mystery and enchantment surrounding the "roof of the world." Moreover, mountaineering can be viewed as the earliest form of tourism in the Himalayas (Johnston and Edwards 1994, 462), laying the foundations for the adventure, cultural, and spiritual tourism the region is known and marketed for today. As the sport grew and every significant European mountain had been "conquered" during the "golden age of alpinism" in the 1850s, mountaineers began setting their sights on peaks in reach of colonial outposts, most notably the Himalayas (Unsworth 1994, 66). It is worth noting that the British, who by virtue of occupying India had the easiest access to the Himalayas, were also the most enthusiastic about mountaineering at the time.

While eighteenth-century mountaineering was part of the Enlightenment project of amassing scientific knowledge, which included mapping and measuring the Earth's surface, Unsworth (1994, 53) notes the scientific element declined fairly rapidly in the nineteenth century. Instead, a sense of individual and nationalistic competition and the desire to be the first to summit the highest Alpine peaks became central to early mountaineering. As Unsworth observes, the rapid growth of British mountaineering by the mid-nineteenth century was part of a broader expansionist mood: "Imperialism and exploration went hand in hand and national self-confidence was such that nothing seemed impossible" (54). By 1877, all of the European Alps first ascents had been claimed (80 percent of which by British climbers), though it was not until 1892 that the first Himalayan expedition took place (Johnson and Edwards 1994, 462). Compared to European mountains, the Himalayas posed a whole new set of challenges for climbers, with 14 peaks reaching higher than 8,000 meters—more than double Europe's highest point, Mount Blanc (4,807 meters). Moreover, while railroads in Europe allowed for increasing speed and accessibility to mountains, leading to what Simmel (1997, 219) called the "wholesale opening of the Alps," it took between three and four weeks on foot to reach the base of most of the Himalayan peaks. It was both the literal heights of these mountains and their distance from Western civilization that allowed the Himalayas to symbolize a frontier of adventure par excellence.

Adventure and escape: twentieth-century Himalayan travel

Ortner (2001) observes that mountaineering is both a "game of masculinity" and a "game of adventure" representing escape from the perceived pitfalls of modern life.

> The kind of adventure involved in mountaineering—a basically twentieth-century phenomenon—is very specifically constructed in relation to issues of "modernity"; one drops out of the continuity of (modern) life because one finds it lacking—lacking in "adventure," among other things. A discourse in which the adventure-ness of mountaineering is linked with a critique of modernity is perhaps the dominant (though not the only) discourse of the sport.
>
> (Ortner 2001, 36)

The point for many climbers, Ortner explains, is to overcome the boredom, lack of challenge, and materialism of modern life and to experience a spirituality that cannot be realized on the flatlands of modern society. While a counter-modern discourse of Himalayan mountaineering runs through the entire twentieth century, Ortner chronicles how it undergoes significant variations in different eras. In the early period, occurring between the 1920s and 1940s, mountaineering was heavily influenced by a romantic ethos that was "organised around the desire to transcend the limits of the self" (2001, 39). Ortner notes how many early mountaineers incorporated elements of asceticism, mysticism, and/or moralism into their Himalayan adventures. Even mountaineers who did not speak of their journeys as "mystical" often displayed anti-modern ideology by denouncing the use of oxygen and other modern technology in climbing. Such technologies ran counter to the idea of mountaineering being about pushing limits and actualizing an authentic self.

The 1950s and 1960s were characterized, on the other hand, by what Ortner calls the "hyper-masculine sahib [Westerner]," epitomized by New Zealander Edmund Hillary, who along with Tenzin Norgay was the first to reach the summit of Mount Everest in 1953. Upon their descent, Hillary is famously quoted as saying, "We knocked the bastard off"; while Norgay gave thanks to the gods who spared their lives. As every Himalayan peak came to be "knocked off" by the early 1970s and the region saw more and more climbing teams every year, mountaineering paved the way for tourism. Johnston and Edwards (1994, 462–463) observe how increasingly visitor numbers occurred in conjunction with improvements in technology, including climbing equipment, more accurate maps, access to weather reports, and improved communications between climbers and base camps. Moreover, the establishment of airstrips in remote regions, together with the general lowered cost of air travel, led to "dramatic increases in the number of people traveling up and down the mountains," and to the region in general. As Zafren (2007, 279) notes, the opening of Nepal and Tibet to foreign tourism gave rise to a "modern generation of secular pilgrims," who included not only mountain climbers, but the "flower children" of the 1960s and 1970s.

The hippie trail and the emergence of the Himalayas as a mass tourist destination

By the late 1960s, Nepal was arguably the biggest magnet in the world for the counterculture, with "hippies from virtually every nation flocked there for the cheap living, 'Eastern religion' and legal marijuana'" (Ortner, 2001, 186). Only opening its doors to outsiders in the 1950s, by 1961 a little over 4,000 tourists had traveled to Nepal. Even then, these early visitors were mostly limited to Kathmandu, owing to the lack of transportation and communication facilities outside the valley. Fifteen years later, more than 100,000 tourists had visited the country and the tourism revenue had quickly become a central facet of Nepal's economy (Nepal 2000, 663). Ortner, who first conducted fieldwork in the Solu-Khumbu region around Mount Everest in the early 1960s and had only encountered a handful of foreigners, revisits her original field notes from the mid-1970s: "There were a *million* tourists. Both the trail and Lukla itself were *awful*, essentially *polluted* with tourists" (Ortner, 2001, 199, emphasis original).

Many came from Europe, having traveled overland on what became known as the Hippie Trail. As the "end of the road," Nepal became a secular pilgrimage centre for those seeking alternative lifestyles. In the new millennium, British author Rory Maclean (2006) remade the journey, chronicling the changes it has undergone over the last four decades. In Kathmandu, he speaks with Desmond, a long-term resident of Nepal who arrived from Ireland in the late 1960s. Describing how he had fallen in love with the mountain landscapes and spiritual traditions of the region, like Ortner he laments the gradual overrun of global tourism:

> Kathmandu's full of people reading the Lonely Planet guide to Vietnam. They sit in cafés sending each other text messages. I mean, at their age we wanted to get into each other and society, not to live in a meltdown world. We didn't have guidebooks; we didn't even know the *name* of the next country. "What's this place called? Bhutan? Where the hell is Bhutan?" he shouts, his voice filled with angry energy.
>
> (Maclean 2006, 271)

In his collection of travel essays, *Video Night in Kathmandu: And Other Reports from the Not-So-Far East*, Pico Iyer ([1989] 2001, 83) similarly depicts the changes Nepal has undergone amidst the boom of global tourism by the late 1980s:

> For twenty years now, it [Nepal] had cashed in on being the closest place on earth to the remotest place on earth, a country just around the corner from Shangri-La. And if Tibet's charm lay in its remoteness, Nepal's lay in its availability; a veteran of the mystic market, it knew exactly how to sell itself as a wholesale, second-hand Tibet. Thus the magic title of the Forbidden Land found its way into every single brand name: local stores were stocked not just with handcrafts, but with Tibetan handcrafts, Tibetan paintings, Tibetan bells, Tibetan scarves, Tibetan pizzas. Nepal had the additional advantage of being on the fringes of India . . . still the biggest spiritual department store in the East, an economy-sized convenience store with many of the same goods at even better prices.

Employing the irony characteristic of postmodern travel writers, Iyer and Maclean trace the trajectory of a "mystical East" fallen prey to the nascent culture of consumer capitalism. Beneath the humor and irony of these texts lies the critical recognition that, paradoxically, tourism both produces and destroys its object of desire (MacCannell 1976; Shepherd 2002; Van den Berghe 1994). Travel writing such as Iyer's also points to the idea that, in tourism encounters, both groups, but particularly the one in the supposedly "dominated" position undergo the more significant transformations (see Bruner 1991). In her ethnography *Tigers of the Snow* (1996), Adams for instance argues that Sherpas interpret themselves according to Western textual representations and in doing so take on a virtual identity (18–19). Drawing upon Ricouer's notion of *mimesis* as an inter-active, dialogical process of identity construction, she writes how "Sherpas engage with Western representations . . . in a manner which opens up and discloses ways of being Sherpa—ways created in the Western imagination but based, after all, on real experiences of Sherpa life" (18). This reflects two earlier points raised in this chapter: first, that representations are not static, but actually fuse with perception and continuous experience; second, that representations are always prefigured by power relations.

Conclusions

This chapter has discussed the Himalayas as an evolving symbolic landscape. Himalayan representations first came from wandering holy men who ventured forth and consecrated sites amidst the high altitude mountain-scapes. Mythical tales of "walking with the gods" spread as Himalayan pilgrimages gradually became an institution of Hinduism and Buddhism. I have tried to demonstrate that certain continuities appear to exist between early local representations of the Himalayas and those later deployed by the West; namely the geography of the region as a peripheral frontier and especially the symbolic significance of mountains. Emerging under very different historical and cultural circumstances, however, Western representations of the region and approaches to travel there reflect some particularly modern themes. Modern Himalayan journeys appear to express desires for not only adventure and escape, but also for stability and timelessness in a world of flux and change. Located literally and symbolically above and beyond the profane world with its dulling routines and social demands, as in Mann's *Magic Mountain*, the Himalayas represent a place up there and out of time.

In late modernity, the growth of global tourism, transnational mobilities, and the commodification of Himalayan nature-cultures means this image is contested. As Bishop (2000, 15) observes, "Shangri-La seems to be caught between being a utopia and a brand name, between being a tourist destination and the goal of spiritual pilgrimage, between a museum or a theme park and a political and spiritual summons." In 2011, when the author carried out fieldwork in the region, the Nepal government has launched a campaign that aimed to attract over one million foreign visitors to the country, with the mission statement: "to establish Nepal as a choice of premier holiday destination with a definite brand image" (Ministry of Tourism 2011). From my readings of Himalayan representations, this "brand image" is one based on the promise of adventure and spirituality as a kind of utopian antidote to modernity and its discontents. As part of the discourse of the "mystical East," the Himalayas as both a symbolic and therapeutic landscape signifies a meta-cultural critique and expression of transnational modernity. At the same time, the symbolism of remote frontiers, wilderness areas and high places is not just a uniquely modern phenomenon, but reflects a deeper religious or utopian impulse (Bloch 1986), as part of the human condition.

Note

1 Reproduced with permission from Lonely Planet © 2016 Lonely Planet.

References

Adams, Vincanne. 1996. *Tigers of the Snow and Other Virtual Sherpas: An Ethnography of Himalayan Encounters*. Princeton, NJ: Princeton University Press.

Appadurai, Arjun. 2010. "How Histories Make Geographies: Circulation and Context in a Global Perspective." *Transcultural Studies* 1(1): 4–10.

Arellano, Alexandra. 2004. "Bodies, Spirits, and Incas: Performing Machu Picchu," in Mina Sheller and John Urry (eds.), *Tourism Mobilities: Places to Play, Places in Play*. London: Routledge: 67–76.

Armstrong, Bill. 2006. *Musings from the Mountaintop*. Saint Louis: Xulon Press.

Beedie, Paul and Hudson, Simon. 2003. "Emergence of Mountain-based Adventure Tourism." *Annals of Tourism Research* 30(3): 625–643.

Bellah, Robert N. 2006. *The Robert Bellah Reader*. Durham: Duke University Press.

Berger, Arthur Asa. 2004. *Deconstructing Travel: Cultural Perspectives on Tourism*. Walnut Creek, CA: Alta Mira Press.

Bhardwaj, Surinder Mohan. 1973. *Hindu Places of Pilgrimage in India*. Berkeley: University of California Press.

Bishop, Peter. 1989. *The Myth of Shangri-La: Tibet, Travel Writing and the Western Creation of Sacred Landscape*. Berkeley: University of California Press.

——. 2000. "The Death of Shangri-La: The Utopian Imagination and the Dialectics of Hope." *Psychology at the Threshold*. University of Santa Barbara, California. http://www.pinnigerclinic.com/events/BISHOPThe__utopian_imagination.pdf.

Bloch, Ernest. 1986. *The Principle of Hope*, 3 vols. Cambridge, MA: MIT Press.

Bohm, David. 1996. *On Dialogue*. New York: Routledge.

Bruner, Edward M. 1991. "Transformation of Self in Tourism." *Annals of Tourism Research* 18(2): 238–250.

Clarke, J.J. 1997. *Oriental Enlightenment: The Encounter between Asian and Western Thought*. New York: Routledge.

Cooper, Adrian. 1997. *Sacred Mountains: Ancient Wisdom and Modern Meanings*. London: Floris Books.

Eliade, Mircea. 1991. *Images and Symbols: Studies in Religious Symbolism*. Princeton, NJ: Princeton University Press.

Fabian, Johannes. 2002. *Time and the Other: How Anthropology Makes Its Object*. New York: Columbia University Press.

Frost, Warwick. 2010. "Life Changing Experiences: Film and Tourists in the Australian Outback." *Annals of Tourism Research* 37(3): 707–726.

Goethe, Johann Wolfgang von. 1892. *Travels in Italy*. London: Bell.

Graburn, Nelson and Salazar, Noel B. 2014. "Introduction: Towards an anthropology of Tourism Imaginaries," in Nelson Graburn and Noel B. Salazar (eds.), *Tourism Imaginaries: Anthropological Approaches*. London: Berghahn Books: 1–28.

Howard, Christopher A. 2012, "A Horizon of Possibilities: The *Telos* of Contemporary Himalayan Travel." *Literature & Aesthetics* 22(1): 131–155.

Hoyez, Anne-Cécile. 2007. "The 'World of Yoga': The Production and Reproduction of Therapeutic Landscapes." *Social Science & Medicine* 65(1): 112–124.

Iyer, Pico. 2001. *Video Night in Kathmandu: And Other Reports from the Not-So-Far East*. London: Bloomsbury.

Jacobsen, Knut A. 2013. *Pilgrimage in the Hindu Tradition: Salvic Space*. New York: Routledge.

Johnston, Barbara R. and Edwards, Ted. 1994. "The Commodification of Mountaineering." *Annals of Tourism Research* 21(3): 459–78.

Kolås, Åshild. 2004. "Tourism and the Making of Place in Shangri-La." *Tourism Geographies* 6(3): 262–278.

Lim, Francis Khek Gee. 2008. "Of Revervie and Emplacement: Spatial Imaginings and Tourism Encounters in Nepal Himalaya." *Inter-Asia Cultural Studies* 9(3): 375–394.

MacCannell, Dean. 1976. *The Tourist: A New Theory of the Leisure Class*. London: Macmillan.

Maclean, Rory. 2006. *Magic Bus: On the Hippie Trail from Istanbul to India*. London: Viking.

Mathieu, Jon. 2009. "The Sacralization of Mountains in Europe during the Modern Age." *Mountain Research and Development* 26(4): 343–349.

Mayhew, Bradley. 1999. *Lonely Planet Tibet*. Hawthorn, Vic., AU: Lonely Planet Publications.

Mercille, Julien. 2005. "Media Effects on Image: The Case of Tibet." *Annals of Tourism Research* 32(4): 1039–1055.

Ministry of Tourism, Nepal. "Nepal Tourism Year 2011." http://www.tourism.gov.np/%3E (accessed February 12, 2011).

Nepal, Sanjay K. 2000. "Tourism in Protected Areas: The Nepalese Himalaya." *Annals of Tourism Research* 27(3): 661–681.

Ortner, Sherry B. 2001. *Life and Death on Mt. Everest: Sherpas and Himalayan Mountaineering*. Princeton, NJ: Princeton University Press.

Picard, David and Giovine, Michael A. Di (eds.). 2014. *Tourism and the Power of Otherness: Seductions of Difference*. Bristol: Channel View.

Picard, David and Zuev, Dennis. 2014. "The Tourist Plot: Antarctica and the Modernity of Nature." *Annals of Tourism Research* 45(1): 102–115.

Preston, James. 1992. "Spiritual Magnetism: An Organizing Principle for the Study of Pilgrimage," in Alan Morinis (ed.), *Sacred Journeys: The Anthropology of Pilgrimage*. Westport, CT: Greenwood Press: 31–46.

Rancière, Jacques. 2007. *The Future of the Image*. London: Verso.

Ricœur, Paul. 1985. *Time and Narrative*, 2 vols. Chicago: University of Chicago Press.

Rountree, Kathryn. 2002. "Goddess Pilgrims as Tourists: Inscribing the Body through Sacred Travel." *Sociology of Religion* 63(4): 475–96.

Ruskin, John. 2010. *Frondes Agrestes*. New York: The Mershon Company.

Said, Edward. 1978. *Orientalism*. New York: Pantheon Books.

Schell, Orville. 2000. *Virtual Tibet Searching for Shangri-la from the Himalayas to Hollywood*. New York: Holt.

Shepherd, Robert. 2002. "Commodification, Culture and Tourism." *Tourist Studies* 2(2): 183–201.

Simmel, Georg. 1997. *Simmel on Culture: Selected Writings.* Edited by David Frisby and Mike Featherstone. London: SAGE.

Singh, Salani. 2004. "Religion, Heritage and Travel: Case References from the Indian Himalayas." *Current Issues in Tourism* 7(1): 44–65.

——. 2005. "Secular Pilgrimages and Sacred Tourism in the Indian Himalayas." *GeoJournal* 64: 215–223.

Singh, T. V. 1992. "Development of Tourism in the Himalayan Environment: The Problem of Sustainability." *IE/PAC UNEP* 15(3–4): 22–27.

Taylor, Charles. 2007. *A Secular Age*. Cambridge, MA: Belknap Press.

Tuan, Yi-fu. 2001. *Space and Place: The Perspective of Experience*. Minneapolis: University of Minnesota Press.

Turner, L. and Ash, J. 1975. *The Golden Hordes: International Tourism and the Pleasure Periphery*. London: Constable.

Unsworth, Walt. 1994. *Hold the Heights: The Foundations of Mountaineering*. Seattle: The Mountaineers Press.

Urry, John and Larsen, Johnas. 2012. *The Tourist Gaze 3.0*. London: Sage Publications.

Van den Berghe, Pierre L. 1994. *The Quest for the Other*. Seattle: University of Washington Press.

Vidal Clarmonte, Carmen Africa. 2005. "Re-presenting the 'Real': Pierre Bourdieu and Legal Translation." *The Translator* 11(2): 259–275.

Wang, N. (2000). *Tourism and Modernity: A Sociological Analysis*. New York: Pergamon.

Zafren, Ken. 2007. "Pilgrimages in the High Himalayas," in Wilder-Smith Annelies, Schwartz Eli, and Shaw Marc (eds.), *Travel Medicine*. Oxford: Elsevier: 279–283.

10 Constructing and celebrating a national object of desire

The Amazonian Oriente frontier and Ecuadorian society, 1900–1946

William T. Fischer

Introduction

The Oriente, Ecuador's vast Amazonian frontier lying to the east of the Andes (see Figure 10.1), received minimal attention from the state for nearly a century after Ecuador's independence in 1830. Despite this neglect, by the end of the nineteenth century the Oriente occupied an important place in Ecuadorian politics and national discourse. A component of this was the "elaboration of a national imaginary about the Oriente," a key part of the "symbolic incorporation [of the Oriente] to the national State" (Esvertit 2008, 264). The continuation of this process in the twentieth century resulted in the construction of Ecuador's Amazonian frontier as a national object of desire. As such, the region was both an ideological tool for Ecuadorian state-makers who argued for greater attention to be paid to the region, and a patriotic yardstick to measure Ecuador's international prestige. These discourses prepared the region in the minds of citizens for its use as a stage on which Ecuadorian history would unfold. This creation of "stage space," as described by historian Raymond Craib, extends the reach of the state and nation; it allows for certain actions and kinds of agency and permits the operation of "science, statecraft, and political economy" (Craib 2004, 5–6). As the anthropologist Anne-Christine Taylor has written, the Oriente's contribution to the national economy was sporadic until the middle of the twentieth century. However, waves of immigration to the region beginning in early the twentieth century eventually led to the "definitive and irreversible" incorporation of the region and the indigenous people who lived there (Taylor 1994, 60). The elaboration of "stage space" was a prerequisite for this development, in this author's view. Jean-Paul Deler has described how Ecuador came to control its territory by elaborating the "rules of the game" between different regions and socioeconomic groups (Deler 2007, 13). For Deler, the Oriente remained marginal for much of the republican period due to the region's lack of an "effective economic interest" (Deler 2007, 175). However, as this chapter argues, Ecuador's Amazon was anything but marginal in cultural and nationalist terms. A significant portion of the Oriente's history, then, took place not inside the Amazonian region itself, but in the press and other publications that were available to citizens in Ecuador's capital of Quito and other principal cities. These publications modified the "national imaginary" about the

Oriente to more fully include the Ecuadorian citizens who extended the nation to the Amazon. Also important was the intensified popularization of the Oriente as a component of civic pride and patriotism, thanks to its increasing prominence in national publications and in public celebration. The *Día del Oriente Ecuatoriano* (Day of the Ecuadorian Oriente) in 1939 and subsequent years offered a condensed view of the Amazon's symbolic incorporation into the national imaginary and furthered its construction as a national object of desire. Cultural identities are in a constant state of production and are never an "accomplished fact," as Stuart Hall has stated (Hall 1990, 222). Ecuadorian national identity therefore had to be produced over the course of the nation's existence; the incorporation of the Amazonian frontier into this cultural identity has a history of its own. As was true for other South American republics, such as Colombia, the frontier occupied an important position in political discourse that shaped frontier policy (Rausch 1993; Rausch 1999). This chapter calls attention to the important role that frontiers play in a nation's history not just because of the social and political developments in the frontier space, but also because of the cultural importance of the frontier as an object of desire that exists throughout the nation.

Additionally, the constant tension with Peru over territory, the 1941 war that removed a large portion of Ecuador's Amazon, and the acute feeling of loss that followed, all affected patriotic discourse about the Oriente. As a national object, the Oriente was tied to the international prestige of Ecuador. Given the role that competition with Peru played in how the Oriente was viewed, it must be understood that a component of Ecuador's national identity was inherently transnational. Ecuador's self-esteem and its future prosperity were both tied to this region's relationship to its neighbors (especially Peru), which was constantly at issue. The 1941 war and the soul-searching that followed ensured that the transnational aspects of the Oriente's frontier existence remained a vital component of Ecuadorian national identity. Therefore, the study of the cultural significance of the Ecuadorian frontier has transnational importance, for as Akira Iriye explained, there is an "intricate interrelationship between nations and transnational existence, between national preoccupations and transnational agendas." This study contributes to our understanding both of nations and also their role "in contributing to defining the world at a given moment in time" (Iriye 2013, 15).

The Oriente in Ecuador's national history

In order to examine the Oriente as a national object in the time before the war with Peru, we must first understand the state of the national imaginary about the Oriente at the turn of the twentieth century. An important component of discourse about the Oriente was the justification of the region's deep, purely Ecuadorian past. Because Ecuador's neighbors were encroaching on its Amazonian territory, Ecuadorian authors and statesmen needed to write a place for the disputed territory in Ecuador's history, and indeed, pre-history, to justify their territorial claims. Like other postcolonial nations, Ecuador aimed to give a deep national past to a political entity of recent creation. Mark Thurner has discussed how Peruvian

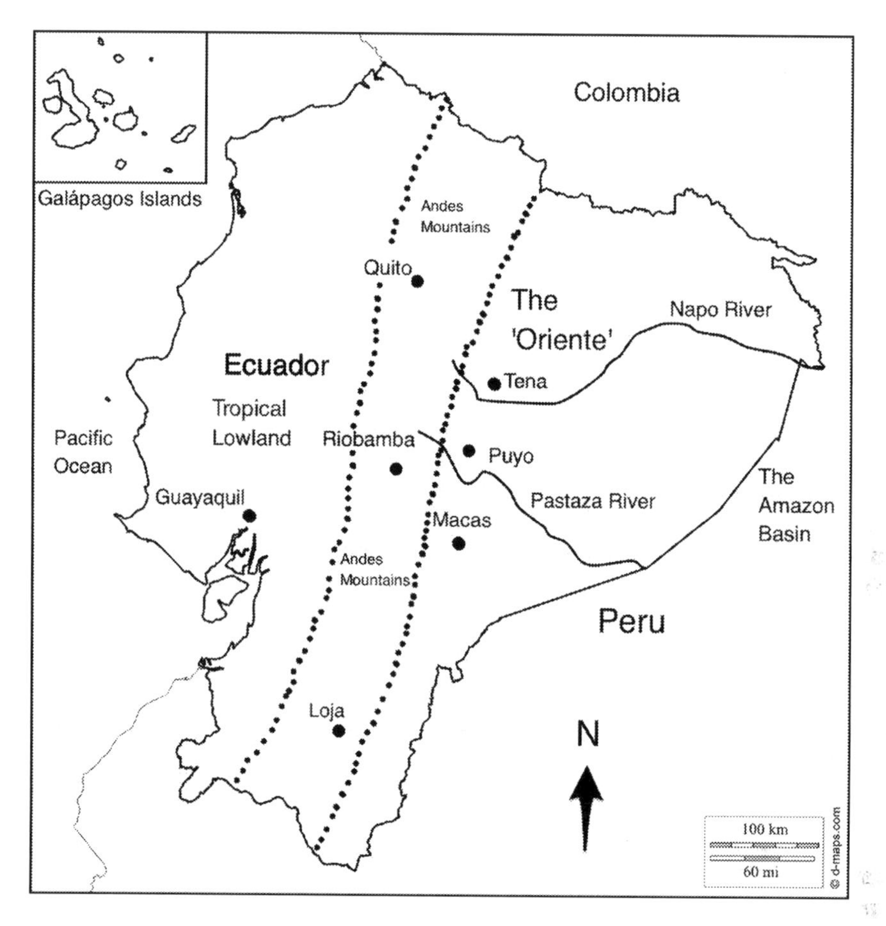

Figure 10.1 Map of Ecuador and the Oriente, east of the Andes.

Source: D-maps.com. Edited by the author.

historiography in the nineteenth century sought to establish a "transhistorical subject 'us'" of Peruvians that was tied to "a timeless place with a proper name" (Thurner 2003, 142). In Peru, this involved recovering the Inca past as part of national narrative. For Ecuador's Oriente, however, it required the nationalization of Spanish colonial antecedents. The 1541 expedition by Spanish conquistadors Gonzalo Pizarro and Francisco de Orellana, which was the first European discovery of the western Amazon River, departed from Quito. These explorers and later colonial officials were described as national heroes, despite the fact that Ecuador, the independent republic, would not exist for centuries. Ecuador broke free of Spanish colonial rule in 1822, first as part of *Gran Colombia*, which included what is now Venezuela and Colombia. Sectional strife led political leaders in Quito to break free of this larger entity in 1830. Nevertheless, Ecuadorians rhetorically bridged the gap between sixteenth-century events and their own national existence. The border between Ecuador and Peru was based on the colonial-era jurisdictions of the Real Audiencia of Quito and the Viceroyalty of Peru. The two nations disagreed, however, on how much of the upper Amazon basin legitimately pertained to each republic. In a pamphlet addressed to the Congress of 1894, one promoter of colonization in the Oriente argued that the bases for territorial legitimacy between Ecuador and its neighbors were the sixteenth- and seventeenth-century explorations of conquistadors and the subsequent political and juridical administration of the Real Audiencia of Quito (Mora López 1894, 11–12). These lands represented a vital source of wealth for Ecuador's future, according to the pamphleteer, and their proximity to Quito made them governable, while their distance from Lima, capital of Peru, made that nation's claims untenable (Mora López 1894, 18). Famed Ecuadorian geographer and mapmaker Enrique Vacas Galindo gave a speech at the University of Quito in 1905 promoting the construction of a railway to the Oriente. This project's purpose was to take advantage of lands that derived their legitimacy, as Vacas explained, from the discoveries of Francisco de Orellana and others, and from the pacification of Amazonian tribes by missionaries from Quito throughout the colonial period. By this logic, Ecuadorian national dominion could be traced back nearly 400 years (Vacas Galindo 1905, 2–3). Sixteen years later, a speechmaker at a university in Loja reinforced this idea by discussing the colonial-era efforts in the Amazon by missionaries and soldiers from Quito in terms of the suffering and sacrifice of "our" national heroes (Eguiguren 1922, 12–13). In 1923, a brief geographic history for young Ecuadorians presented the exploration of the Amazon as an Ecuadorian achievement thanks to the importance of Quito and Guayaquil in the explorers' efforts (Veintemilla 1923, 26–28). These narrations of the territorial history of Ecuador before its existence mirrored the efforts of "official nationalists," as described by Benedict Anderson, to represent "specific, tightly bounded territorial units" in historical maps. As Anderson argues, geographical representation is an important component of an imagined national community (Anderson 1991, 174–175). Veintemilla's and similar publications represent an effort to convince Ecuadorians that the Oriente was a fully national space; if part of the nation's past had been performed on that stage, then so could the nation's future.

The Amazon as international measure of prestige

During the twentieth century, Quito's press presented the Oriente's progress as a question of national prestige, especially in contrast to Peru's efforts in its Amazon and in lands claimed by Ecuador. Peru, and Bolivia also, pursued policies of Amazonian integration that relied upon the use of missionaries and military forces to strategically occupy the region as well as to Christianize and "civilize" the indigenous people that lived there (García Jordán 2001, 17). Ecuadorians watched these efforts and compared them to their own. In May of 1924, Quito's daily newspaper *El Comercio* published an article concerning Peru's plans to colonize its Amazon with 80,000 European immigrants (No Author 1924b). *El Comercio's* editors opined that this plan should serve as a wake-up call; it should push Ecuador's own government and citizens to be more decisive in seeking progress in the Oriente (No Author 1924a). In March of 1925, *El Comercio* lamented the abandonment of the Oriente and the stagnation of government efforts toward the region. But a letter written from the southern Oriente argued that, although the government might not be praiseworthy, Ecuadorian citizens in the region were. The letter detailed the works by "landholding colonists" using their own capital in the face of Peruvian encroachments. "They are," stated the author, "truly patriotic people" who would soon see the benefits of their toil in the exuberant forests (No Author 1925a). In October of that year, *El Comercio* published articles titled "Propaganda Oriental," arguing that preoccupation with the Oriente was a patriotic duty. One contribution to this series was addressed to the Ecuadorian proletariat, exhorting this class to fly the Ecuadorian flag in all national territory. According to the author, those who became "Orientalists" would be "true patriots" and "martyrs of democracy" that would save the Homeland (No Author 1925b). Their efforts were not in vain; the 1920s were a critical decade for the Oriente's integration into the wider republic. This is the time when "Amazonia finally began to be integrated into the political-administrative structure of the Ecuadorian state" thanks to new legislation passed in Quito and the increasing presence of Catholic missionaries from the highlands (Little 2001, 46–47).

Those who contributed to the Oriente's development were acting in the nation's best interests. A letter published in *El Comercio* from the president and secretary of a new pro-Oriente committee in the northern highlands stated that their project to colonize the Oriente would be an honorable example of true patriotism for any "communists" who stirred up trouble by complaining of unequal distribution of lands; in the Oriente, lands were available and abundant (Maldonado 1926). The same committee later stated their intention to "serve as sentinels of national integrity" that would pursue economic development (Maldonado 1924). The word "sentinel" implies that the Oriente's colonists were especially patriotic for their quasi-militaristic attitude toward the region. It was not just a fully national region; it was filled with citizens who had taken on a special mission.

The campaign to make the Oriente a component of Ecuadorian patriotism and national identity was given another outlet when *El Oriente Dominicano* was founded in 1927. Published by Dominican priests, it served to report the progress of the Dominican missionary settlements in the central Oriente. The magazine commented in 1933 while armed conflict loomed between Colombia and Peru over disputed territory in the Amazon. José María Vargas, the magazine's director, argued that Ecuador "is under the patriotic obligation" to direct its efforts toward the Oriente rather than remain a mere spectator of the ongoing international conflict. Ecuador's national prestige depended on progress in such factors as road construction to the Oriente and agricultural development, both of which promised to employ out-of-work Ecuadorians and to increase national wealth. Vargas stated that because all the Dominican missionaries were Ecuadorian by nationality, they were natural defenders of the territory. They worked not for personal interest, but for the Homeland and its future (Vargas 1933, 1–2).

The magazine also stressed the importance of the Oriente to all Ecuadorians, not just missionaries or colonists. A typical example from 1935 argued that patriotism demanded familiarity with the borders and history of all the land that was Ecuador's national inheritance, that the land be loved like a prized possession, and that it be made to flourish through Ecuadorians' labor. In short, the nation's efforts toward the region should not cease until all citizens have easy access to the Oriente to make "that green region the true *hacienda* of the Homeland" (No Author 1935, 65–66). The following year, José María Vargas reiterated that all Ecuadorians should be familiar with the Oriente's borders. Vargas reviewed the various nineteenth-century treaties that established Ecuador's right to the Oriente but that Peru refused to respect during its frequent incursions into the territory. The only solution would consist of Ecuador's cultivation of every last section of the disputed territory. Doing so, he argued, would provide economic benefit and a navigable passage to the Atlantic for the whole country (Vargas 1936, 281–282). Again, developing the Oriente was the solution to myriad problems: the international conflict with Peru on the one hand, and economic shortfalls and unemployment on the other.

Another important publication about the Oriente was the secular magazine *Miscelánea*, which entered its fourth year of publication in 1935 with the promise to "intensify our propaganda in favor of the Oriente, which is . . . threatened or invaded by . . . envious neighbors." In the magazine's view, the people and the government both had the obligation to "affirm, with material and moral reality, the Ecuadorianness of the Oriental Region, which belongs to us without doubt" (No Author 1934a, 1–2). In a subsequent editorial, *Miscelánea* cited the fact that the Spanish discoverers of the Amazon River were guided by Indians from Quito as justification for Ecuador's claim to the Oriente. That the Oriente was the future of the nation was well known; *Miscelánea* exhorted the new government of José María Velasco Ibarra to follow through on this vision for the future (No Author 1934b, 1–3).

El Comercio continued throughout the 1930s to print articles about the patriotic effort to develop the Oriente. A January 1933 article by an author with 25 years' experience in the Oriente asserted that the region's abundant wealth, especially the gold in its rivers, "offers to all poor men . . . that love work" the chance to "shake off their misery" and "conquer for themselves life and independence" (Romero Terán 1933). The paper addressed the region's more symbolic importance by publishing a manifesto from university students addressing the looming border conflict in the Amazonian region between Peru and Colombia. The students repeated the reasons for Ecuador's claim to Amazonian territory. In their view, Ecuador's right to river passage to the Atlantic was "inviolable and undeniable" and dated back to the "first footstep" of "civilized man" in the region (Yepes et al. 1933).

Civic celebration of a national object of desire

Perhaps no effort to make the Oriente a part of the Ecuadorian nation was so clear and concerted as the creation of a holiday to celebrate the region, the *Día del Oriente Ecuatoriano*, which in 1939 was officially designated for the 27th of February

(Chiriboga 1939). Civic holidays have various meanings; one of their purposes is, in the words of historian Thomas Benjamin, to "strengthen patriotism and social solidarity [and] enhance the legitimacy of the state" (Benjamin 2000, 99). The Día del Oriente certainly promoted patriotism and Ecuadorian unity; furthermore, it celebrated missionaries, private citizens, and agricultural colonists that dared to make their way in the region despite poor communications, illness, and inclement weather. Many of the speeches and publications around the holiday argued that the day was an occasion for regular Ecuadorians to look to themselves as potential redeemers of the nation through action in favor of the Oriente.

El Comercio's editors did not miss a chance to explain the significance of this new holiday, which was intended to raise the patriotic spirit of Ecuadorians by describing the Oriente's potential wealth, and to demonstrate the damages caused by Peruvian incursions. Through the use of conferences, speeches, and articles in the press, the holiday would seek to gain the support of all citizens in addressing the region's enormous needs. As *El Comercio* stated, "it is the duty of each Ecuadorian to cooperate with the State in a generous form: the 27th of February will be a day for a national contribution for the Oriente." The date chosen was also significant: the 27th of February, 1829, saw the victory of Ecuadorian troops over Peruvian invaders at the Battle of Tarqui, an event that helped define the borders between the new republics (No Author 1939a). So the Día del Oriente fulfilled another purpose of civic celebrations, which includes the "transfiguration" and "transmitting" of collective memory "so that the past can better sanction the particular . . . arrangements of the present" (Benjamin 2000, 99). Like many decades' worth of Ecuadorian publications about the Oriente, the celebration of the holiday was tied directly to the nation's past events. Remembering these events in a public, officially sanctioned manner helped cement the essential nationality of the Oriente and Ecuador's territorial rights.

Although the holiday's planners made no claim that celebrating the Oriente in this fashion was a longstanding tradition, the holiday nevertheless exemplifies the "invented traditions" discussed by Eric Hobsbawm that are essential components of nationalism. The holiday's celebration included "a set of practices . . . which seek to inculcate certain values and norms of behaviour by repetition, which automatically implies continuity with the past" (Hobsbawm 1983, 1). Choosing the 27th of February was one clear link to the past; invocations of heroic forebears in the Oriente were others. Invented traditions tend to come in a time of need or crisis (Hobsbawm 1983, 4); for the Ecuadorian planners of the Día del Oriente, Peru's threat to territorial integrity and national prestige provided the necessity for an "invented tradition" to reinforce Ecuador's claim to the Amazon and cement it as a component of national identity. The Día del Oriente celebrations also conformed to the typology of traditions Hobsbawm describes; it promoted "social cohesion" and "the membership of groups" by presenting the residents of the Oriente as full and laudable members of the national community. It "legitimiz[ed]" relations of authority by placing Ecuador's territorial claims above those of Peru. And it "inculcat[ed] . . . beliefs, value systems and conventions of behavior" by promoting the notion that it was incumbent upon Ecuadorian citizens to work for progress in the Oriente (Hobsbawm 1983, 9). Although when the holiday was first celebrated the presence of the state was weak and the degree of incorporation of the region into the nation at large was low, the Día del Oriente can be taken as "evidence" of "developments which are otherwise difficult to identify and to date" (Hobsbawm 1983, 12). That is, the civic celebration showed that the Oriente had become part of Ecuadorian national identity as well as a component of the imagined future of the state and economy.

On the day of the celebration, *El Comercio* published articles reviewing the history of the Amazon. The Orientalist committees in Quito put on a program that began with a display at the monument to founding father Antonio José de Sucre, one of Simón Bolívar's lieutenants and a military leader instrumental to the victory of patriot armies over Spanish forces in the areas around Quito during the war for independence. In the afternoon, public speeches about the Oriente included such speakers as the President of the Comité Orientalista Nacional, and the Chief of the Fourth Military Department, which was in charge of the Oriente's administration. Throughout the city, in each public office, one employee was put in charge of collecting donations for the Oriente and encouraging his colleagues to wear the national tricolor (No Author 1939b).

Some commenters viewed celebrating the Oriente as a way to strengthen the nation as a whole. One article in *Miscelánea* argued that February 27th represented a vow before the nation to "sustain and defend the sovereignty of the Oriente in all its vast extension, whatever it costs." Strength would come in unity, in his view: "if here we sacrifice ourselves in blood and fire among brothers for internal political questions, how will we be lions or Spartans before the eternal and usurping enemy?" It was time for Ecuadorians to make themselves respected again, as they once had at the Battle of Tarqui (Gabela 1939, 25–26). The celebration of the holiday, then, had transnational significance as it was in part about pitting Ecuador against those neighbors, especially Peru, that had designs on its Amazon territory.

The celebration of the Día del Oriente in 1940 was grander still, probably due to the greater time available for preparation. For one thing, the President of the Republic, Andrés Córdova, offered government funds for the festival. In addition the Asociación Deportiva del Napo (The Napo Sports Association) was preparing a group of athletes to travel to the capital from the Oriente, including "authentic Indians skilled in the sport of swimming, of blowguns, and arrows" (No Author 1940). A contributor wrote later in *El Comercio* that there was much enthusiasm for sports in the Oriente:

> it causes true pleasure to see how our brothers that live in the Oriente's jungles show themselves to be so enthusiastic for all that is progress, both cultural and material, of our un-renounceable land that sits on the other side of the eastern cordillera and that is now and always very much ours.
>
> (Gelves 1940b)

That year, then, Ecuadorians in the capital would have the opportunity to see their Amazonian compatriots in person. This included the region's indigenous people, for among them would be some true "Jívaros," "who with so much tenacity and patriotism defend and are advanced sentinels of our frontiers. . . . They will let us know that civilization has also arrived to where they live" (Gelves 1940a). In this view, the Oriente's belonging to the nation was without question, as even the fierce, warlike Jívaros (as they were so often described) were not savages but "sentinels" for Ecuador. During this time period, the region's indigenous people were frequently driven into debt peonage by white entrepreneurs and landowners who coveted their labor (Muratorio 1991, 146–147; Moreno Tejada 2015, 104–109). Their rhetorical value within the framework of the Oriente as a national object of desire was irresistible, however. Expositions of Oriente products, parades, theatrical performances, sporting events, and the exaltation of "enlightened Ecuadorians who have labored in and for the Oriente" rounded out the 1940 celebrations (No Author 1940).

The importance of the Día del Oriente and its implications for all Ecuadorians were felt in the highland provincial capital of Riobamba, where a secondary school marked the occasion with a speech by the student Maria Antonieta Gallegos. This young woman reviewed the historical events and treaties that gave Ecuador its right to the region, and she described the lamentable state of Peruvian intrusion and the road system that lacked development. It was high time, in her view, to leave political ambitions aside, tap the wealth of the Oriente, and thereby elevate Ecuador among the prosperous American nations (Gallegos 1940, 5).

Rallying around the Oriente during wartime and grappling with defeat

The long-predicted loss of Oriente territory to Peru, when it finally came, provoked a great deal of soul-searching among Ecuadorian writers in the press and other publications. This conflict brought the two nations' disagreements over their Amazonian border to a head. Oil was the catalyst for this war; Peru's government, under pressure from Standard Oil, was intent upon acquiring territory claimed by Ecuador in order to begin oil drilling (Ayala Mora 1990, 107–108). The war and eventual defeat helped cement the view of the Oriente as an "integral part of national territory" among Ecuadorian state-makers, according to historian Marco Restrepo. It spurred new legislation for the region as well as new road construction projects to connect the Oriente with the highlands (Restrepo 1993, 160).

Fighting with Peru broke out first on Ecuador's southern border in July of 1941. There were reactionary shows of patriotism in the Republic's central provinces and the Oriente. In Tena, a patriotic parade involving all local citizens took place, demonstrating that "in the Oriente shines the sun of love and faith in national honor" (Corresponsal 1941). Puyo held a large patriotic demonstration, and its residents saw their feelings stirred by two telegrams that circulated among the populace. The first, from a local resident to the Senator for an Oriente province, stated that "here all our wills are united, ready to do what we can for our beloved Homeland." The second, from the Senator to Puyo's local administrators and citizens, praised their "true indignation over the outrages committed by the invader" and their readiness to punish them (D. E. P. 1941). Later in the month, another public event decrying Peruvian aggression was held in Tena. Writing in *El Comercio*, a resident there described his neighbors as the "vanguard of Ecuador in the jungle" and praised their "virile and Spartan" souls (Peláez 1941).

El Oriente Dominicano began its 1942 issues with an article titled "Lessons of Time" in reaction to the Rio de Janeiro Protocol that formalized Ecuador's loss of territory to Peru. The Ecuadorian statesmen who acceded to the loss of territory in signing the Protocol were pressured to do so by international arbitrators in favor of "hemispheric unity" in the war against Japan (Ayala Mora 1990, 108). For many years, the Dominican editors wrote, the government had ignored Peruvian encroachment. Ecuador's defeat, however, was through no fault of the missionaries, who made the only efforts to maintain control over the territory. The proper reaction to this defeat should be a fuller understanding of patriotic duty among all Ecuadorians; the need to "sacrifice themselves for the Homeland" should be imparted to all. As the authors stated, "the history of these days is making us see how the idea of a defined nationality polishes over party or ideological roughness to realize the wonder of the union of all citizens" (No Author 1942b, 1). This editorial accomplished two things: it lauded the work of the missionaries at the expense of the state and private citizens, implying their inadequacy, while also portraying defeat in the Oriente as a reason to make a renewed commitment to the nation and, specifically, the Oriente as a part of it. The defeat added new significance to the Amazonian frontier as a region in need of redemption. As Wolfgang Schivelbusch has noted, national defeat can quickly turn into a "kind of victory" as commenters diagnose the problems that led to the defeat and identify a better way forward (Schivelbusch 2001, 10). This article by Dominican missionaries was like many others that did precisely that.

Reeling from national embarrassment, the Comité Orientalista Nacional did not organize any activities for the Día del Oriente in 1942. The President of the Ladies' Orientalist Committee wrote in *El Comercio* that her committee planned to continue its activities to aid "those brave sentinels of our sovereignty" in the Oriente, and she urged all Ecuadorians to "continue to defend that which remains to us in our Oriente" (Campi de Yoder 1942). Later, *El Oriente Dominicano* commented, "THE *ORIENTE*, before its horrendous disgrace, should it remain squalid, livid, gaunt, trembling, and fearful? No, never again! In no way!" Puyo, as the magazine pointed out, did not lack for brave, warlike, and defiant men and gracious women. The people of the area, from schoolchildren to military officers, celebrated the holiday with speeches and athletics; they swore "that, happily, we are all Ecuadorians and we promise to love her and defend her even with the sacrifice of our lives" (No Author 1942a). If nothing more, the Día del Oriente in 1942 offered an occasion to mourn the recent dismemberment of the Ecuadorian Amazon and also to re-affirm citizens' commitment to the region.

Later, *El Oriente Dominicano* reprinted an article from the *Revista Universitaria* that lamented the Rio de Janeiro Protocol but praised the missionaries and issued a strong call to arms. "The Ecuadorian Oriente, since the Rio de Janeiro treaty, as all know," the author argued, "has been converted into an unredeemed land of martyrs." However, the missionaries would continue to work there until "the aurora of a sun of justice for vilified America finally dawns." The nation found itself in a "profound stupor" after its sovereignty was "dismembered" in part thanks to the "oligarchic" governments that had neglected the Oriente and the threat posed by Peru (No Author 1944c). Now, however, "we must recover the feeling of patriotism beginning with the reconstruction of the Homeland," the missionaries wrote; "only then will our Nation be reborn to inscribe on its flag this grandiose motto: 'Ecuador defends itself'" (No Author 1944c). Again, the loss to Peru conforms to the phenomenon described by Wolfgang Schivelbusch in which the defeated nation experiences a profound feeling of rebirth or of having been freed from "a former life of sin." The loss was viewed as a "purifying and renewing force" that allowed Ecuadorians to identify past problems and how to fix them (Schivelbusch 2001, 29).

Miscelánea also lamented defeat and rallied patriotism. A 1944 editorial in the magazine argued that the loss of territory could have been avoided if the government had, over the years, given more attention to the Oriente. However, Ecuador made a good showing in the war, as it took many thousands of Peruvians to defeat just a few hundred Ecuadorians. The duty of Ecuador now lay in multiplying the number of roads and schools, planting colonists throughout the region, and promoting industry, commerce, and agriculture. *Miscelánea* called on the new government of José María Velasco Ibarra to improve the Oriente, "which constitutes for the Ecuadorian Homeland its greatest reservoir and future" (No Author 1944b). More than just the government would need to intervene, however; the magazine believed that private citizens had a "supreme obligation" to "place at least a grain of sand toward the erection of the national monument there in the jungles . . . that indicates our direction, our dominion, our property." "Let us carry in our souls the vision of the Oriente," the author exhorted in closing, "so we may never forget that a part of the Homeland extends there and is defended with the arm of the Ecuadorians and even with their life" (No Author 1944a).

Improving the Oriente required the efforts of the whole nation, but it offered much in return. *El Oriente Dominicano* repeated the decades-old notion of the Oriente as a solution to Ecuador's problems in 1946, when the magazine's editors reminded their readers that Ecuador's cities were full of unemployed men who did not realize that "extensive and varied uncultivated lands" awaited them. These lands meant that "unemployment" and "class struggle" should not exist in Ecuador. "May the exploited peasant, the unhappy worker, the poor employee, all come to search for the fortune that only the vigorous are given," they exhorted (No Author 1946). The Oriente promised, then, not only to create material wealth, but also to smooth over social and political discontent. This view could be repeated even in those years after such an immense portion of Ecuador's Amazon was transferred to Peru through war and international arbitration. The cultural importance of the frontier as an object of desire throughout Ecuador made this possible.

Conclusions

Beyond the social and political developments in the frontier space itself, Ecuador's Oriente frontier existed throughout Quito and other major cities as an object of desire. The region entered national myth through civic celebration, inculcating new generations with notions about its importance. Akira Iriye has stated that national memory may seem, at first glance, to be a poor subject for transnational history; however, he notes that the memory of war can be shared transnationally (Iriye 2013, 71). The preoccupation with the Amazonian frontier, as this chapter shows, shaped Ecuador's national identity. However, Ecuador was not the only nation to have such a preoccupation, as tension with Peru was an important shaping factor for decades. The transnational history of Amazonian mythmaking, territorial fixation, and economic development would be a rich exploration of social and political developments in the Amazon itself. It would also illuminate transnational concerns related to the frontier territory as a desired object, such as environmental protection, anthropological studies, and indigenous movements.

References

Primary sources

Campi de Yoder, María Elvira. 1942. "Actividades de un Comité Orientalista de Señoras con motivo de la conmemoración del 'Día del Oriente.'" *El Comercio*, March 1.

Chiriboga M., Aurelio. 1939. "Comité Orientalista Nacional." *El Comercio*, February 5.

Corresponsal. 1941. "En Tena se realizó un desfile patriótico en el que se protestó por traidora agresión peruana." *El Comercio*, July 14.

D. E. P. 1941. "Una grandiosa manifestación patriótica se realizó en Puyo (Oriente)." *El Comercio*, July 18.

Eguiguren, Joaquín S. 1922. "Algo sobre límites y Oriente." *Crisálida: Ciencia-Literaria, Órgano del Centro Local de la Universidad de Loja* 1: 12–13.

Gabela, Luis F. 1939. "Ecos sobre el Comité Orientalista." *Miscelánea* 9: 25–27.

Gallegos, Maria Antonieta. 1940. *En el Día del Oriente Ecuatoriano*. Riobamba: Ediciones Siembra.

Gelves. 1940a. "El Oriente y los Deportes." *El Comercio*, January 19.

——. 1940b. "Los deportes en la Región Oriental." *El Comercio*, January 15.

Maldonado, Heriberto. 1924. "Nota que el Comité Pro-Oriente dirige al Comité Central de Colonización." *El Comercio*, April 9.

Maldonado, Humberto. 1926. "El Comité Pro-Oriente de Cayambe." *El Comercio*, April 7.

Mora López, José. 1894. *Límites del Ecuador y Proyecto de Colonización*. Cuenca: Imprenta del Autor.

No Author. 1924a. "La Colonización de Huallaga." *El Comercio*, June 1.

——. 1924b. "La Publicación Sobre Concesiones Peruanas a un Sindicato Americano." *El Comercio*, May 22.

——. 1925a. No Title. *El Comercio*, March 21.

——. 1925b. "Propaganda Oriental." *El Comercio*, October 18.

——. 1934a. "Al comenzar un año más." *Miscelánea* 4: 1–2.

——. 1934b. "La Defensa Amazónica." *Miscelánea* 4: 1–3.

——. 1935. "Patriotismo Oriental Ecuatoriano." *El Oriente Dominicano* 8: 65–66.

——. 1939a. "El Día del Oriente." *El Comercio*, February 19.
——. 1939b. "Hárase colecta de fondos para la apertura de caminos al Oriente." *El Comercio*, February 26.
——. 1940. "Programa General de los Festejos con que el Comité Orientalista Celebrará el Día del Oriente." *Selva* 1: 38.
——. 1942a. "El Día del Oriente en El Puyo." *El Oriente Dominicano* 15: 76.
——. 1942b. "Lecciones del Tiempo." *El Oriente Dominicano* 15: 1–2.
——. 1944a. "La Visión del Oriente." *Miscelánea* 14: 8–9.
——. 1944b. "Por el Futuro de nuestro Oriente." *Miscelánea* 14: 7.
——. 1944c. "Por el Oriente y las Misiones." *El Oriente Dominicano* 17: 97–98, 133.
——. 1946. "La Voz de Nuestro Oriente." *El Oriente Dominicano* 19: 57.
Peláez, José. 1941. "Los colonos de Tena y Napo contribuyen para la Defensa Nacional." *El Comercio*, August 4.
Romero Terán, Domingo. 1933. "Ofrece una ventajosa colocación a cuantos quieran explotar riquezas del Oriente." *El Comercio*, January 3.
Vacas Galindo, Enrique. 1905. *Conferencia acerca de la Importancia del Ferrocaril del Oriente*. Quito: Imprenta Nacional.
Vargas, José María. 1933. "El Patriotismo Ecuatoriano hacia el Oriente." *El Oriente Dominicano* 6: 1–2.
——. 1936. "Nuestra Cuestión Limítrofe." *El Oriente Dominicano* 9: 281–282.
Veintemilla M., Julio. 1923. *Geografía del Ecuador: Apuntes sobre el Oriente Ecuatoriano y sus Límites*. Quito: Tipografía de la "Prensa Católica."
Yepes, Jorge Luna et al. 1933. "El Ecuador es País Amazónico – Manifiesto de los Estudiantes del Grupo Orto, de la Universidad Central del Ecuador." *El Comercio*, March 23.

Secondary sources

Anderson, Benedict. 1991. *Imagined Communities: Reflections on the Origin and Spread of Nationalism*. Revised edition. London and New York: Verso.
Ayala Mora, Enrique. 1990. "El arroísmo y el conflicto territorial." In *Nueva Historia del Ecuador*, Volume 10, edited by Enrique Ayala Mora, 107–108. Quito: Corporación Editora Nacional.
Benjamin, Thomas. 2000. *La Revolución: Mexico's Great Revolution as Memory, Myth, and History*. Austin: University of Texas Press.
Craib, Raymond. 2004. *Cartographic Mexico: A History of State Fixations and Fugitive Landscapes*. Durham and London: Duke University Press.
Deler, Jean-Paul. 2007. *Ecuador: del espacio al Estado nacional*. Quito: Corporación Editora Nacional.
Esvertit Cobes, Natalia. 2008. *La Incipiente Provincia. Amazonía y Estado ecuatoriano en el siglo XIX*. Quito: Corporación Editora Nacional.
García Jordán, Pilar. 2001. *Cruz y arado, fusiles y discursos. La construcción de los Orientes en el Perú y Bolivia, 1820–1940*. Lima: IEP.
Hall, Stuart. 1990. "Cultural Identity and Diaspora." In *Identity: Community, Culture, Difference*, edited by Jonathan Rutherford, 222–237. London: Lawrence & Wishart.
Hobsbawm, Eric. 1983. "Introduction: Inventing Traditions." In *The Invention of Tradition*, edited by Eric Hobsbawm and Terence Ranger, 1–14. New York: Cambridge University Press.
Iriye, Akira. 2013. *Global and Transnational History: The Past, Present, and Future*. London and New York: Palgrave Macmillan.

Little, Paul E. 2001. *Amazonia: Territorial Struggles on Perennial Frontiers*. Baltimore and London: The Johns Hopkins University Press.

Moreno Tejada, Jaime. 2015. "Castles in the Air: Rise and Fall of the *Hacienda* System in the Ecuadorian Amazon, 1910–1940." *Canadian Journal of Latin American and Caribbean Studies* 40(1): 97–115.

Muratorio, Blanca. 1991. *The Life and Times of Grandfather Alonso: Culture and History in the Upper Amazon*. New Brunswick: Rutgers University Press.

Rausch, Jane M. 1993. *The Llanos Frontier in Colombian History, 1830–1930*. Albuquerque: University of New Mexico Press.

——. 1999. *Colombia: Territorial Rule and the Llanos Frontier*. Gainesville: University Press of Florida.

Restrepo, Marco G. 1993. "El Proceso de acumulación en la Amazonía ecuatoriana (Una breve visión histórica)." In *Amazonía nuestra: Una visión alternativa*, edited by Lucy Ruiz, 125–148. Quito: CEDIME.

Schivelbusch, Wolfgang. 2001. *The Culture of Defeat: On National Trauma, Mourning, and Recovery*. Translated by Jefferson Chase. New York: Picador.

Taylor, Anne-Christine. 1994. "El Oriente ecuatoriano en el siglo XIX: 'el otro litoral.'" In *Historia y región en el Ecuador: 1830–1930*, edited by Juan Maiguashca, 75–108. Quito: Corporación Editora Nacional.

Thurner, Mark. 2003. "Peruvian Genealogies of History and Nation." In *After Spanish Rule: Postcolonial Predicaments of the Americas*, edited by Mark Thurner and Andrés Guerrero. Durham and London: Duke University Press.

Part VI

Ethnographies

11 Frontier Bali

Local scales and levels of global processes

Graeme MacRae

Introduction

Bali is justly famous for a number of things, most of them revolving around a heady nexus of international relationships (tourist and expatriate, artistic and academic) with a supposedly timeless traditional culture, always threatened but always surviving, indeed thriving on its encounter with global cosmopolitanism. What it appears to have largely escaped is the frontier violence of national and transnational regimes of resource extraction that have laid waste to indigenous cultures and environments over much of Asia and Latin America (Sawyer and Gomez 2012). Balinese culture itself, enduring, traditionalist and resilient in the face of change, seems likewise the antithesis of the unstable dynamics of frontier process. There is some truth in these impressions but also a degree of illusion.

The development and investment fueled by the long tourism/expatriate boom, all feeding off the primary resource of "local culture," may be usefully seen as a kind of frontier in itself, in the broad conceptual sense of "an imaginative project . . . molding both places and processes . . . in the shifting terrain between legality and illegality, public and private ownership, . . . violence and law, restoration and extermination" (Tsing 2005, 33); but also in the more literal and concrete sense of a geographic locale characterized by the combination of abundant resources and relative freedom from regulation, first identified in the Turner–Webb Thesis of the expansion of European populations across the north American continent (Barbier 2011, 9).

The metaphor of the frontier has reappeared in recent years in critical response to the rapid and largely unchecked spread of neoliberal capitalism and resource extraction since the 1980s in terms both concrete and conceptual, such as "relational zones of economy, nature and society; spaces of capitalist transition, where new forms of social property relations and systems of legality are rapidly established in response to market imperatives" (Barney 2009, 46). In Bali such radical regimes of environmentally and socially violent resource extraction have not occurred, and development has appeared relatively benign. But between the sheer mass of tourism, the increasing economic and cultural influence of the expatriate community, the financial dominance of outside investment capital, and the social and environmental impacts of building projects, elements of classic frontier process are becoming evident. So also are the traditional frontier side-effects of environmental damage, unsustainable use of resources, social dislocation and violence, and marginalization of local communities.

This island-wide frontier is driven by wider national, regional, and global processes, but it manifests locally in the form of multiple smaller sub-frontiers based on a range of sub-resources. The aim of this chapter is first to show how frontier processes have long been part of a "hidden history" of Bali; second to map the broad contours of the contemporary island-wide meta-frontier; third to document small parts of this frontier; and finally to argue that the frontier processes unfolding in Bali are best understood as contemporary local nodes in a global frontier process of long standing dynamics.[1]

Frontier Bali: a hidden history

The well-known Chilean film *La Frontera* (1991) is about the exile of a dissident intellectual to the remote south of the country. The title ("the frontier") needs little explanation to Chileans—it refers to a national cultural geography created historically by the relentless southward expansion of Spanish colonization, driving back of the indigenous Mapuche people to a region of rivers, lakes and forests where they remained invincible until the late nineteenth century. To this day *la frontera* refers both to a recognizable landscape, in which the majority of Mapuche still live, and to a sense of the "wild south" of the country, where anything is possible, as the exiled intellectual discovers. Few frontiers remain so embedded in national consciousness and landscape, but they were a dynamic feature of most settler societies into the twentieth century, immortalized in fiction and convincingly argued in the writings of Frederick Turner (1920) and Walter Webb (1952).

This spectacular and often romanticized frontierism typical of settler societies is less evident in Balinese history or national identity, but neither is it entirely absent. There is no Balinese word equivalent to frontier, but boundaries and borders (*batasan*) are important in Balinese culture. A primary one, common to many Southeast Asian cultures, is between cultivated land and the wild domain of forest, and their social correlates, human civilization and a natural/spiritual wilderness of (non-human) animals and unruly spirits. These distinctions remain pervasive in Balinese thinking about space and place, and are marked materially by walls around domestic and sacred spaces. Villages are also bounded spiritually, by temples located (at least notionally) at their uphill and downhill boundaries, and (sideways) along contour lines by deep ravines in which both wild forest and powerful spirits remain. Cycles of seasonal ritual protect these boundaries against incursions of all kinds, from dangerous spirits from across the sea to contemporary material equivalents such as terrorist attacks.

The macro-ambitions of the micro-polities of precolonial Bali, celebrated in Clifford Geertz's *Negara* (1980) were likewise constrained by landscape, shaped by the relative ease of uphill/downhill movement and difficulty of cross-slope movement, into long thin wedges, sloping from the mountains down to the sea, an axis mirrored in Balinese cultural and spiritual notions of landscape

(MacRae 2006, 84). Within the conservative, traditionalist self-representations of Balinese culture, lies a deep sense of borders and boundaries. But there is also a hidden sub-history of frontier expansions.

Oral histories from all over the island record the founding of villages, somewhat in the manner of Macondo in Gabriel Garcia Marquez's *One Hundred Years of Solitude,* by migrants and refugees from another part of the island, looking for land to cultivate and/or freedom from the oppression of predatory rulers or troublesome neighbors (Howe 1980; MacRae 1997, 288; Schulte-Nordholt 1996, 56). These stories begin typically with the discovery of an auspicious place/space with abundant resources, the physical clearing of forest and the spiritual pacification of deities of the local landscape: frontier process and frontier thinking.

The biggest of these frontier stories, the meta-story, is that of Rsi Markandeya, a holy man of Javanese and/or Indian origin who led an expedition to Bali (in the distant pre-historical period, perhaps the sixth century). His followers died from diseases and attacks of wild animals and he was forced to abandon his quest. Somewhat chastened, he meditated and realized that his failure was due to his neglect of proper respect for the powerful invisible inhabitants of the frontier. So he tried again, fortified with better spiritual armaments, and this time he was able to subdue the wilderness and establish the primary institutions of Balinese culture: the foundations of civilization (MacRae 1997, 231–235; Stuart-Fox 2002, 260–262).

The Markandeya story reads like a metaphorical correlate of the mythologized history of the later migration and survival of Javanese Majapahit culture in Bali. Majapahit was the greatest of the classical Indianized Javanese empires, based in East Java, but dominating much of what is now modern Indonesia from the late thirteenth to late fifteenth centuries. It invaded Bali in the mid-fourteenth century and retained political control through strategic marriage links with Balinese rulers. When its base in East Java was finally overrun by coastal Islamic states in 1478, its surviving aristocracy fled to Bali, where they became a new cultural and political elite, especially over the southern plains and foothills, while the older indigenous culture remained dominant in mountains (Reuter 2002, 265–267; Hobart 1996, 35). The extent to which modern Balinese culture is an historical continuation of Majapahit culture is debated, but the point for our purposes is that this widespread mytho-historical understanding is essentially one of cultural frontier-making[2].

Colonial frontier

The Dutch colonial enterprise in the East Indies began in the seventeenth century as state-sponsored mercantile capitalism but developed by the nineteenth century into a system of ruthless extraction on two frontiers: internally into under-utilized resources of land and labor in the "civilized" heartland of Java, but also outward into the relatively unexploited territories to the west, north and especially east, which in the mid-nineteenth century, was still "the least known part of the world" (Wallace 2000, 1; see also Locher-Scholten 1994, 97–98; Ricklefs 2008, 161).

Through most of this period of expansion, Bali, although immediately east of Java, was largely bypassed by this multi-faced frontier, protected by difficult coastline as well as the renowned ferocity of its inhabitants (Boon 1976; Ricklefs 2008, 164; Vickers 1989, 16–20). In the mid-nineteenth century the Dutch managed to establish a foothold on the north coast by exploiting the shifting rivalries between multiple small Balinese kingdoms. While these kingdoms were largely engaged in contests for control over established populations in Bali, one of them, Karangasem at the eastern end of the island, had embarked in the late eighteenth century on its own frontier project—expanding into the neighboring island of Lombok, less populated and fertile than Bali, but productive and strategically valuable nevertheless. The Dutch conquest of Lombok in 1894 put an end to Balinese frontier ambitions and provided a platform for their assault on south Bali.

When the Dutch finally achieved control of the south of Bali in the first years of the twentieth century, colonial attitudes had changed and the ruthless exploitation of earlier times had been replaced (at least notionally) by one of "ethical" development (Ricklefs 2008, 183; Vickers 1989, 92–93). The resource frontiers typical of colonial development elsewhere did not happen. Instead, the primary resource identified was Balinese culture itself, and this became the basis of a distinctive kind of development—a frontier of a rather different kind. In the year of their takeover of south Bali, the Dutch established a system of official government-sponsored tours of their colonial empire and Bali soon became one of its centrepieces.

In 1929, Hickman Powell, an adventurer on another cultural frontier, of elite escape from between-the-wars Europe and America (Fussell 1980), published a book entitled *The Last Paradise* (Powell 1985) embedding forever the core metaphorical resource on which the tourism/expatriate development of Bali has depended ever since (Vickers 1989). The search for the next "paradise" (because the previous one has been "spoiled") continues to be one of the most durable frontiers of the tourist industry.

Indonesia became independent from Dutch control after the Second World War. The high hopes of the new nation were undermined by economic and political instability and, by the early 1960s, President Sukarno was casting around desperately for sources of economic development. One of them was tourism; Balinese culture was the resource. Sukarno enlarged the airport, built the first large hotel and invited the international association of travel agents to hold a conference in Bali (Picard 1996, 44).

But before Sukarno was able to realize his dream he was deposed in a confused coup in Jakarta in 1965 which led to nationwide massacres and the establishment of a regime known as the New Order which ruled Indonesia until 1998. The New Order opened the country to foreign investment and created new frontiers exploiting resources ranging from forestry and mining to cheap labor for the factories of multinational corporations.

Again Bali was spared most of these kinds of developments. Instead Suharto followed Sukarno's lead by seeking World Bank funding and commissioning international tourism consultants. Their advice reflected classical frontier-thinking—to open up Bali to mass tourism based on the resource of traditional culture. But they also predicted frontier side-effects of destruction of the resource and collateral damage which they recommended managing by restricting tourism to the southern corner of the island (Picard 1996, 45–49).

Through the New Order period, tourist numbers in Bali grew from around 6,000 in 1968 to over a million by 1994 (Picard 1996, 52). Tourism became a major source of national income and turned Bali into a cash-cow for central government but also an investment frontier for capital from Jakarta and beyond (Aditjondro 1995; Warren 1998).

At the lower end of the economy, the long boom in Bali also created an employment frontier drawing migrants from less affluent parts of the country, especially East Java, where employment was scarce and incomes were only a fraction of what could be earned in Bali. Massive migration led to enclaves of (mostly Javanese) immigrants who were seen by Balinese both as a breeding ground for petty crime and as the fifth-column frontier of an Islamic plot to undermine and ultimately destroy and replace Balinese Hinduism. The bombings by "Islamist" terrorists in 2002 and 2005 served to reinforce these suspicions (Hitchcock and Darma Putra 2007, 120–121; MacRae 2010, 23; Schulte-Nordholt 2007).

Whatever the motivations of the bombings, both tourism and economic development came to an abrupt halt, recovery from which remained only gradual until 2007 when they took off again and have continued to increase every year since. These raw figures tell part of the story, but the other part is a set of qualitative changes in the shape of tourism itself and of development generally, which may be seen as an ongoing frontier process, consisting of both an expansion of existing sub-frontiers and opening up of new ones. I will illustrate this with a set of brief accounts of three of these sub-frontiers.

New frontiers

Sub-frontier 1: tourism/expatriate nexus

In 2010, the magazine *Condé Nast Traveller* named as the "Best City in Asia" the small town of Ubud in Bali. To describe it as a city, let alone the best in Asia, is patently absurd to anyone who knows Ubud as a village morphing into a town. Regardless, the award reflects an emergent cosmopolitanism, not only of tourism fashions, but of the style of Ubud and the way it is being positioned in the global tourism market. It has been a centre for tourism for nearly a century, based largely on its deserved reputation for traditional culture and arts. Over the same period it has also been home to a series of foreign expatriates, many of them artists, musicians or scholars, or simply *aficionados* of Balinese culture.

But over the past decade, this has changed, at first gradually and now more rapidly. Locally owned businesses, especially restaurants, are being replaced by increasingly sophisticated and expensive foreign-owned ones, with which locals find it hard to compete. Tourist numbers have tripled since 2010 (from around

70,000 in 2010 to over 245,000 in 2013),[3] but there has also been a change in the style of tourism. One of the more visible changes is the daily arrival of several tour buses full of Chinese visitors. Another is what locals call *tamu yoga* (spiritual tourists)—hundreds of young people from all over the world for whom Ubud has become a node in a global network of centres for yoga, world music, "alternative" therapies, organic/raw/vegan food and other New Age practices (Citrayanthi 2015). The catalyst for this development is the annual Bali Spirit Festival, organized by a local-foreign couple. It attracts thousands of people, many of whom stay for weeks, months, or even years. Some support themselves by selling various goods and services within their own sub-economy. But this, like the new retail and restaurant frontier, is increasingly at the expense of local businesses and based on exploiting the gap between foreign and local costs and incomes.

Some of these people, as well as others less "spiritually" oriented, support themselves by working online, as graphic or web designers, advertising copywriters, journalists, video editors, or online marketers. These people are from all over the world—they could be living anywhere—their work and lifestyles are "location independent." They have chosen to live in Ubud (at least for the time being)—but their reasons for being there are to do with the pleasant environment, good facilities, including restaurants and international schools, fairly high speed internet and what they call "community"—by which they mean a lot of people like themselves—certainly hundreds, probably thousands. Many work and interact via Hubud (Hub-in-Ubud), a digital "co-working space" or "post-corporate, post-institutional" workspace. While Hubud is physically located in the middle of Ubud, it is better understood as an outpost of a cyber-digital frontier spreading across the world from an epicentre in Silicon Valley.

Hubud was founded by parents of children at the Green School, an innovative environmentally-oriented international school near Ubud. This school, with a roll of some 250 students, has attracted dozens of middle-class families from all over the world. They have added to the expatriate population, pushing up the local real estate market and added a distinctive global environmentalist voice to expatriate culture. Ubud is now on the frontiers of global environmental and international education movements.

The other major component of the international demographic frontier in Ubud is the growth of "later life migration" (that is, retirement migration). There are reportedly some ten thousand Australian retirees living in Bali and a conservative estimate would amount to at least one thousand in Bali, plus at least another thousand from other parts of the world. These people too have effects on local real estate, labor and retail markets (Green 2014).

Taken together, there are now several thousand of these people, in a small town of some 11,000 (and wider sub-district of about 42,000) where their numbers alone have an inevitable effect.[4] But more importantly for our argument here, their choice to live in a linguistically and culturally separate community, combined with the development of a "dual economy" in some ways respects similar to colonial ones (Boeke 1953), may be seen as a frontier of the fast-changing mobility of global tourist/expatriate development.[5]

Sub-frontier 2: cultural heritage

In 2012, the UNESCO World Heritage Committee (WHC) formally recognized and added to its World Heritage List the "Cultural Landscape of Bali Province: the subak system as a manifestation of the Tri Hita Karana Philosophy." Barely a year later the WHC had become sufficiently concerned about reports of uncontrolled development resulting from the listing, that it advised the Indonesian government of these concerns and especially that the statutory body supposed to be managing the heritage estate had not yet been set up (Sutika 2014b). Since then it has been pointed out should this not be addressed in an adequate and timely manner, it could lead to a nationally embarrassing de-listing by WHC (Bali Post 2013a, Sutika 2014b).

The main purpose of World Heritage listings is "collective protection of the cultural and natural heritage of outstanding universal value" (UNESCO 1972)—in effect a global system of protection of heritage against what is often the collateral damage of processes of frontier development. However, as Jan Turtinen points out, "World Heritage is constructed in an institutional system of formalized routines, beliefs and practices, and in a centre/periphery relationship . . . [part of] a larger socialising and civilising project" (2012, 3). He goes on to analyze this system as one of transnational government and ordering, and in this sense it is the opposite of and a corrective to frontier processes. What he does not mention is the way in which the WH system has expanded to become a significant global industry managing a frontier-like process of expansion and claiming of new territories and resources. In the process it has the local side-effect of creating a new resource of "heritage capital," a cultural or symbolic resource which, as Bourdieu insisted of all capital, is inherently exchangeable for other forms of capital (1994, 173).

In the case of Bali, what the listing was seeking to protect was a system of management of irrigation water, decentralized among hundreds of local communities by way of cycles of ritual linked through a hierarchy of temples. The listing was intended to protect against the appropriation of water resources and farmland driven by the tourism/expatriate frontier described above by "provid[ing] support to sustain the traditional systems and to provide benefits that will allow farmers to stay on the land" (UNESCO 2012, 190). However, almost immediately after the WH listing, in a particularly beautiful mountain valley in the heritage estate called Jatiluwih, outside investors began buying land and building hotels and villas. At the same time, small operators began setting up roadside stalls. Land values began to rise steeply. Farmers sold dry-field land for development, which in turn affected the integrity of the very landscape the WH listing was supposed to protect. The government charges visitors an entry fee, but at least one enterprising division of the subak began charging an additional fee for viewing and photographing their fields. Conflicts began to arise between subaks and between farmers and tourism operators over differential benefits from the new situation. Within six months of the listing, the head of the district parliament, was publicly lamenting an "invasion" (*serbuan*) of investors who he believed were threatening the environment and the livelihoods of farmers and could even result in the status of the WH listing being revoked (*Bali Post* 2013b; Kompas 2013). An unintended consequence of the listing was the creation of new heritage capital with considerable exchange value in the tourism and real estate economies, thus opening new frontiers of these markets (Reeves and Long 2011). As research in other WH sites suggests, "the enhanced profile and opportunities for marketing were just too good to miss" for various "stakeholders" (Fyall and Rakic 2006, 161).

In late 2013, in response to these frontier side-effects, the district (*kabupaten*) government established a management body for Jatiluwih. However, the name of this organization (Jatiluwih Tourism Area Management Body) reflects its focus on tourism development rather than heritage protection, let alone conservation or agriculture. As Bodley observes of classical frontiers, government intervention is usually triggered by conflict, but such interventions tend to favor the interests of the invaders over those of local community or environment (1999, 46–48).

As this heritage frontier unfolded, or perhaps folded in on itself, another was unfolding. In 2012, UNESCO approved an application for membership of Mount. Batur in the Global Geoparks Network (GGN). Mount Batur is an active volcanic cone sitting alongside a lake within the crater of its former, much larger self. Like Jatiluwih, it is a spectacular and beautiful site, which has attracted tourists for decades, as well as being one of the most powerful places in the ritual landscape of Bali and a kind of control centre of much of the irrigation system recognized by the WH listing (Lansing 1991, 74). A geopark is "a geographical area where geological heritage sites are part of a holistic concept of protection, education and sustainable development" (GGN 2010). This listing was another significant step in Indonesia's recognition in the global heritage and conservation system.

Again, within days of the declaration, investors began applying to the district administration for developments ranging from a golf course to bamboo farming (Erviani 2012), reflecting the value added by the GGN certification. These applications also immediately raised the possibility of the kinds of contradictions that occurred in the wake of the WH listing.

The WH and Geopark listings are outposts on the frontier of an expanding global system of heritage management. This universal system comes down to ground in specific locations it encounters, like all global/universal systems, the "frictions" of local culture, history, and economy (Tsing 2005). In this case, among the products of this friction are new resources with new value in tourism and real-estate economies, which has led to frontier development of entrepreneurs seeking to exploit these resources. This has led in turn to the frontier side-effects of unregulated tourism development, land-grabbing, and competition for tourism money. These have begun to marginalize local communities, damage local ecosystems and create conflicts among different interest groups. Government has intervened, but to date mostly in classic frontier mode, on behalf of the interests of the "frontiersmen" rather than local communities.

Sub-frontier 3: black sand

Every day, from before dawn till after dusk, at least 1500 (a rough but conservative estimate) trucks grind their way painfully out of the crater of Mount Batur.

They are loaded with black sand, gravel and rock, and at the crater rim they turn and speed away downhill to deliver their cargoes to construction sites in the tourism and urban centres on the plains and coast.

Down in the caldera, a narrow road circles the caldera floor. Alongside it are piles of black gravel or rocks and makeshift shelters under which men or women shovel gravel through large sieves into piles of finer sand. When it is done, they flag down a truck and load it by hand. Roadside signs invite trucks into a hinterland of even narrower dirt tracks where more piles are waiting. Each hamlet the trucks pass through shares in the boom by levying its own little toll.

Some of this material is dug by hand from private or public land, but here and there this landscape of backyard industry expands into something bigger. Tracks lead down into pits 20 meters or more deep, and up to 100 meters wide, at the bottom of which large diggers drop buckets of gravel onto monster versions of the roadside sieves, and into waiting trucks. These pits spread until they threaten to undermine houses, villages, schools, even temples, which now sit precariously atop cliff-faces.

The villages strung along the roadsides are black and dusty, bleak and treeless—no tourist comes here and the only water arrives by truck. But between them, in places close enough for water to be pumped from lake or spring, are vegetable gardens or orchards, and off-road in the labyrinth of tracks, fields of tomatoes, cabbages, onions and enormous orange pumpkins.

But even this vision is deceptive. Because horticulture is the other viable economic option for a rapidly growing population, forest regrowth is being cut on the crater walls to clear more fields. The eroded topsoil is silting up the lake and raising water levels. At the same time runoffs of petrochemical fertilizers, fungicides and pesticides find their way into the lake, which is lined with fish farms using synthesized food laced with antibiotics. Together, these have increased nutrient loads in the water which feed an invasion of water hyacinth, while DDT and other toxins have caused deformities in fish.

There are two main communities in the crater: Batur, whose village and main temple was shifted from the floor to the rim of the crater after an eruption in 1926, but which retains traditional ritual jurisdiction over the western half of the crater; and Songan, a larger village by the eastern end of the lake, which has, until recently, had less resources or economic development than Batur. But the new resource rush has changed that, with most of the quarrying and trucking being done by people from Songan. Most, if not all of it is illegal. Not everybody in Batur is happy with this new wildcat economy, but nobody has authority to stop it.

Responsibility for environmental protection and development lies with the district government, which levies an entry fee on tourists. When I asked (in mid-2014) what the money is used for, staff recited a litany of environmental and infrastructure problems that government was clearly not using these revenues to remedy. Likewise, the local police and army post say they are prepared to stop the quarrying and trucking, but only at the direction of government. The lake and its temple are part of the UNESCO World Cultural Heritage listing, and the whole crater is part of a UNESCO Global Geopark. Both are likewise supposed to be managed by the district government. To date the governing bodies required by these designations have not even been established.

But there are other sources of authority in Batur. In Batur, a temple called Puru Ulun Danu is generally recognized as one of the main temples of Bali and a kind of control centre for the irrigation water for most of the island. Its high priest, Jro Gede, has used his spiritual authority to encourage and facilitate environmental awareness and conservation around the crater. Any quarrying or forest cutting visible from the temple results in a polite delegation from the temple and few would dare continue in the face of disapproval of the high priest. At the same time he has restored and revitalized a circle of temples at strategic ecological points on the crater floor that likewise few violate. On his own land near the crater wall, the high priest has planted gardens above which nobody has yet been prepared to cut forest. These and other strategic interventions have limited the damage, but anywhere out of direct sight of the temple remains fair game and at the other end of the crater his jurisdiction gives way the that of Songan, where different values prevail and support for the present district government is high. However the authority of Jro Gede stems ultimately from yet a higher source.

According Jro Gede, Batur is "not just a lake, but the garden of Dewi Danu (the Goddess of the lake)" and the god Wisnu, on whom the water and ecosystem of the whole of Bali depends. In mid-2011, the water of the lake suddenly changed color and soon after thousands of the farmed fish died. The official explanation was the release of a plume of sulphur from an underwater volcanic vent. Others say that it was the goddess cleansing the lake of pollutions.

A few months later, in the wet season of 2011–12, heavy rains caused landslides of deforested slopes at the edge of the crater. The lake level rose, exacerbated by the silting, and on the eve of Nyepi, the Balinese day of silence, Songan began to flood. Most of the inhabitants evacuated up to Penelokan on the crater rim but returned when the rains abated next morning, to observe Nyepi in the seclusion of their sodden homes. Since then, there has been a noticeable increase in ritual activity and religious fervor throughout the crater community. But the trucks still run.

Conclusion: frontier Bali

What is happening in Batur is an unprecedented rush for resources reminiscent of the classic frontiers of New World settler societies or, closer to home, the environmental destruction on the contemporary resource frontier in southeast Kalimantan, documented by Anna Tsing (2005). What is unusual about Batur is that it is not distant corporations that are stripping the modest resources, but local people competing to grab what they can before it all runs out. Authorities are often quick to lay blame on local people, and indeed to define such mini-frontiers as local problems. But this too can be deceptive, and has many precedents in frontier history, including Kalimantan where local people "tap the slender ends of arteries flowing with capital from rich urban entrepreneurs [and] conglomerates . . . branching in thinner and thinner capillaries out into the forest" (Tsing 2005, 34).

The household enterprises in the dusty crater of Batur are "capillaries" of a system of circulation whose heart lies far away. An hour or so downhill from Batur is another frontier where centuries-old rice fields are being converted by the hour into a sea of villas, hotels and resorts. The sand and rock frontier in Batur is one of the upstream ends of the many supply chains feeding this construction boom. The chronically underemployed labor landscape of East Java is another. On the other side of the construction frontier are national and international demands for villas and hotels as well as the fuel supply lines—of investment money (legal or otherwise) flowing from Jakarta and overseas.

The new tourism/expatriate nexus in Ubud is another node in this system, at once a cultural frontier of global modernity, and a minor pump driving the investment/construction frontier that sees the streets of Ubud lined with small piles of black Batur sand. Likewise the "heritage frontier," is on one hand a far "capillary" along which cultural capital flows from the global heritage system, to national, provincial and district levels of government, but it also creates frontier processes and side-effects at the local level.

While resource frontiers such as Batur and Jatiluwih take the shape of failures of local behavior, they are the inevitable downstream ends of money being made somewhere else. Frontiers are driven by demand for resources but can only happen with the complicity of government. Government inaction, in centres half an hour downhill from Batur in Bangli and Jatiluwih, is an integral part of those frontiers.

What I have tried to show here is the way in which, behind the smiling faces of "cultural tourism" in Bali, there is a multi-faceted process of economic development and environmental consequences. Elements of this process display, in various ways, features recognizable as those of frontiers in other times and places. Taken together, they constitute an island-wide pattern that is in itself a local branch of a global frontier process that can be traced back to sources of power and wealth in faraway metropolitan centres. What Bali also shows us is the ways in which local processes take different forms and occur at different levels, which tend to obscure their linkages into greater frontiers.

Postscript

In early 2016, the district government, in the face of mounting pressure from various quarters, did finally restrict the movement of trucks out of the crater and the sand frontier has almost closed, as fast as it opened. Local residents who depended on it for their livelihoods are now shifting to (chemically-driven) horticulture and hoping that the government promises about the Geopark will create a new tourism frontier.

Notes

1 Much of the empirical material reported here was collected during eight weeks of ethnographic research in mid-2014, but the real background is over two decades of ethnographic research in Bali. I am grateful to many colleagues and friends in Bali including W. Alit Arthawiguna, J. Stephen Lansing, Sarita Newson, Kadek Adhidarma, Trishna Newson, Carol Warren, Diana Darling, Rio Helmi and Sandy Cuthbertson. I am

grateful also to The Indonesian Institute of Sciences (LIPI) and Udayana University for their sponsorship of my initial research in Bali and to Auckland and Massey Universities for contributions to funding of my research.

2 Majapahit itself is well documented. Its relationship with Bali is widely known, but documented less systematically in works such as Creese (1997) and Hobart et al. (1996, 34–36).

3 These figures are of visitors registered with the local police as staying one or more nights in the sub-district (*kecamatan*) of Ubud.

4 Accurate figures are impossible to obtain: first because Indonesian statistics are notoriously unreliable, second because there is no discrete category of "foreign residents" in the way statistics are recorded, and third because foreigners reside in Bali under a range of legal statuses and visa types.

5 These developments are discussed in more detail in MacRae (2016).

References

Aditjondro, George J. 1995. "Bali, Jakarta's colony: Social and Ecological Impacts of Jakarta-Based Conglomerates in Bali's Tourism Industry."—Asia Research Centre. Murdoch University. Perth. Australia.

Bali Post. 2013a. "Diserbu Investor, Status WBD Jatiluwih Terancam." 30 January.

——. 2013b. "Hindari Serbuan Investor: Jatiluwih Ditetapkan Jalur Hijau." 10 December.

Barbier, Edward, 2011. *Scarcity and Frontiers: How Economies Have Developed through Natural Resource Exploitation*. Cambridge: Cambridge University Press.

Barney, Keith. 2009. "Laos and the Making of a 'Relational' Resource Frontier." *The Geographical Journal* 175(2): 46–159.

Bodley, John H. 1999. *Victims of Progress*. Mountain View, CA: Mayfield.

Boeke, Julius Herman. 1953. *Economics and Economic Policy of Dual Societies, as Exemplified by Indonesia*. New York: Institute of Pacific Relations.

Boon, James. 1976. "The Birth of the Idea of Bali." *Indonesia* 22: 70–83.

Bourdieu, Pierre. 1994. "Structures, Habitus, Power: Basis for a Theory of Symbolic Power." In *Culture/Power/History: A Reader in Contemporary Social Theory*, edited by Nicholas B. Dirks, Geoff Eley, and Sherry B. Ortner, 154–199. Princeton: Princeton University Press.

Citrayanthi, Made. 2015. "'Ubud is Calling Me': Spiritual Tourism and Development in Vibrant Ubud." Unpublished M.A. thesis. Leiden University.

Creese, Helen. 1997. "In Search of Majapahit: The Transformation of Balinese Identities." *Working Paper 101*. Clayton (Australia): Monash Asia Institute.

Erviani, Komang. 2012. "Investors Race to Develop Batur Geopark." *Bali Daily*, 29 November.

Fussell, Paul. 1980. *Abroad: British Literary Traveling between the Wars*. Oxford: Oxford University Press.

Fyall, Alan and Rakic, Tijana. 2006. "The Future Market for World Heritage Sites." In *Managing World Heritage Sites*, edited by Anna Leask and Alan Fyall, 159–176. Oxford: Butterworth-Heinemann.

Geertz, Clifford. 1980. *Negara: The Theatre-state in Nineteenth-Century Bali*. Princeton: Princeton University Press.

Global Geoparks Network. 2010. "Guidelines and Criteria for National Geoparks Seeking UNESCO's Assistance to Join the Global Geoparks Network (GGN) (April 2010)." UNESCO.

Green, Paul. 2014. "Contested Realities and Economic Circumstances: British Later-Life Migrants in Malaysia." In *Contested Spatialities, Lifestyle Migration and Residential Tourism*, edited by Michael Janoschka and Heiko Haas, 145–157. London and New York: Routledge.

Hitchcock, Michael and I. Nyoman Darma Putra. 2007. *Tourism, Development and Terrorism in Bali.* Aldershot: Ashgate.

Hobart, Angela, Urs Ramseyer, and Albert Leemann. 1996. *The Peoples of Bali.* London: Blackwell.

Howe, Leopold E. A. 1980. "Pujung: The Foundations of Balinese Culture." Ph.D. diss. University of Edinburgh.

Kompas. 2013. "Menjaga Jatiluwih untuk Pariwisata Bali." 4 November.

Lansing, J. Stephen. 1991. *Priests and Programmers: Technologies of Power in the Engineered Landscape of Bali.* Princeton: Princeton University Press.

Locher-Scholten, Elsbeth. 1994. "Dutch Expansion in the Indonesian Archipelago around 1900 and the Imperialism Debate." *Journal of Southeast Asian Studies* 24(4): 91–111.

MacRae, Graeme. 1997. "Economy, Ritual and History in a Balinese Tourist Town." Ph.D. thesis. University of Auckland.

——. 2006. "*Banua* or *Negara*? The Culture of Land in South Bali." In *Sharing the Earth, Dividing the Land: Land and Territory in the Austronesian World*, edited by Thomas Reuter, 83–112. Canberra: ANU E-Press.

——. 2010. "If Indonesia is too Hard to Understand, Let's Start with Bali." *Journal of Indonesian Social Sciences and Humanities* 3: 11–36.

——. 2016. "Community and Cosmopolitanism in the New Ubud." *Annals of Tourism Research* 59: 16–29. http://dx.doi.org/10.1016/j.annals.2016.03.005.

Picard Michel. 1996. *Bali: Cultural Tourism and Touristic Culture.* Singapore: Archipelago Press.

Powell, Hickman. 1985. *The Last Paradise.* Oxford: Oxford University Press.

Reeves, Keir and Colin Long. 2011. "Unbearable Pressures on Paradise?" *Critical Asian Studies* 43(1): 3–22.

Reuter, Thomas. 2002. *Custodians of the Sacred Mountains: Culture and Society in the Highlands of Bali.* Honolulu: University of Hawaii Press.

Ricklefs, Merle. 2008. *A History of Modern Indonesia since c.1200.* Basingstoke: Palgrave Macmillan.

Sawyer, Suzana and Edmund Terence Gomez (eds.) 2012. *The Politics of Resource Extraction: Indigenous Peoples, Multinational Corporations, and the State.* New York: Palgrave Macmillan.

Schulte Nordholt, Henk. 2007. *Bali, an Open Fortress, 1995–2005: Regional Autonomy, Electoral Democracy and Entrenched Identities.* Singapore: NUS Press.

Sutika, I. Ketut. 2014a. "Pengurus Subak Mulai Putus Asa." *Antara News.* 28 May.

——. 2014b. "WBD Subak di Bali Tanpa Rencana Aksi." *Antara News.* 29 May.

Schulte-Nordholt, Henk. 1996. *The Spell of Power: A History of Balinese Politics.* Leiden. KITLV.

Stuart-Fox, David. 2002. *Pura Besakih: Temple, Religion and Society in Bali.* Leiden. KITLV.

Tabanan. 2014. "Bupati Tabanan Resmikan Badan Pengelola DTW Jatiluwih." 14 February. http://www.tabanankab.go.id/berita/pemerintahan/2048-bupati-tabanan-resmikan-badan-pengelola-dtw-jatiluwih.

Thorburn, Craig. 2002. "Regime Change—Prospects for Community-Based Resource Management in Post-New Order Indonesia." *Society & Natural Resources: An International Journal* 15(7): 617–628.

Tsing, Anna. 2005. *Friction: An Ethnography of Global Connection.* Princeton: Princeton University Press.

Turner, Frederick J. 1920. *The Frontier in American History.* New York: Henry Holt.

Turtinen, Jan. 2000. "Globalising Heritage—On UNESCO and the Transnational Construction of a World Heritage" (SCORE Rapportserie 2000: 12). Stockholm Center for Organizational Research.

UNESCO (World Heritage Committee). 1972. "Convention Concerning the Protection of the World Cultural and Natural Heritage." http://whc.unesco.org/en/conventiontext/.

UNESCO. 2012. "Convention Concerning the protection of the World Cultural and Natural Heritage." World Heritage Committee. Thirty-sixth Session, Saint Petersburg, Russian Federation. 24 June–6 July 2012 (WHC-12/36.COM/19).

Vickers, Adrian. 1989. *Bali: A Paradise Created.* Ringwood: Penguin.

Wallace, Alfred R. 2000. *The Malay Archipelago.* Boston: Periplus.

Warren, Carol. 1998. "The Cultural and Environmental Politics of Resort Development in Bali." In *The Politics of Environment in Southeast Asia Resources and Resistance*, edited by Philip Hirsch and Carol Warren, 229–261. London: Routledge.

Webb, Walter P. 1952. *The Great Frontier.* Boston: Houghton Mifflin.

Wiener, Margaret. 1996. *Visible and Invisible Realms: Power, Magic and Colonial Conquest in Bali.* Chicago: University of Chicago Press.

Windia, Wayan and Wayan Alit Artha Wiguna. 2013. *Subak: Warisan Budaya Dunia.* Denpasar: Udayana University Press.

12 An ambivalent nation

Ch'orti' in eastern Guatemala and western Honduras

Brent Metz

Introduction

Nations are imagined and constructed, not transhistorical and given in the nature of things, as Anderson (1983) established for Latin America. In the process of nation building, as with the formation of ethnic groups (Barth et al. 1969), distinction and sometimes opposition are accentuated between those within and those outside the nation, thus separating people by geographical and identity borders. In the case of indigenous national identities, an oft-used colonial strategy has been to divide and conquer by imposing such borders on groups. Whether this was intentional for the Ch'orti' Maya or not, the results of separating the group into the three provinces of Guatemala, Honduras, and El Salvador during the colonial period and corresponding nation-states thereafter served to break any unity that might have existed before. Moreover, various postcolonial ethnocidal practices such as land privatization and invasion, prejudicial education, military campaigns, and everyday discrimination has dissolved much distinctive Ch'orti' identity and culture. However, after five corrosive centuries, "the Ch'orti' Maya" are now reemerging as part of a transnational indigenous identity movement with the potential to (re)unite people long divided by nation-state borders.

When Hondurans, Guatemalans, and Salvadorans who had publicly shunned indigenous identities formed Ch'orti' movements in the early 1990s, I began a longitudinal historical investigation into the extent of the former Ch'orti'-speaking area, under the questionable assumption that language is the greatest unifying cultural factor upon which identity is based, and an ethnographic survey to ascertain the cultural and political bases of the movements and the possibility that other people may join the movements by no longer repressing their common cultural and identity characteristics. Could latent identities come to the fore among more people? By naming "the Ch'orti' Area" and investigating the contemporary commonalities of the people therein, including language, culture, and political position, I found myself to be inescapably enmeshed in transnational, counter-nation-state political construction. I have embraced this contribution because of its potential to reduce Ch'orti' marginality and thereby improve their quality of life, while privileging historical and social scientific accuracy and recognizing the dangers of national divisiveness.

Any analysis of the concept of a Ch'orti' Maya nation or people must first confront what the corresponding territory would be. Was there, is there, or could there ever be a Ch'orti' Maya territory? I myself have contributed to the construct of the Ch'orti' Maya "area" and must admit that the more we know, the more it dissolves. Greater historical knowledge of *any* "culture area" or "territory" tends to complicate its unity. Still, if one remembers that "peoples" or "nations" and their "territories" are constructs, they serve as useful conceptual devices, especially when the region shares some common factors. When constructing indigenous territories, the most straightforward foundation for determining commonality is language, something routinely and even subconsciously practiced by scholars and indigenous people alike in Mesoamerica. An "area" or "territory" is determined by the predominant indigenous language spoken there. However, even something as simple as mapping language to "territory" for just a single point in time is often problematic because of multilingualism due to trade, conquest, and intermarriage. For "the Ch'orti'," the area of Central America where the language was spoken during the Spanish invasion seems indeed to have been multilingual. As we will see, though, the construct of a "Ch'orti' area" is not completely arbitrary, and an investigation of contemporary peasant culture and identification there provides a basis for recognizing some ongoing unity, but one severely weakened by modern state formation.

Mapping "the Ch'orti' area": historical sources

Scholars (e.g. Robertson et al. 2010; Thompson 1975) who have studied colonial and linguistic records suggest that Ch'olti' was spoken roughly in the area below, and that Ch'orti', then known as Apay, was a dialect of Ch'olti' spoken in the tri-border area of Guatemala-Honduras-El Salvador in the Spanish colonial period (1524–1821). In fact, during and after the invasion, survivors in the Apay area fled north to join the Manché Chol beyond Lake Izabal, while the Torquegua and Chontales to the east, whose identities are disputed, eventually disappeared as unique groups after being brutally conquered by the Spanish and their Tlaxcalan allies.

However, existing archeological and colonial sources are not clear where Ch'orti'-speakers and polities were. The people of the region referred to their language as *Apay* or *Apayac*, and their Nahuatl-speaking Pipil neighbors called them *chontales*, a label commonly applied to "foreigners" (García de Palacios 1985 [1576]: 11–13, fn. 3; 46–48, 51–52; Newson 1986: 35). They called their collective territory over which they held dominion *Chiquimula*, and their Pipil neighbors referred to it as *Payaquí.* It included 12 Apay (presumably Chorti) chiefdoms and city-states that may well have included Alaguilac-, Nahua-, and Poqomam-speakers; cf. Brasseur de Bourbourg 1966 in García de Palacios 1985: 51–52; Brewer 2002; Brewer 2009: 138–139; Carmack 1981; Kramer 1994: 38; Orduna 1530; Walters and Feldman 1982: 595–596). Chronicler Fuentes y Guzmán (1933 [1699]: 119–122, 169–182, 204–209) was unequivocal that the region united to resist the Spanish invasion from 1524–1532. The word *chorti*, meaning "the language of the corn farmers or country folk" (Wisdom 1940: 6;

Girard 1962: 1; Morán 1935 [1695]; cf. Metz 2009: 2; Robertson et al. 2010: 9), was first recorded by Cortés y Larraz in 1770 in several towns in eastern Guatemala and northwestern El Salvador.

The Ch'olti'-speakers virtually disappeared in the north by the early 1700s due to epidemics, displacements, and predations by outsiders, and the Apay area was gradually reduced as well. By the late 1500s the Spanish displaced indigenous populations in the Motagua Valley as it became a principal trading route valuable for the production of sugarcane, indigo, cocoa, and mule and horse haciendas (plantations and commercial ranches). Already traditional society was breaking down, with Indian chiefs now acting as intermediaries for the Spanish, and thus losing their moral authority; many Ch'orti's were already learning Spanish (Terga 1980: 45, 56, 69–70, 73; Metz 2006: 37–56; Brewer 2009: 141; MacLeod 1973: 291; Little-Siebold 1995: 68). By the mid-1600s, the Ch'orti's had lost about 90 percent of their population due to epidemics, warfare, starvation, displacement, and flight (Pérez 1997), and the Honduran territory seems to have become largely depopulated. By the second half of the 1700s, some Ch'orti's began escaping their heavy communal tribute and forced labor obligations, as well as fleeing the devastating epidemic diseases, by taking up residence as peons on the haciendas, joining newly formed communities of *ladinos* (Hispanicized Indians, mestizos, mulattos, poor Spaniards, and freed Blacks) in the interstices of haciendas and Indian communal lands; others moved towards the center of the territory, away from the large Spanish-dominated towns of Chiquimula and Esquipulas (Brewer 2002: 83, 159–160; Cortés y Larraz 1768–1770 in Brewer 2002: 135; Dary 1995 in Girón 2007: 312–313; Feldman 1985; Fry 1988: 32–33, 38–39, 40, 52; Girón 2007: 312–313; Terga 1980: 26, 42, 43, 61–64, 75–76). In El Salvador, populations shrank throughout the colonial period, and in the major center of Tejutla nearly everyone spoke Spanish as well as Ch'orti'. Although Ch'orti's benefitted to some degree by their own indigo production and cattle herding for the leather to package the dye, ladino land pressure in El Salvador reduced Indians to occupy ever shrinking territories (Lauria-Santiago 1999a: 18–21, 29, 32; Browning 1975: 124–142).

The provincial boundaries of colonial Central America, which included Chiapas (in contemporary Mexico), were set early. Over the course of 300 years of colonization, the Crown exercised control over the northern portion of the Central American isthmus via Mexico City, and then made Antigua, Guatemala (Santiago de los Cabelleros de Guatemala) into a more autonomous control center. Under the Bourbon Reforms in the late 1700s, greater bureaucratic autonomy was given to what would become the Central American states, as more control devolved from central Guatemala to the provincial capitals of future Central American countries. This final fragmentation would be the basis for the formation of independent Central American countries.

As power gradually dispersed and borders hardened, the Ch'orti' area fragmented. Early in the colonial period, all "Indians" were "reduced" into legally protected but tribute-paying communities, thus breaking indigenous regional polities and the maximal level leadership positions that went with them, but indigenous people in eastern Guatemala continued regional trade and interacted with each other in forced labor (*repartimiento*) obligations (Feldman 1985). Elites in Santiago, Guatemala dominated the export economy of Central America, which emphasized indigo dye production for the growing textile and ink industries in proto-Industrial Revolution Europe. Spanish ladino entrepreneurs sought Indian labor to process indigo, which involved soaking leaves in fetid water swarming with flies and mosquitos. Because of the epidemics rife in indigo workshops, the Crown prohibited the use of Indian labor in them on multiple occasions, and the entrepreneurs turned to African slave and ladino labor when officials were paying attention. These same laborers would seek to get their own land, including illegally squatting on Indian communal lands. Indians were also forced to work in small mines in all three countries of the region, as well as in Caribbean ports as porters and soldiers against pirates, all of which in the second half of the eighteenth century led to a consequent population decline (Feldman 2009; Metz 2006: 49–53).

After independence in 1821, Central America became a unified federation (1823–1841), but poor communications between regions, sharp divides between social classes and ethnic groups, lack of common identification, lack of transportation infrastructure, vicious battling between conservatives and liberals, and meddling by the Catholic Church ensured its demise. Chronic and devastating warfare between Guatemalan, Honduran, and Salvadoran factions took place in the circum-Ch'orti' area, which in turn hardened emerging national identities of the respective populations and divided an already weakened Ch'orti' regional identification. Local identities were subverted further by land privatizations in the three countries, starting in the 1870s. All three sought to modernize the economy by privatizing Indian and Church corporate lands. The Indians could buy their own lands, but it was not uncommon for the process to be corrupted, and often the Indians who secured individual titles to plots were robbed by ladinos or their own leaders, or else forced to sell the land at extremely low prices (see Lauria-Santiago 1999b). In Ocotepeque, Honduras, and Chalatenango, El Salvador, some communities struggled valiantly with legal battles, knowing that to lose communal title to land was to lose the community, but they finally lost (Lauria-Santiago 1999a: 201–204). The Guatemalan and Salvadoran states added insult to injury by implementing forced labor laws, including "vagrancy" laws for peasants without a certain minimum amount of land because it had been expropriated in the "reform" process.

All the while, the overall population in the three countries climbed as a result of both natural population growth and ladino immigration. By the 1800s Ch'orti's were in the minority in all three countries, and by mid-twentieth century no records of Indians can be found for El Salvador, in part because in the aftermath of the genocide of thousands of Indians in the Pipil area to the south in 1932, after which the government stopped taking ethnic censuses (Tilley 2006: 80–81, 169, 179–180, 187, 218). The brutal Salvadoran Civil War of 1979–1992 was especially harsh in Chalatenango in the former Ch'orti' area, tearing to shreds what fabric of peasant life remained. In Honduras, Guatemalan Ch'orti's began immigrating to the Copán area at least as early as the 1810s, from which they slowly spread to the far north and east, especially in the twentieth century, and as they emigrated they tended to abandon their indigenous communal cultures. When a border dispute was settled between Guatemala and Honduras in the early 1930s, and Copán Ruinas definitively became Honduran, wealthy Honduran landowners rushed in to title the lands long occupied by the Ch'orti's and made them hacienda peons for tobacco labor (Loker 2005). They were severely impoverished and overworked, and largely abandoned most of their spiritual traditions and Ch'orti' identification. When the Honduran military government finally allowed for and even encouraged indigenous ethnic recovery in the 1980s, no self-identified Ch'orti's were recognized. In Guatemala, the Ch'orti' language was abandoned in all townships by the twentieth century except for contiguous Jocotán and Olopa (see Figure 12.1). In many non-Ch'orti'-speaking towns and villages, however, complex Maya rituals involving sacrifice of fowl to bring the rains and fertility were continued by *cofradías* (religious brotherhoods). Modernization in eastern Guatemala was violent and arguably lasted from the 1930s to 1990s, during which military governments forced Ch'orti's to work without pay in the 1930s and 1940s, curtailed or rescinded democracy regularly, and marginalized Ch'orti's from state education, healthcare, and transportation services. Internally, the last 18 Ch'orti'-speaking communities suffered from overpopulation on the poor lands to which they were relegated, and as population continued to increase, thousands emigrated to northern Guatemala and Honduras.

Transnational persistence of Ch'orti' Maya culture in Central America

By the 1990s when I began my research, it seemed that, after nearly five centuries of colonization and two of modern state formation, Ch'orti' identification, language, and culture were finally destined for complete abandonment. Even my Guatemalan Ch'orti' language instructor was embarrassed to speak the language in public. Indigenous markers like carrying loads on one's back by headstrap (*mecapal*) were seen as embarrassments except in a reduced area of Guatemala and a sliver of Honduras. Subsistence traditions were viewed as means of survival at best. Honduras and El Salvador showed no signs of Ch'orti' identification whatsoever. Then, the Latin American indigenous movement exploded in the early 1990s.

In Guatemala, it arrived in the Ch'orti'-speaking area via the Maya Movement in 1992, the year of the Columbian Quincentennial, and within a couple of years *campesinos* (peasants) and even town-dwelling mestizos began to identify as Ch'orti's in opposition to the racist and ethnocentric state and society. Initial public forums run by western Guatemalan Maya organizations transformed into lists of demands, and eventually resulted in the creation of permanent Ch'orti' satellite offices. While the demands addressed just about everything dissatisfying about contemporary life, from gender discrimination to lack of modern services to disrespect for religion, the weight of the Ch'orti'-run programs revolved around preserving and recovering the language, resulting in several rural language

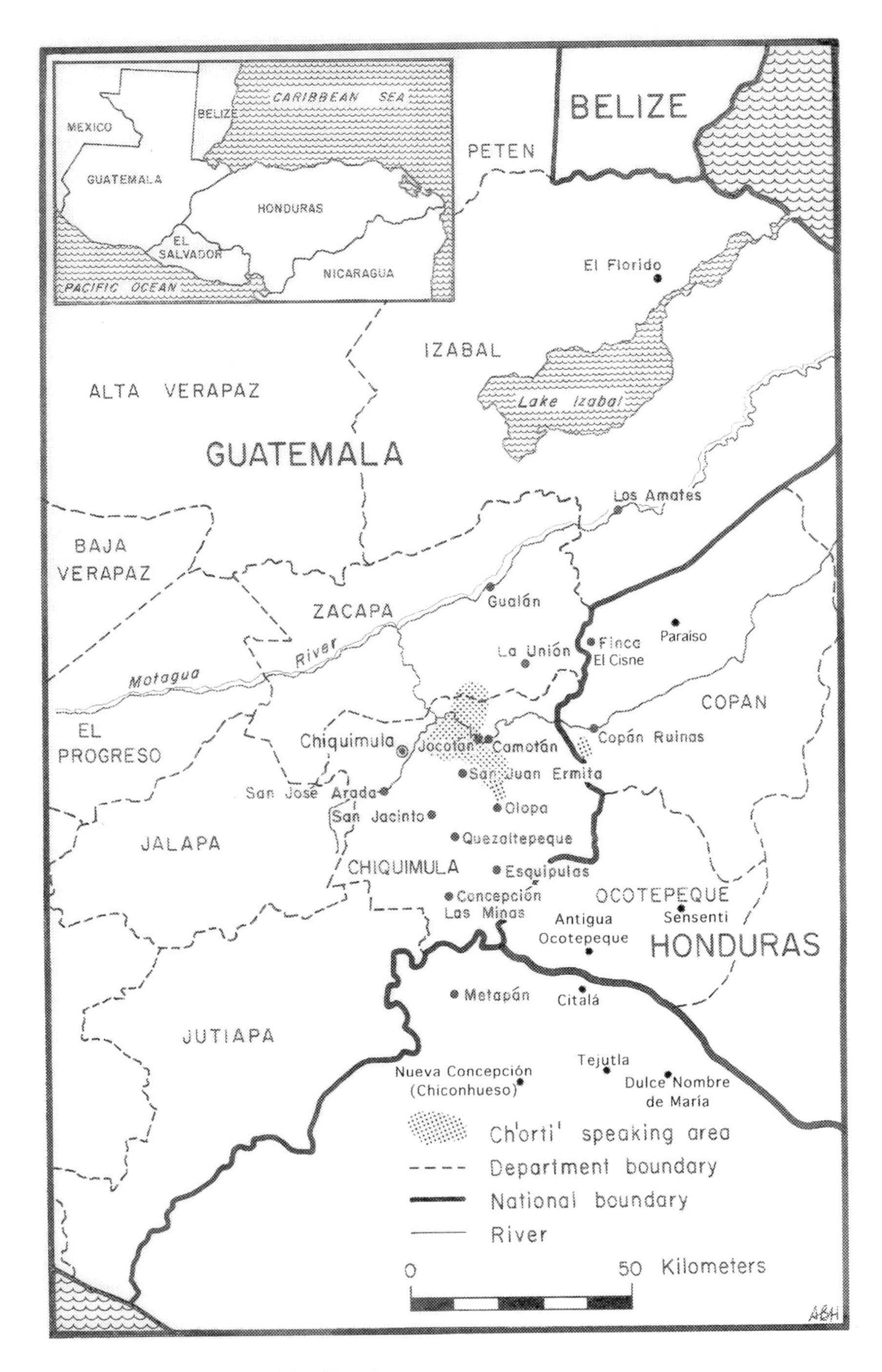

Figure 12.1 Map of the Ch'orti' region.

programs and a Ch'orti' public junior high and teachers' school. The institutionalization of the Ch'orti' language received a major boost with the 1996 Peace Accords, about half of the provisions for which concerned indigenous–state relations. While some academics might criticize language programs as state palliatives to distract from political and economic reforms, the reversal in ethnic shame, motivation, and confidence to pursue modernization and education is palpable.

In Honduras, official programs to strengthen indigenous languages and cultures in the 1980s were appropriated by peasants-turned-indigenous activists in the 1990s. Some Lenca activists, including academics, began recruiting in Copán and Ocotepeque in 1994 and could find only six campesinos in the least bit sympathetic with the movement. Slowly but surely, as the benefits of Honduras' signing of the ILO Convention 169 on indigenous rights to land were explained under the threat of landlord harassment and intimidation, the peasants realized its revolutionary potential. When the Ch'orti' leader and tour guide of Copán's archeological park, Cándido Amador, was murdered by landlords in 1997, the boil of landlord oppression was lanced. Several hundred re-identified Ch'orti's marched on the capital Tegucigalpa and demanded justice, land, language education, development, and proceeds from the archeological park. The Honduran state began a pattern of capitulation, delay, and recapitulation, while the Ch'orti's maintained the pressure with marches, roadblocks, and archeological park occupations. In 2000, the army, police, and Chamber of Commerce attacked the Ch'orti's who were blocking Copán's archeological park (and thus hurting tourism), using clubs and dispensing tear gas from helicopters. One woman from Ocotepeque eventually died from her injuries, another lost her fetus, several suffered permanent injuries, and over 150 were injured. The Ch'orti's eventually won justice against the murderers of Amador, but the state has been very slow to fulfill the promise of about 20,000 hectares of land. Land continues to be the principal motivating force of the movement, and the state has slowly ceded land or has intransigently refused to do so, arguing that it is too expensive, or else insinuating that the claimants are "not indigenous enough" (Metz 2010). As a result, some communities have formed new groups (for example, the Maya Ch'orti' Indigenous Communal Council, and the Chorti Indigenous Community of Antigua Ocotepeque), while others have left the movement altogether.

In El Salvador, as the civil war of 1979–1992 came to a stalemate, indigenous groups found a political opening when the state and the U.S. military were preoccupied with class warfare. In fact, El Salvador's recognition of indigenous groups and foundation of the local Houses of Culture in the 1980s with funding by foreign donors, was a play for national reconciliation (Tilley 2006: 24, 33, 39–41, 222–224, 230). The Pipils of southwestern El Salvador, who had submerged their identity and culture following the 1932 genocidal massacre, were at the forefront of the new National Association of Salvadoran Indians (ANIS). With "Indians" coming out of the closet after decades of state exclusion and social discrimination, Chapin (1989) argued that as many as 500,000 "Indians" could be living as a hidden social class in the country.

The re-emergence of Ch'orti's in Honduras, and the possibility of such in El Salvador inspired my historical and ethnographic analysis of the entire former Ch'orti'-speaking region. I collected data from 2003–2015 with the support of a Fulbright-Hays grant. The primary questions to be investigated were:

- Do the people (re)claiming to be Ch'orti' have ancestors who self-identified as Ch'orti'?
- Could more people "come out" as Ch'orti', particularly in El Salvador?
- What is the likelihood that re-identified Ch'orti's would unite in a transborder movement?

I approached these questions through historical research and ethnography. To build and strengthen any ethnic movement, including indigenous ones, unity must be built in part around a vision of a common history and at least some aspects of current culture. Visions and conscious memories (as opposed to embodied ones) need not be accurate, but I argue that greater accuracy means a greater *probability* of a stronger, more cohesive, effective identification. This is due in part to challenges from enemies. However, even more important for the re-emergence of Ch'orti' identification are shared interests. People may recognize a common history and culture and never identify if there is no occasion of opportunism or defense which calls for it. In such cases, indigenous identification lies latent (e.g. Canessa 2000; Cepek 2008, Gabbert 2006; Mayberry-Lewis 1997; Metz 2010; Tilley 2006). I researched, therefore, history as recorded by scholars and chroniclers, memories both imagined and embodied, common culture, and shared interests that could unite Ch'orti's within and across nation-states.

I visited 31 *municipios* (each a little smaller than a U.S. county) from 2003 to 2015 to explore memories, culture, common interests, and ultimately the possibility of latent Ch'orti' identification. Such an exploration is trickier that it might seem, since recording Ch'orti' identification is not simply a matter of observing distinctive dress, which has almost disappeared, noting language use, which many self-proclaimed Ch'orti's have abandoned, or flatly asking their identity, which they often repress due to social stigma. The Ch'orti'-speakers in Guatemala I knew best were often embarrassed to have their identity openly raised, and the (re)emergence of Ch'orti' identification in Honduras and possibly El Salvador suggested potentially the same. Ch'orti' rituals in particular have been attacked, largely due to Christian missionization campaigns, and from the 1960s to 1980s, in Guatemala and Honduras, armies and paramilitary groups equated saint cults and Ch'orti' identity with subversion. Deepening poverty has also added to the stigma of being a failed people. Hence, I approached identification throughout the region through circuitous questioning about historical memory, ancestry, and discrimination. Fortunately, interested officials, activists, and amateur anthropologists and historians provided me with contacts in communities thought to be the most indigenous, and accompanying Guatemalan Ch'orti' friends opened doors as well. I used my knowledge from Charles Wisdom's encyclopedic 1940 ethnography and my previous research living among Ch'orti'-speakers of eastern

Guatemala as bases for questioning, and I was attentive to any distinctive culture and identification.

At the start of the project, the characteristics (besides language) that distinguished Ch'orti'-speakers from others included a wide range of social practices: the integration of subsistence agriculture with observation of the phases of the moon for fertility; having large families; the sharing of food as a moral imperative; sacrifices or "payments" offered at planting time; food offerings to the ancestors in the month of the dead, or *Tz'ikin*; the linkage of saints with natural forces; utilitarian production of handmade crafts such as *petate* mats (*pojp*), hammocks, ropes, brooms, soap, wooden pipes, ceramics, and more; folklore about female water and male mountain spirits; beliefs in *nawales* or animal spirit companions, especially for sorcerers; phenotypic appearance that is predominantly Amerindian; the suffering of ethnic discrimination and its emotional effects; and the utilization of unique recipes for corn drinks and herbal remedies. To be clear, these traits are not comprehensive or *inherently* or primordially Ch'orti', but in combination they become distinguishing cultural features of contemporary Ch'orti'-speaking communities.

In Guatemala, I found many of these same features or memories of them in the non-Ch'orti'-speaking municipios of Camotán, San Juan Ermita, Chiquimula, Quezaltepeque, La Unión, San Jacinto, and Esquipulas, where people also remembered their ancestors speaking Ch'orti'. Campesinos and town-dwelling ladinos alike use Ch'orti' words like *ixchoko* (child, from Ch. *ijchok*, "girl"), *tz'ikin*, *kume* (youngest child, from Ch. *ku'm*), and *akukuch* (carrying with headstrap), but more ubiquitous are Nahuatl terms, especially for fruits, vegetables, and other food traditions (for example, *achiote, chilacayote, zapote, guisgil* or *chayote, milpa, chucte*, and many more). Particularly remarkable was non-Ch'orti'-speakers performing richly symbolic sacrifices of turkeys to bring the rains in Quezaltepeque, Chiquimula, and San Jacinto, as described by Girard (1949, 1962). Campesinos throughout the region, including mestizos from Honduras, travel to Quezaltepeque to pay respect and provide offerings to procure rain and fertility. Handmade crafts are rarely made beyond Jocotán, Camotán, and Olopa, partly because of a lack of materials and partly for the low market prices. In other words, the extreme poverty in these townships contributes to the maintenance of distinctive subsistence traditions. In most townships, except for the low-lying areas of Esquipulas, I detected a strong sense that peasants (*campesinos*), whether with indigenous identification or not, were discriminated by more powerful, well-connected, and corrupt townspeople, even though the latter also have Amerindian, as well as African, and European heritage.

In Honduras, I visited three populations active in the Ch'orti' movement. In and around Copán Ruinas, including parts of Santa Rita and Cabañas, many campesinos are nearly indistinguishable culturally from the aforementioned Guatemalan population, and in fact most emigrated from Guatemala in the nineteenth and twentieth centuries, as mentioned above. While only a few political refugees speak the language, save for a new generation of children now learning it in schools, traces of Ch'orti' vocabulary can be heard in everyday speech, place names, and rituals. They even have traditions abandoned by Guatemalan Ch'orti'-speakers, such as the spinning of cotton thread (Lincoln Vaughn, personal communication 2004). The sense of ethnic discrimination is entrenched, as is their extreme poverty.

In the northern extremes of Copán Ruinas and the municipio of El Paraíso, the population of fairly recent immigrants is more culturally and biologically mixed. I interviewed over a dozen families in eight communities in El Paraíso, and the oldest lineages date to only the late 1940s. Most immigrants arrived from Guatemala (with families often spending a generation of more in the mountains of Copán Ruinas first) or from the interior of Honduras, and many moved farther east to central Honduras after fertility declined and well-connected ladinos moved in to title the land for cattle ranching and sugarcane. The degree of cultural uniqueness depends on their original communities of origin, with those coming from other parts of Honduras showing little distinctiveness and more European ancestry, even if they are ardent members of the Ch'orti' movement. Those coming from Guatemala and Copán continue traditions of subsistence agriculture, herbal remedies, and the use of Ch'orti' words popular in Guatemala that set some of them apart from the general Honduran population, such as the use of *ixchoko* instead of *sipote* for children. Membership in the movement throughout Honduras connotes indigenousness, regardless of one's heritage, and when I asked whether a given community was indigenous, they often consulted with each other about whether they were in the movement ("organized" or not). Some in Paraíso were quite frank that they joined the Ch'orti' movement as a strategy to get land and aid, because their efforts to organize as campesinos had been ineffective. The exception is a core of communities in Copán Ruinas near the Guatemalan border, who once welcomed into the movement anyone claiming to be Ch'orti'; however, they have come to view Ocotepeque in particular with a critical, competitive eye regarding authenticity.

The self-identified population in Ocotepeque, including the municipios of Nuevo Ocotepeque, Dolores Merendón, Santa Fe, San Marcos, and Sensenti, form a separate group with little connection to the Guatemalan Ch'orti's. Except for some groups adjacent to the Guatemalan border of Esquipulas and in the high remote mountains of Dolores Merendón, their physical features show a range of mixing, from Amerindian, Spanish, and even African. They used no Ch'orti' words. In Antigua Ocotepeque, the heart of Ch'orti' movement, instead of emphasizing the planting ceremonies, they direct their attention to the harvest ceremony of *Los Parales*; however, they have no concept of the harvest-time *Tz'ikin* month of the dead. Some of the festive revelry organized by a loosely organized *cofradía* of San Andrés involves medieval Spanish images known as *corasquines* and *ujiganga*, and a burlesque representation of "Indians." When I showed some of them photos of Ch'orti's from Guatemala, they considered the people to be exotic and authentic. Another unique tradition includes the masked dance drama of Moors and Christians, something commonly introduced by priests to Indian communities in the colonial period. Other legends revolve around local landmarks, such as enchanted springs and a local peak called Cayahuanca, the name of which is not a Ch'orti' derivative. It may well be, as stated before, that the indigenous roots in Ocotepeque are more

Nahua or Lenca than Ch'orti', or a mix of all of these. Like those in Paraíso, the Ocotepeque Ch'orti's are very enthusiastic and nostalgic in hopes to discover that any of their contemporary practices or local sites have Ch'orti' roots; in this manner, they are unlike the Ch'orti's of Guatemala and Copán who are more leery of insinuations of their Indianness.

In El Salvador, people have "moved on" from any indigenous identities and Ch'orti's are spoke of as people of the past, although most people clearly have indigenous ancestry with little mixing, particularly in La Palma and Concepción Quezaltepeque. As Chapin (1989) explained, this does not mean that the concept "Indian" is not in use. The label in El Salvador as well as Honduras is applied to anyone, regardless of physical features, who lives a "backwards," peasant lifestyle. The main axis of identification is not ethnic but which side one had taken during the civil war. Some said the last remnants of peasant culture were destroyed during the war; this was when the landscape became disenchanted. No one spoke of racism or even discrimination between towns and the country folk, and people spoke of indigeneity without hesitation. Those aligned with the guerrillas were even excited to hear about possible indigenous roots. Fewer signs of Ch'orti' influence can be found, but some words of Ch'orti' origin remain in use in Tejutla, San Fernando, and Dulce Nombre de María. What the Salvadorans share with everyone in the region are pan-Mesoamerican and archaic Spanish practices, such as beliefs in the power of the moon in affecting fertility, the practice of calling cats by "meesh, meesh, meesh" (a pan-Mesoamericanism from the Nahuatl *mixtl*, puma), beliefs in bewitching water spirits, and other folklore, such as *cadejo* dog spirits. Some recounted folklore seemingly unique to El Salvador, such as tales of female hurricane spirits. In Concepción Quezaltepeque, people have maintained a hammock-making cottage industry, and now are using small motors and plastic twine instead of maguey fiber. My Guatemalan Ch'orti' companions felt the most affinity for the campesinos in the mountains near the Honduran border, where indigenous speakers, whether Lenca or Ch'orti', were said to have lived three or four generations ago.

As for common interests, Guatemalans have united around language rights, whereas Hondurans primarily have united around land but also language rights, and Salvadorans in the Ch'orti' area have not truly organized at all around ethnic issues. Attempts have been made to unite peasants of the three countries. From 2000 to 2003 a European-funded project based in Jocotán, Guatemala attempted to create a binational Ch'orti' territory but stalled when the Ch'orti's in Guatemala refused to even have their communal lands legally demarcated, which would have been the first step towards territorialization. Most Guatemalan Ch'orti's have no legal title to their lands and are essentially squatting on municipal lands since time immemorial, so they feel especially vulnerable to the issue of legalization. They are also concerned that legally titling land would mean paying taxes every year. Thus, the uniting of land is problematic in Guatemala itself. In Honduras, where CONIMCHH[1] has had some success in procuring land from the government, land has been ceded to communities as members of the organization, not to the organization as an "indigenous nation" on the whole. This has created conflicts in that some *aldeas* (villages) upon receiving their land, have lost interest in paying CONIMCHH dues and attending meetings on behalf of other communities that have yet to receive their land. The slow compliance of the governments in meeting promises from the 1990s has complicated the dynamic of frustration and impatience, but it is remarkable that many communities, even after getting their land, have remained in solidarity with those who have not, which has occurred mostly in the central Copán Ruinas area. Nonetheless, entirely new Ch'orti' organizations have formed and weakened the unity of the movement, not to mention decreasing the possibility of cross-border organizing. When I helped to organize a meeting of the three most active indigenous/ethnic organizations in each country in late 2004 (COMACH in Guatemala, CONIMCHH in Honduras, and RAIS in El Salvador), the effort was largely wasted. The date was changed at the last minute which made attendance difficult, and COMACH, and to a lesser extent CONIMCHH, saw RAIS as an organization that could assist, rather than partner with them on an equal basis.

This year (2015) seven communities in Jocotán, with the help of a human rights NGO called Nuevo Día, made the great stride of forming Indigenous Councils, which was the first and most momentous step in titling their land as "Indigenous Communal." This designation wrests some control away from municipal governments and was done in defense of threatening megaprojects in which they have received little consultation from the mayor, including a proposed private superhighway or "dry canal" project, a hydroelectric dam project, and mining. At meetings of potentially affected communities, many spoke of how "we Indians are always pushed around by the rich, and no one cares whether we live or die." Similar threats from development are occurring in Honduras and El Salvador, and self-defense organization against such neoliberal projects is likely to provide the most potential for a tri-country Ch'orti' indigenous reclamation.

Conclusion

If we are to make the concept of indigeneity workable in social science without falling into a misguided primordialism (i.e. the myth of transhistorical cultures and territories, in which peoples/traditions/identity are conflated; see Gupta and Ferguson 1992), then we must attend to historical and ethnographic detail to examine the bases of any indigenous group's unity. Applying such methods to the various Ch'orti' cases demonstrates why indigenous attempts to unite across borders can be very challenging. The farther back one goes in time, and the longer states have been involved, the more trajectories and bases for common identification diverge.

Identification requires commonality and distinction from others. In Latin America, the distinction has come to indigenous people, rather than being sought by them; it has come by way of invasion and colonization by others who have treated them as an inferior class of "Indians." In periods of harsh discrimination or lack of incentives for open identification as an indigenous group, identification has lain latent. Latent indigenous identification becomes manifest not by "natural affinity" but by the political conditions of the moment. People recognize their commonalities, recall and recognize common historical memories and contemporary culture. The people of the circum-Copán Ruinas and Guatemalan Ch'orti' areas have a stronger basis for building such self-recognition and identification with each other than people of other parts of Honduras and El Salvador do with them. In the latter two, cultural mixing, separation, and distinctive national dynamics have long been sending populations on diverging paths. All of this could be overcome for the entire former Ch'orti'-speaker area with the most important ingredient of indigenous movements, and of ethnic movements in general, which is a set of strong common interests.

Currently, such interests in all three countries exist to a weak degree, and include a desire to relearn Ch'orti' for reasons of nostalgia, historical vindication, strengthening of claims to indigeneity, and to strengthen the potential for building cross-border alliances against megaprojects. The megaprojects are taken by the indigenous as a sign of discrimination against them, and thus have begun to raise their consciousness and self-recognition as a defensive measure.

Seen from a wider lens, the Ch'orti's may not be "indigenous enough" for scholars of western Guatemala, where distinctive language and a particular kind of dress (that Ch'orti's never wore) are expected. Consequently, edited volumes on "Guatemala" often exclude the east, and foreign ethnographic research in other Central American countries has been scarcer still (Tilley 2006). Within the former Ch'orti'-speaking region, indigenous dynamics are of a different order, but this has not prevented western Guatemalan Mayanist activists from coming to redeem their wayward cousins—recognizable to them by their language, subsistence traditions, physical features, and proximity to the ancient site of Copán—with re-Mayanization via greater autochthonous spirituality and community organization. They recognize not only common roots but also common political interests.

Note

1 CONIMCHH is the Honduran Ch'orti' Maya Indigenous Council, and was formed in 1994. Two other organizations mentioned below are: COMACH, the Ch'orti' Maya Council, formed in Jocotán, Guatemala in 1994; and RAIS, the Network of Indigenous Associations of El Salvador that began operating in Chalatenango in the early 2000s.

References

Anderson, Benedict. 1983. *Imagined Communities: Reflections on the Origin and Spread of Nationalism*. London: Verso.

Barth, Fredrik, ed. 1969. *Ethnic Groups and Boundaries: The Social Organization of Culture Difference.* Long Grove, IL: Waveland Press.

Brewer, Stewart W. 2002. *Spanish Imperialism and the Ch'orti' Maya, (1524–1700): Institutional Effects of the Spanish Colonial System on Spanish and Indian Communities in the Corregimiento of Chiquimula de la Sierra, Guatemala.* Dissertation, University at Albany.

——. 2009. "The Ch'orti' Maya of Eastern Guatemala under Imperial Spain." In *The Ch'orti' Area, Past and Present.* Brent Metz, Cameron L. McNeil, and Kerry Hull, eds. Pp. 137–148. Gainesville: University Press of Florida.

Browning, David. 1975. *El Salvador: La tierra y el hombre.* San Salvador: Ministerio de Educación.

Canessa, Andrew. 2000. "Contesting Hybridity: Evangelistas and Kataristas in Highland Bolivia." *Journal of Latin American Studies*, 32: 115–114.

Carmack, Robert. 1981. *The Quiche Mayas of Utatlán: The Evolution of a Highland Guatemala Kingdom.* Norman: University of Oklahoma Press.

Cepek, Michael L. 2008. "Bold Jaguars and Unsuspecting Monkeys: The Value of Fearlessness in Cofán Politics." *Journal of the Royal Anthropological Institute*, 14: 334–352.

Chapin, M. 1989. "The 500,000 Invisible Indians of El Salvador." *Cultural Survival Quarterly*, 13(3): 11–16.

Cortés y Larraz, Pedro. 1958 [1768]. *Descripción geográfico-moral de la diócesis de Goathemala.* Vol. 20. Guatemala City: Sociedad de Geografía e Historia.

Dary, Claudia. 1995. *Ch'orti's, negros y ladinos in San Miguel, Gualán, Zacapa: Una perspectiva etnohistórica. La Tradición Popular no. 103.* Guatemala: CEFOL-USAC.

Feldman, Lawrence H. 2009. "Some Data and Reflections on the Demographic Dynamism and Continuity of the Colonial Ch'orti' Population: The Many Copáns and San Juan Ermita." In *The Ch'orti' Area, Past and Present.* Brent Metz, Cameron L. McNeil and Kerry Hull, eds. Pp. 148–156. Gainesville: University Press of Florida.

——. 1983. "Distribución prehistórica e histórica de los pipiles." *Mesoamérica*, 4(6): 348–372.

——. 1985. *A Tumpline Economy.* Culver City, CA: Labyrinthos.

Fry, Michael Forrest. 1988. "Agrarian Society in the Guatemalan Montana, 1700–1840." Ph.D. diss., Tulane University.

Fuentes y Guzmán, Francisco Antonio De. 1933 [1699]. *Recordación florida: Discurso historial y demostración natural, material, militar y política del reino de Guatemala.* 3 vols. Guatemala: Biblioteca "Goathemala."

Gabbert, Wolfgang. 2006. "Concepts of Ethnicity." *Latin American and Caribbean Ethnic Studies*, 1(1): 85–103.

García de Palacio, Diego. 1985 [1576]. *Letter to the King of Spain.* Trans. and with notes by Ephraim G. Squier [1859], with additional notes by Alexander von Frantzius and Frank E. Comparato. Culver City, CA: Labyrinthos Press.

Girard, Rafael. 1949. *Los chortis ante el problema maya: Historia de las culturas indígenas de América, desde su origen hasta hoy.* 5 vols. Mexico City: Antigua Librería Robredo.

——. 1962. *Los Maya Eternos.* Mexico: Libro Mex.

Girón, Felipe. 2007. "Significados étnicos, sentidos locales: Dinámicas socioeconómicas y discursos identitarios en Huité." In *Mayanización y vida cotidiana: La ideología multicultural en la sociedad guatemalteca*, Vol. 2. Santiago Bastos and Aura Cumes, eds. Pp. 307–45. Guatemala City: FLACSO, CIRMA, and Cholsamaj.

Gupta, Akhil and James Ferguson. 1992. "Beyond 'Culture': Space, Identity, and the Politics of Difference." *Cultural Anthropology*, 7(1): 6–23.

Kramer, Wendy. 1994. *Encomienda Politics in Early Colonial Guatemala, 1524–1544: Dividing the Spoils.* Boulder, CO: Westview.

Lauria-Santiago, Aldo A. 1999a. *An Agrarian Republic: Commercial Agriculture and the Politics of Peasant Communities in El Salvador, 1823–1914.* Pittsburgh: University of Pittsburgh Press.

——. 1999b. "Land, Community, and Revolt in Late-Nineteenth-Century Indian Izalco, El Salvador." *The Hispanic American Historical Review*, 79(3): 495–534.

Little-Siebold, Todd R. 1995. "Guatemala and the Dream of a Nation: National Policy and Regional Practice in the Liberal Era, 1871–1945." Ph.D. diss., Tulane University.

Loker, William M. 2005. "The Rise and Fall of Flue-Cured Tobacco in the Copán Valley and Its Environmental and Social Consequences." *Human Ecology*, 33(3): 299–327.

Macleod, Murdo. 1973. *Spanish Central America: A Socioeconomic History, 1520–1720*. Berkeley: University of California Press.

Mayberry-Lewis, David. 1997. *Indigenous Peoples, Ethnic Groups, and the State*. Needham Heights, MA: Allyn & Bacon.

Metz, Brent. 2006. *Ch'orti' Maya Survival in Eastern Guatemala: Indigeneity in Transition*. Albuquerque: University of New Mexico Press.

——. 2009. "The 'Ch'orti' Area.'" In *The Ch'orti' Maya Area, Past and Present*. Brent E. Metz, Cameron L. McNeil, and Kerry M. Hull, eds. Pp. 1–14. Gainesville: University Press of Florida.

——. 2010. "Honduran Chortís and the Inherent Tension of Generalized Indigeneity." *Journal of Latin American and Caribbean Anthropology*, 15(2): 289–316.

Morán, Francisco. 1935 [1695]. *Arte en lengua cholti, que quiere decir lengua de milperos*. New Orleans: Tulane University.

Newson, Linda. 1986. *The Cost of Conquest: Indian Decline in Honduras under Spanish Colonial Rule*. Boulder: Westview.

Orduna, Francisco De. 1530. *Información del señor capitán e juez de residencia, Francisco de Orduna. Actas de Cabildo de Guatemala, tomo 1*. Archivo General de Centroamerica A1.2.2–11763–1768.

Pérez, Hector. 1997. "Estimates of the Indigenous Population of Central America (16th to 20th Centuries)." In *Demographic Diversity and Change in the Central American Isthmus*. Anne R. Pebley and Luis Rosero-Bixby, eds. Pp. 97–115. Santa Monica, CA: Rand.

Robertson, John S., Danny Law, and Robbie A. Haertel. 2010. *Colonial Ch'orti': The Seventeenth-Century Morán Manuscript*. Norman: University of Oklahoma Press.

Terga, Ricardo. 1980. "El valle bañado por el río de plata." *Guatemala Indígena*, 15(1–2): 1–100.

Thompson, J. Eric S. 1975. *Historia y religión de los mayas*. Mexico: Siglo Veintiuno.

Tilley, Virginia Q. 2006. *Seeing Indians: A Study of Race, Nation, and Power in El Salvador*. Albuquereque: University of New Mexico Press.

Walters, Garry Rex and Lawrence H. Feldman. 1982. "On Change and Stability in Eastern Guatemala." *Current Anthropology*, 23: 591–604.

Wisdom, Charles. 1940. *The Chorti Indians of Guatemala*. Chicago: University of Chicago Press.

Ximenez, F. 1710 [1932]. *Historia de la Provincia de San Vicente de Chiapa y Guatemala de la Orden de Predicadores*. 2nd ed., Vol. 5. Guatemala: Biblioteca Goathemala.

Part VII

Entangled histories

13 Nation-state building and transnationalism

Central American connected histories

Luis Roniger

Introduction: Central America in transnational perspective

Geopolitically, Central America is a small region uniting the continental masses of North and South America and an isthmus that separates the Pacific Ocean from the Caribbean Sea (see Figure 13.1). In spite of its small dimensions when compared to other regions of the globe, it is a highly differentiated area, host to multiple societies and cultures, as well as the seat of seven sovereign countries that have maintained close yet often tension-ridden relationships since the attainment of political independence in the early nineteenth century. Drawing attention to how the process of nation-state building has been long embedded there in undercurrents connecting these societies to one another, this chapter embraces the analytical perspective of connected histories and transnationalism. It suggests that, following independence, the societies of this region were unable to completely disengage from transnational forces that pulled them together time and again.

The perspectives of connected histories and transnationalism, stemming from the earlier moves to world history and the criticisms such move has generated among supporters of historical distinctiveness, has opened the way for bringing comparative analysis and transfer studies closer to the analysis of transnational interactions (Subrahmanyam 1997, 2005; Werner and Zimmermann 2002, 2006; Sznajder and Roniger 2009; Preuss 2011; Roniger 2011, 2014; Blumenthal 2013). This is a turn to analyses neither fully determined by developed countries' geopolitical priorities and visions, nor driven fully by globalization. It shifts attention to inquiries into frontiers, cross-border practices, networks and movement of ideas and peoples as bridges between societies. Accordingly, this perspective suggests being aware of how processes and reflexivity about them are socially structured, reflecting particular positions in competition or power struggles. It stresses mutual impacts, resistances, inertias, new combinations and transformations "that can both result from and develop themselves in the process of crossing"; and conceives the transnational level as

> a level that exists in interaction with the others, producing its own logics with feedback effects upon other space-structuring logics. Far from being limited to a macroscopic reduction, the study of the transnational level reveals

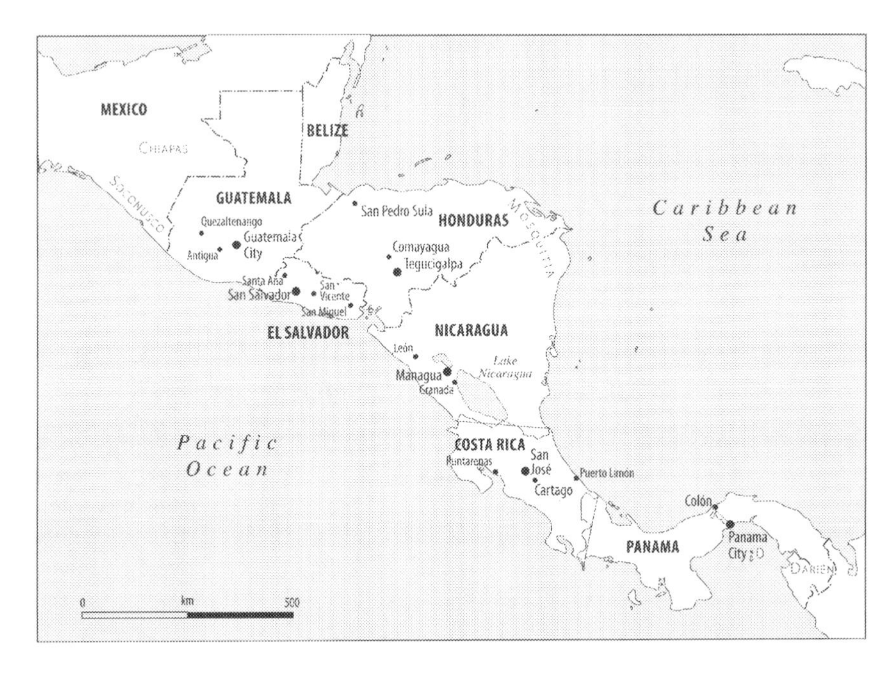

Figure 13.1 Map of transnational Central America.

Source: Map by Tamar Sofer © Luis Roniger.

> a network of dynamic interrelations whose components are in part defined through the links they maintain among themselves and the articulations structuring their positions.
>
> (Werner and Zimmermann 2006, 43).

Such a transnational turn has been shaped as well by anti-colonial and post-colonial scholarship, contributing to the analysis of "units that spill over and seep through national borders, units both greater and smaller than the nation-state" (Seigel 2005).

From the perspective of our times, the societies of Central America constitute an ideal setting for reflecting on the complex dynamics of connected histories and transnationalism cutting across distinct nation-state identities, without one axis fully replacing the other. An analysis of how the societies under consideration led a Janus-faced dynamic embedded interface of nation-state building and transnationalism is particularly relevant as research demystifies a lineal reading of history that had once canonized the hegemony of the nation-state and now seems to promulgate its decline under the aegis of globalization.

The interface of nation-state building and transnationalism

Societies construct collective identities in diverse ways, involving a process of symbolic creation and reconstruction crafted to denote the shared experience and orientations that may unite the members and set them apart from others. The images societies construct and the stories they tell about themselves are crucial in representing who they are and in catalyzing feelings and emotions (Geertz 1973). They are equally crucial in crystallizing the group's consciousness and motivating individuals to identify with collective visions and aims. Condensed in key symbols and artifacts—such as flags, shields and national anthems—these images acquire a presence of their own and a representational power, which can be projected across time, shaping the visions and collective memory of future generations, albeit with an enormous variability among and within nations and groups (Roniger 2008; and see Ortner 1973; Smith 1996; Geisler 2005).

Different societies have undergone the process of nation-building through a variety of historical paths and contingent developments (Herb and Kaplan 2008). Shmuel N. Eisenstadt and Berhard Giesen analyzed how in Germany since the eighteenth century, the nation was represented as a *kulturnation*, the locus of a transcendental realm of sublime essences and forces of history or as a sublime operator of historical action, in romantic ways that predated the process of state construction and nation-building undertaken by Otto von Bismarck in the late nineteenth century. By then, the romantic idealization had led to a blood-and-soil-related nationality discourse, which later on, under Nazism, assumed extreme primordial overtones leading to genocide and the Holocaust (Eisenstadt and Giesen 1995, 85–93). Contrastingly, in Japan, primordiality was given priority and was almost divinized, although not in terms of a transcendental mission as was the case in the monotheistic religions and civilizations (Eisenstadt and Giesen 1995, 93–102).

The case of Central America differs from either one of these cases. The region has been seat of sovereign countries that have maintained close yet often tension-ridden relationships since their political independence. The political and intellectual elites—with the possible exception of Costa Rica—could not easily romanticize their historical origins, in which power competition, domination, and fragmentation were typical developments. Nor could the elites assume primordiality for constructing collective identities, since their hierarchical social visions predicated cultural subjugation of ethnic and cultural groups that were bearers of distinctive primordial identities.

In regional terms, Central America stands out comparatively since most states in the region were born out of a shared colonial administrative jurisdiction and a short-lived attempt at unification following independence, with long-lasting effects on the ways these countries have strived to construct their national identities and idiosyncrasy, develop their distinctiveness, while at the same time being unable to completely disengage themselves from the sister republics of the isthmus.

In Central America, as elsewhere in Spanish America, most states emerged as a consequence of imperial disintegration, in the great majority of the cases structured initially on the basis of previous colonial administrative jurisdictions. But unlike South America, independence in the isthmus was attained without major heroic sacrifices and "hallowed grounds" that could be celebrated in poetry, ballads, holidays, monuments, or pantheons for fallen martyrs. States were destined to eventually create nations, attempting to "render them real" through the use of official accounts and rituals, the elaboration of hegemonic views and symbolic practices, and the structuring of images of people-hood, connected to spatial and temporal boundaries. Such strategies of nation-building involved the partitioning of territories that once belonged to the same political entity; the formation of confined membership and the delineation of borders, organized according to principles of national sovereignty. The creation of nations also implicated systems of cultural representation that legitimized or de-legitimized different sectors' and people's access to the resources of the nation-state (Alonso 1994; Mande 2000).

This process of nation-building, shared in its generic traits by many states worldwide, became convoluted and protracted in Central America, as the new states could hardly elicit in the population a sense of being part of what Benedict Anderson (1991) called an "imagined community." The new states had to define and create national membership and boundaries, which implied recognizing certain categories of citizenship as paramount, while replacing, ignoring or denying—without fully eradicating—earlier forms of identification, including the pan-isthmian identity, and subsuming more localized and ethnic identities.

The complex and prolonged process of nation-building in Central America was a result of many factors, with the following as most prominent: the shared colonial and early post-colonial identity; a strategy stressing the political structuring of their countries and denying distinctive primordial identities to build separate nationalities; the improvisational character of the states representing cities interested in dominating other urban centers and enlarging the hinterland they controlled; and, as stressed by Robert Holden (2004), the contested nature of factional and clientelistic politics, which fueled public violence and did not allow the early consolidation of political centers with authority over the inhabitants of each state.

For decades after their separation, the states of Central America could not consolidate their boundaries and seclude themselves from a dynamic of regional intervention. The interference came from factional armies and clientelistic entourages, driven by the prospects of taking power in their own home region or another region, disregarding borders, and state jurisdictions. Rebels in one area were supported by allies in the neighboring states, willing to topple those in power and facilitate the rise to power of political forces sympathetic to their own designs.

What in contemporary views could be interpreted as "invasions" were at that time considered as mere advances of forces willing to change constellations of power and in some cases, define state boundaries anew. The wars that ensued were not seen as "national" wars or "anti-imperialist" wars. All political forces shared the understanding that these were internal, fratricidal wars. It will take external threats and interventions to generate a "national" interpretation of the struggle for independence. Yet, as we shall see below, initially, even the sense of national struggle was in fact embedded in the transnational resistance to external intervention and threats.

Moreover, overshadowing the construction of sovereign realms and separate identities were the common origins, which left a legacy of cross-national networks of kinship, economic, social and political ties, and an image of an alternative project of regional nation-building. Individuals could rely on such image when relocating to sister-countries or challenging current institutional arrangements and political divisions. From the perspective of the symbolic enactment of separate national identities, primordiality—in the form of ethnicity or race—was secondary to the political and civic strategies adopted while constructing nationhood. From early on, elites were fully aware that local identifications existed, but at the same time, they were also conscious that there were no strong lines separating republics from one another or portraying the others as unalterably different in an incommensurable manner. Moreover, the way in which these states declared independence implied that they could not envision their collective identity as naturally given, but rather as a civic accomplishment.

The construction of nation-states had to embody institutional arrangements resulting from the promulgation of agreed-upon documents representing the will of the *pueblos* (communities) of the isthmus. The institutional arrangements depended on the continued will of the members of these communities and particularly on the will of leading elites to respect the promulgated charters and civic rules. Once routinized, these arrangements could be endowed gradually with transcendental meaning, through the use of myths of historical origins (e.g. Cruz 2000), the cult of founding fathers and heroes, commemorative rituals, school textbooks, and recurrent practices and symbols aimed at stressing the distinctive virtues of those belonging to the national core (Cruz 2000; Taracena and Piel 1995).

In the case of the republics of the isthmus, this process was complicated due to their shared origins, the complex process of promulgation of independence and the protracted mutual involvement of each state in the affairs of its neighbors. This set of factors made it hard to even promulgate clear-cut, joint or separate, independence dates. Accordingly, even establishing the date of independence has often been a source of dissent in the isthmus. As the process of construction of distinct collective ("national") identities gained momentum in the mid to late nineteenth century, it was intertwined in transnational connections that superseded and in fact even allowed such construction.

Connected histories and the incomplete national configuration of space and identity

Hindsight into such connected histories shaping the inability of complete disengagement from the transnational framework of the region can be provided by reviewing some of the main processes that Central America witnessed under external threats and interventions. A major such external intervention was the case of the entire region facing the rise of William Walker and his private army of North American freebooters taking control of Nicaragua in the 1850s.

Walker's intervention led to a war that was fought by what today we would define as a "transnational" alliance of nationals of various isthmian countries against Walker, while paradoxically—or perhaps not, since it fitted the logic of state claims, advanced in parallel by several of the isthmian states—such war became known in isthmian historiography as the National War. The transnational dimension of the struggle was so evident that its symbolic appropriation in terms of the emerging images of nationhood was done in a plural fashion and could not completely obliterate the transnational angle as an underlining current of narrative and symbolic representation. In such manner, the war only rekindled once again the tension-ridden process of nation-states' construction embedded within the persisting transnational dimension of their existence in the isthmus.

After gaining independence, Nicaragua was involved in endless fighting between Liberals, with a stronghold in the city of León, and Conservatives, with their center in Granada. Facing a rising tide of Conservatism, led by Rafael Carrera in Guatemala, Nicaraguan Liberals called in North American adventurers to their cause, promising them generous land grants. Under contract with the Liberals, William Walker and 57 Californians arrived in the region, and were reinforced by 170 locals and about 100 Americans upon landing. Initially, Walker suffered defeat by forces led by a Conservative commander from Honduras, but due to parallel defeats of other Liberal commanders, he managed to rise as commanding officer, taking over the city of Granada. This opened the door to Walker's recognition as chief of the armed forces and eventually led to his becoming head of state of Nicaragua. By executing and exiling opponents, he controlled the country and was elected president in June 1856. Walker's agents recruited over a thousand American and European men to sail to the region and fight for the conquest of the other four Central American republics (Leonard 1993, 6).

This move would galvanize the resistance of Central American states. Moreover, by aligning himself with railroad and shipping capitalists Cornelius Garrison and Charles Morgan in their competition with Commodore Cornelius Vanderbilt for control of the company that secured the only major trade route from New York to San Francisco through Nicaragua, Walker made an enemy of Vanderbilt. Vanderbilt supported the military coalition of Central American states, led by Costa Rica; prevented supplies and men from reaching Walker; pressured the U.S. government to withdraw his recognition as the Nicaraguan power holder; and gave North American defectors free passage to the United States.

Walker managed to gain control of Nicaragua by the mid-1850s, but he was soon thrown from power in a struggle that would become immortalized also as the "War of Independence" in Nicaraguan and Central American historiography, serving as a substitute for the war of independence that Central America had never experienced (Domville-Fife 1913). By May 1857, Walker had surrendered to the U.S. navy and was repatriated. Subsequent attempts by him to regain power failed and led to his imprisonment by the British who controlled the Caribbean coast of Honduras. The British handed Walker over to the Honduran government, which executed him in September 1860.

The fight to oust Walker became a source of collective pride and was portrayed as a cornerstone of national identity in more than one country of the region. The struggle also carried transnational connotations which emphasized Central American solidarity and patriotism. The built-in tension between nation-building and transnationalism was thus encoded symbolically in the recreation and commemorations of this historical event, instrumental in the construction of collective identity.

In Costa Rica, April 11th became a key national holiday, commemorating the battle of Rivas in 1856, in which a makeshift army of Costa Rican peasants chasing Walker defeated him at the locality of Rivas, Nicaragua, thus beginning the decline of Walker's power which culminated in his eventual surrender to the U.S. navy. Juan Santamaría, an illegitimate mulatto drummer boy from Alajuela, Costa Rica, volunteered to set fire to the wooden fort of Rivas where Walker and his men had taken refuge, so as to force them out. Although Santamaría, who lost his life as a consequence, was not celebrated at the time, beginning in the late 1880s and continuing officially in 1891 he was held up as a national hero. A statue of Santamaría depicting a strong and handsome soldier carrying a torch was placed in Alajuela, and Nicaraguan poet Ruben Darío immortalized his memory. Today, Santamaría's heroism is taught in schools and his sacrifice is eulogized and celebrated by Costa Rican children in commemorative performances. Unsurprisingly, the story of that war has been equally central to the construction of national identity in Nicaragua. Promoted by historians and politicians, it was portrayed there as a war fought by Nicaraguan patriots joined by the armies of neighboring states, thus giving it a national interpretation without completely obliterating the themes of Central American solidarity and transnational patriotism.

This constructive process glossed over but could not eradicate the tensions existing between national narratives and previous interpretations embedded in a transnational realm. These different interpretations and narratives continued to coexist, albeit with changes operated in their relative saliency in different states and times (on this shifting balance see Acuña Ortega 1995, 535–571 for Costa Rica; and Fumero Vargas 1995 for Nicaragua). The experience of that War was highly important for the process of construction and reshaping of national and transnational identities in the region. Political centers promoted ceremonial traditions, which served both to commemorate the heroic deeds of the recent past as well as to recognize the vision and progress of the countries and in the case of Nicaragua, the military capacity of its armed forces. The Central American experience clearly fits in this regard within Eisenstadt and Giesen's analysis of construction of collective identities, Eric Hobsbawn and Terence Ranger's analysis of invented traditions, and Anthony Smith's studies of naturalization of historical events. That experience has the additional aspect of projecting a persistent tension between the redefinition of state boundaries and the reformulation of meanings of the "national," thus recreating the transnational dimension of politics, now from beyond the crystallization of discrete state boundaries.

The war was rapidly appropriated symbolically as a National War, but in spite of its name, it had been fought by armies from all over the region and across state borders, thus ensconcing persisting transnational threads, even as these nations entered new phases of state consolidation. The very character of the confrontation could not erase the transnational dimension that became projected into the next stages by the very fact that the war could be claimed in parallel by various states, so that its transnational character was projected into the rhetoric and memory of future generations, even when this was officially discouraged.

Transnational social movements and inter-state constructions

The Janus-face character of collective identities and political membership in Central America determined an intermittent renewal of projects of unification drawn by political elites as well as by social movements, and processes of transnational spillover from one society to another. Sustaining the attempts of reunification, and beyond the specifics of each attempt, was an underlying feeling of disappointment with the current achievements of the separate republics, often seen as "sections" of a more substantive Fatherland. The supporters of these initiatives were sometimes motivated by the conviction that a golden age awaited to the strategically located isthmus, provided the countries managed to overcome past divisiveness. In the nineteenth and early twentieth centuries there were at least ten major attempts at reunification (Roniger 2011, chapter 8). In the early twentieth century, even the U.S. government supported such attempts. A General Treaty of Peace and Friendship was signed in Washington D.C. in December 1907, under the sponsorship of the US. Efforts to revive a confederation failed, but the delegates signed eight conventions aimed at stopping transnational fighting as countries agreed not to intervene in the affairs of their neighbors through the manipulation of political exiles and other measures. With the opening of the Panama Canal only years away, the US was strongly interested in the political stability of the region. Within the isthmus, the position of El Salvador was crucial, since its policies of distrust of Guatemala's striving for regional hegemony assured the parallel projection of Nicaraguan influence until 1908. Likewise, Salvadoran opposition to the increasing dependence of Central America on U.S. financial support led to various early transnational initiatives such as the idea of creating a Central American bank in 1912.

An International Central American Office was created, located in Guatemala with the intention of supporting "the peaceful recreation of the Central American Fatherland." The office, in which delegates of the various states were represented, cooperated with the member states in promoting the harmonization of constitutional provisions, the unification of educational contents and custom tariffs, monies, weights and measures, and the adoption of a single shield and flag. The Office published a quarterly journal that promoted the spirit of transnational unity through notes on the common history, national heroes and symbols; reports on regional treaties; and information on the member states. The spirit of transnational mission that pervaded its activities can be assessed by quoting the opening paragraph of its April–June 1920 issue:

> Central America, wake up. The feeling of nationality stands up boldly in each of the five Republics of the Isthmus. Our peoples, melted by misfortune, diminished by animosity, increasingly sense the imperative need of

> transforming into optimal reality those spiritual ties that, albeit in weak form, have maintained the noble and fertile idea of Central American solidarity during the painful phase of our separation. Tying our fraternity within an indissoluble bond and enabling the emergence of a free, great and prosperous nation in the heart of the new world, constitutes the imperative mandate of our peoples. An intensive work of regeneration is needed. After so many years of exhausting experience, it is hard to imagine that the sacred fire of patriotism would still be impotent to melt the ice of indifference that controls minds and of egoism that makes hearts sick.
>
> (*Centro-América*, XII, 2 (1920): 1)

Those committed to a Central American fatherland were encouraged by the experience of Italy and Germany, which were united after centuries of fragmentation. Writing about the Italian experience, they imagined the future of their own fatherland, choosing to stress the voluntary, spiritual commitment of those willing to work for "national unity" beyond "sectionalism." In the 1910s and 1920s, the institutional trend of treaties signed by delegates of the isthmus states was accompanied by renewed transnational activities led by the Unionist movement committed to the idea of a Great Central American fatherland.

The Unionist movement constituted a transnational network of idealist activists, who tried to recreate the project of a Central American nation at the grassroots level. They were convinced that the union of early independence failed due to a combination of factors: the egoistic drive of reactionary elites, their manipulation of ignorant masses, and their cooptation of intellectuals who otherwise should have opposed the dissolution. Accordingly, they envisioned reunification as led by a new cohort of committed intellectuals, who would spread the word and nourish the latent massive support for the Unionist revival, which in their eyes was the "truth of the future," soon to turn into a reality with many benefits for the wellbeing and pride of Central Americans, living in liberty, democracy, justice and the rule of law (Luna 1906, 34; Bermúdez 1912; Serpas 1954).

The core of unionism was constituted by intellectuals, teachers and students, most of upper class background, who were disillusioned with the Liberal and Positivist projects and refused to recognize the true Republic of their dreams in the states that they lived in. The movement's leaders and activists alike envisioned that, in order to prosper, Central America had to unite as a region open to all humankind. Moreover, once Central America was united, both Mexico and South America would be in a stronger position to resist the onslaught of U.S. economic interests. The Unionists wanted to regenerate the nation, promote the spiritual consciousness of a shared destiny among all the inhabitants of the Isthmus and create a just society, in which individuals should enjoy basic rights, irrespective of ethnic, gender or status differences (Casaús Arzú 2006). The unionists were fully aware of the failure of previous projects of Central American union, but they believed in the bottom-up regeneration of the united nation, to be attained by promoting a consciousness of common destiny and building a model of equality, social justice and tolerance, that would encourage the union while respecting the autonomy of the various societies (García Giráldez 2005). Imbued by the vision of Central American regeneration and especially in the years preceding the 100th anniversary of the end of Spanish rule, many of the Unionists wandered through the sister-nations of the region, suffering the perils of exile or living as expatriates, trying to promote enthusiasm for the Unionist creed in their new environments.

From a transnational perspective, probably the most important long-term contribution of the Unionists was the widening of public spheres in the 1920s–1940s through the creation of many publications, along with the opening of spaces of sociability accessible beyond distinctions of gender, nationality and to lesser extent, class and race. This forums and spaces of sociability also expanded public debate on issues such as the incorporation of indigenous and mestizo subaltern sectors and women to full citizenship or the regeneration of society from the bottom up. Through such activism and diffusion of ideas, they influenced ideational paradigms, impacting the thought of thousands of individuals across the isthmus; some of whom—e.g. Juan José Arévalo—would later on take active part in politics and others—the daughters and granddaughters of the women participating in the female circles named after Gabriela Mistral—would attain female rights to vote and stand for elections in Guatemala in 1946.

The spirit of reunification that the transnational movement kept alive combined with the sincere attempts of the International Central American Office to work within the framework of pan-Americanism, publicizing efforts of regional coordination and promoting the sense of Central American solidarity. The 1960s witnessed the creation of Central American institutions, while in the late 1970s and 1980s Central America became a major backyard for military confrontation between East and West, in the last years of the Cold War, producing a spillover of violence across much of the region and shaping one of the most massive legacies of human rights violations in the Americas. Regional efforts finally managed to bring the civil and transnational wars to an end in the 1990s.

In the last decades there has been momentum for an endorsement of regional coordination and the creation of many institutions aimed at confronting shared problems, such as the increasing threat that transnational illicit and illegal networks pose to governability and public security in the entire region. Sociologically, politically and culturally, the region has faced increasing pressures derived from those transnational illicit networks, social violence and persisting socioeconomic gaps. Corruption and administrative inefficiency have added public distrust, while public insecurity has led to citizens' retreat from public spheres, widespread reliance on private security services and a growing fragmentation of civil society. Many have left their countries to escape the vicious circle of unemployment, poverty and lack of prospects of life improvement. Many more had dreamt of taking the road of migration too, at least until the economic crisis hit the US and Mexico in the late 2000s. These have been challenges affecting Central America as a whole and demanding decisions that take into account the transnational dimensions of state and network interactions across the region. Treaties and agreements have been signed which have defined in new forms the need to coordinate actions and achieve higher forms of economic integration, harmonization and regulation. The question is how the regional layer becomes incorporated and supported by the national and sub-national levels and whether the development of non-state civil networks and multi-level governance can be attained (Phillips 2001; Jayasuriya 2008).

Structural constrains lend credibility to the idea of supporting regional coordination and transnational cooperation. Small countries face particularly difficult problems in terms of the limitations of their internal markets, the threat posed by transnational criminal networks, and the lack of resources needed to redefine the public arena in terms of alternative models of developments, public confidence and the emergence of civic accountability and diminishing corruption. Pooling resources and easing the formation of regional transnational strengths thus makes sense. The question is whether the current layer of public agencies at the regional level is working to generate the institutional trust needed to support any further move in the direction of coordinated strategies. Particularly now that these strategies cannot be reduced to economic and trade integration but, to be effective, need to address a broader range of transnational issues, so to support an ethos of increasing democratization, citizen participation and multi-level governance in the region.

Conclusions

This chapter is part of a broader research project that discusses processes shaping collective memories and historical narratives, cross-border practices, social networks, and movements of ideas and peoples (see Roniger 2011, 2014; Sznajder and Roniger 2009). Focusing on Central America, this chapter has more than regional interest. We have followed various aspects interconnecting the states and societies of the isthmus within such a long-term perspective, showing that transnationalism had existed long before the current stages of globalization and beyond the domain of economics. Accordingly, this work suggests that globalization and transnationalism are not coterminous, although they may impinge upon one another, as is the case in the current stage of development.

Analysis has drawn particular attention to the realms of politics, public life and the construction of collective identities. It traced the tension-ridden interplay between the process of constitution and consolidation of states and distinctive national identities, on the one hand, and, on the other, the lingering presence of alternative projects of reconstruction of broader, transnational commitments and identities, entangled with one another in shifting ebbs and flows at different historical moments.

The idea of Central America persisted beyond the short-lived attempt of political unification in the early nineteenth century. For some, that experience became a source of inspiration of later transnational projects, while for others its memory solidified the will to take a distinctive path of disengagement. At times, the region was conceived as part of constructive political formulas designed by elites, while at others, it was the core idea of popular movements, informed by the experiences of exiles, migrants and intellectuals crossing state borders and willing to generate a sense of isthmian brotherhood. There were times when the resistance to foreign interventions triggered a sense of transnational solidarity, while under other circumstances it was the United States as a hegemonic power that supported the idea of regional coordination and integration. In sum, in different periods and under varied circumstances, there were diverse ways of envisaging the future of Central America, many of them geared to the logic of consolidation of nation-states, but others oriented toward projecting transnational connections, as traced in diverse political initiatives, civil society networks, belligerent actions, cross-border coalitions, and transnational practices. Rather than claiming that Central America has existed as an objective entity, we analyzed major developments and forces that promoted transnational commitments or fought them in the last two centuries.

In a broader compass, this work underlines the relevance of focusing attention on the construction of collective identities as an autonomous dimension of social life, connecting visions of history, membership in political communities and future-oriented projects. As this study has been conducted on a region that evolved under the complex dynamics of interconnected histories and transnationalism cutting across distinct nation-state identities yet without fully replacing one another, it has additional relevance for comparative-historical research. It enables a closer understanding of the intertwined geopolitical and cultural history of macro-regions, revealing how collective identities affect power, meaning, and the open-ended character of the social, political and cultural forces, shaping societies connected to one another rather than as mere discrete units evolving separately in a path of multiple modernities.

References

Acuña Ortega, Víctor. 1995. "Autoritarismo y democracia en Centroamérica: La larga duración." In *Nicaragua en busca de su identidad.* Edited by Frances Kinloch Tijerino, 535–571. Managua: Instituto de Historia de Nicaragua, Universidad Centroamericana.

Alonso, Ana María. 1994. "The Politics of Space, Time and Substance: State Formation, Nationalism, and Ethnicity." *Annual Review of Anthropology* 23: 379–405.

Anderson, Benedict. 1991. *Imagined Communities.* London: Verso.

Bermúdez, Alejandro. 1912. *Lucha de razas.* México: Tipografía Económica.

Blumenthal, Edward. 2013. *Exils et constructions nationales en Amérique du Sud: proscrits argentins et chiliens au XIXe siècle.* Ph.D. thesis. Paris: Paris University-Diderot.

Casaús Arzú, Marta. 2006. "Las redes intelectuales centroamericanas y sus imaginarios de nación (1890–1945)." *Circunstancia* 9: 123–205.

Cruz, Consuelo. 2000. "Identity and Persuasion: How Nations Remember Their Pasts and Make Their Futures." *World Politics* 52: 275–312.

Domville-Fife, Charles W. 1913. *Guatemala and the Status of Central America.* London: Francis Griffiths.

Eisenstadt, S. N. and Berhard Giesen. 1995. "The Construction of Collective Identities." *European Journal of Sociology (AES)* 36: 72–102.

Fumero Vargas, Patricia. 1995. "De la iniciativa individual a la cultura oficial: El caso del General José Dolores Estrada, Nicaragua, década de 1870." In *Nicaragua en busca*

de su identidad. Edited by Frances Kinloch Tijerino, 324–325. Managua: Instituto de Historia de Nicaragua, Universidad Centroamericana.
García Giráldez, Teresa. 2005. "La Patria Grande Centroamericana: La elaboración del proyecto nacional por las redes unionistas." In *Las redes intelectuales centroamericanas: Un siglo de imaginarios nacionales (1820–1920)*. Edited by Marta Casaús Arzú and Teresa García Giráldez, 123–205. Guatemala: F & G Editores.
Geisler, Michael E. (ed.). 2005. *National Symbols, Fractured Identities*. Hanover: University Press of New England.
Geertz, Clifford. 1973. *The Interpretation of Cultures*. New York: Basic Books.
Herb, Guntram H. and David H. Kaplan (eds.). 2008. *Nations and Nationalisms in Global Perspective*. 4 vols. Santa Barbara: ABC-CLIO.
Hobsbawn, Eric and Terence Ranger (eds.). 1983. *The Invention of Tradition*. Cambridge: Cambridge University Press.
Holden, Robert. 2004. *Armies without Nations: Public Violence and State Formation in Central America, 1821–1960*. New York: Oxford University Press.
Jayasuriya, Kanishka. 2008. "Regionalising the State: Political Topography of Regulatory Regionalism." *Contemporary Politics* 14(1): 21–35.
Leonard, Thomas M. 1993. "Central America and the United States: Overlooked Foreign Policy Objectives." *The Americas* 50(1): 1–30.
Luna, A. 1906. "San Salvador." *La Quincena*, 1 June: 129–131.
Mande, Anupama. 2000. "Subaltern Studies and the Historiography of the Sandinista-Miskitu Conflict in Nicaragua, 1979–1990." Paper presented at LASA 2000 Congress. lasa.international.pitt.edu/Lasa2000/Mande.PDF (accessed July 11, 2008).
Ortner, Sherry B. 1973. "On Key Symbols." *American Anthropologist* 75: 1338–1346.
Phillips, N. 2001. "Regionalist Governance in the New Political Economy of Development: 'Relaunching' the Mercosur." *Third World Quarterly* 22(4): 565–583.
Preuss, Ori. 2011. *Bridging the Island: Brazilians' Views of Spanish America and Themselves, 1865–1912*. Madrid: Iberoamericana Vervuert.
Roniger, Luis. 2008. "Identidades colectivas: Avances teóricos y desafíos politicos." In *Identidad, sociedad y política*, edited by Judit Bokser Liwerant and Saúl Velasco Cruz, 45–68. Mexico: UNAM.
——. 2011. *Transnational Politics in Central America*. Gainesville: University of Florida Press.
——. 2014. *Destierro y exilio en América Latina. Nuevos estudios y avances teóricos*. Buenos Aires: EUDEBA.
Seigel, Micol. 2005. "Beyond Compare: Comparative Method and the Transnational Turn." *Radical History Review* 91: 62–92.
Serpas, Carlos. 1954. *Diario de Hoy*. San Salvador, 8 May 1954, in Mario Monteforte Toledo, 1967, "Los intelectuales y la integración centroamericana." *Revista Mexicana de Sociología* 29(4): 844–845.
Smith, Anthony. 1996. "The Symbolic Construction of National Identities." Paris: Euroconference on Collective Identity and Symbolic Representation.
Subrahmanyam, Sanjay. 1997. "Connected Histories: Notes towards a Reconfiguration of Early Modern Eurasia." *Modern Asian Studies* 31: 735–762.
——. 2005. *Explorations in Connected History: From the Tagus to the Ganges*. Oxford: Oxford University Press.
Sznajder, Mario and Luis Roniger. 2009. *The Politics of Exile in Latin America*. New York: Cambridge University Press.

Taracena, Arturo and Jean Piel. 1995. *Identidades nacionales y Estado moderno en Centroamérica*. San José: Editorial de la Universidad de Costa Rica.

Werner, Michael and Bénédicte Zimmermann. 2002. "Vergleich, Transfer, Verflechtung: Der Ansatz der *Histoire croisée* und die Herausforderung des Transnationalen." *Geschichte und Gesellschaft* 28: 607–636.

——. 2006. "Beyond Comparison: Histoire croisée and the Challenge of Reflexivity." *History and Theory* 45: 30–50.

14 Infrastructuring the Mekong

Construction of the national border and riverbank development in Vientiane Capital, Lao PDR

Miki Namba

Introduction

"Thailand is not a foreign country [*Thai bor mean than pathet*]," a Lao man in his forties said to me with a rather wry smile. It was an ordinary Sunday afternoon, and we were enjoying a beer at his house in a village located about four kilometers from a border checkpoint in Thanaleng, Vientiane Capital. From his village just across the Thadeua Road, the Mekong River flows and divides the two countries: Laos on the east bank and Thailand on the west bank. Our conversation had begun with his statement, "I have never been to any foreign countries," followed by my question, "Not even Thailand?" He smiled because the answer to my question was obvious to him.

I have heard people in Vientiane make this sort of statement only about Thailand; Cambodia and Vietnam are always regarded as foreign countries. The feeling of proximity between Laos and Thailand is widely shared among Lao people, and has been described in many ways. For example, their languages, which both belong to the Tai language family, share extensive vocabulary and have very similar phonological and grammatical systems (Enfield 1999). Also creating feelings of proximity, there are recent cultural and economic influences that Thailand has had on Laos due to Thai investments, Thai goods, and Thai television programs (Long et al. 2007; Rigg 2005). Finally, and perhaps most importantly, Laos and Thailand share the history of the upper Mekong region in the pre-colonial period, often described with an emphasis on vague territorial borders and characterized as having "galactic" (Tambiah 1977) or "mandala" (Wolters 1982) political structure (Evans 2002; Stuart-Fox 1997). These commonalities are undergirded by a high degree of accessibility: about 1.2 million Lao nationals traveled to Thailand in 2001 by crossing the Friendship Bridge in Vientiane; it was the highest annual number of crossings recorded at the bridge. In relation to this figure, the population of Vientiane was only about 616,000 at that time (Keola 2013).

Although I was aware that the history of the people in this region and their living circumstances required them to be highly mobile, the notion of Thailand not being a foreign country led me to reflect on the border between Vientiane and northeast Thailand, Isan. It is worth exploring *in what sense* people of Laos say that Thailand is not a foreign country, and to discover how this comment indexes the ways in which their lives are *not* territorialized by a political border.

In this chapter, I explore the vulnerability of borders by focusing on the infrastructure development projects on and near the Mekong River between Laos and Thailand, and the way in which a focus on infrastructure might help us to address our concerns with borders and transnational movement. I start with an account of the historical background of the formation of the present border and the city of Vientiane, following the notion of *histoire croisée* (Werner and Zimmermann 2006), which helps us to unlearn the taken-for-grantedness of the presence of the national border and the classical use of analytical categories such as nation. However, such historical constructions of instability do not preclude the possibility of understanding Vientiane residents' feelings about their proximity to Thailand. Rather, the function of a border is sustained not only by metaphysical regulations and laws, but also by physical and tangible actors. Among these non-human actors, this chapter will focus on the First Thai–Lao Friendship Bridge and riverbank protection works in Vientiane, extending the notion of the *histoire croisée* through elaborating upon recent studies on infrastructure.

The central interest of this chapter is the question of how the Mekong River has been transformed into a state border through the construction of physical infrastructure intended to enhance its function, and how these attempts—either by the Lao government, foreign organizations, or foreign business enterprises—create enduring contradictions and controversies regarding the function of the Mekong.

Transforming the Mekong River into the border: the First Friendship Bridge

Since the Lao People's Revolutionary Party (LPRP) came to power in 1975, Laos has maintained a single-party political system, though its economy shifted from a planned economy to a liberalized, market-oriented system after the inauguration of *chintanakan mai* (new thinking), or "new economic mechanism," in 1986. The news quickly spread that a small, landlocked, socialist country was giving up on its centralized planned economy and therefore needed to become more open. This spurred infrastructure development projects and aided Laos in securing financial assistance from foreign donors to further its development. The Greater Mekong Subregion Economic Cooperation Program (also known as the GMS Program) was established in 1992 as an initiative of the Asian Development Bank. The program aimed to expand the cross-border movement of people and goods and to enhance regional integration among countries situated along the Mekong River by facilitating the construction of border infrastructure.

Motivated by the GMS Program, the First Thai–Lao Friendship Bridge was opened in April 1994, funded by Australia. It was the first major bridge across the lower Mekong River, 1,170 meters long, and linked the nearby border town of Nong Khai in Thailand to Thanaleng in Vientiane (see Figure 14.1). The creation of this land transportation link was said to have "symbolized the beginning of a new era of openness to the outside world after a decade and a half of Lao isolation and introspection" (Sage 1998, 1). The somewhat prosaic slogan,

Figure 14.1 Map of Vientiane and northeast Thailand.

Source: Map by the author.

"From landlocked to land-linked country," has been widely publicized since the opening of this bridge.

This representation, "from landlocked to land-linked," implied that Laos had been isolated and marginal prior to connecting to its neighboring countries and that it needed to "open up." Ferguson (1990) observes that it is a common strategy among states and developers to employ "the invention of isolation," as it leads to more development projects. In this myth-making process, the Mekong River becomes the border that must be opened up by developing transportation infrastructure as a part of the "economic corridor" to channel the flows of "regional integration" (cf. Walker 1999). This understanding leads to remarks such as, "The GMS has become the vehicle by which landlockedness can be transformed from a national liability into a national asset" (Jerndal and Rigg 1999, 39).

Grant Evans (1998, 131) put it more cautiously, noting that "[a]lthough there has been continuous contact between Laos and Thailand for eons, popular imagery built up around this event was the 'opening up of Laos to the world,' and speculation about all the changes that would come rushing across the bridge." Indeed, foreigners in the early 1990s widely viewed Laos as an isolated and inherently *laid back* country. However, this image was a result of geopolitical dynamics dating from the pre-colonial period and extending into the post-colonial period.

Formation of a national border and Vientiane as capital

Reconsidering the tendency of historical studies to depict continental Southeast Asian border regions as marginal based on modern state borders, recent studies have focused on the Tai traditional sociopolitical and spatial unit, the *meuang*,[1] and its relation to mandala-esque political power. According to Oliver W. Wolters (1982, 16–17), "[the] mandala represented a particular and often unstable political situation in a vaguely definable geographical area without fixed boundaries and where smaller centers tended to look in all directions for security." In such pre-colonial Tai polities, power was conceptualized as radiating like a candle flame (Thongchai 1994, 100), and the political power map of this area appears as a patchwork of mandalas that often overlap each other. Moreover, the index of power was generally population-based rather than territorial. People could express their displeasure with their leaders by leaving. To avoid the constant possibility of emigration, victors sometimes relocated those they defeated closer to their center of power (Gesick 1976 cited in High 2009, 80).

The space comprising various Lao *meuang*, which were brought together to form the Kingdom of Lan Xang by the mid-fourteenth century, extended across the Mekong into the Khorat Plateau of today's northeast Thailand. The capital city of the Kingdom of Lan Xang was moved from Luang Prabang in the north to Vientiane around the year 1560 to secure the kingdom from the threat of expansionary Burma. Despite the relocation of the capital, Vientiane was constantly threatened and occupied by Burma and later by Siam; hence, Marc Askew (2007, 62) notes that the Mekong did not work as "an effective barrier against attack from the opposite bank." After the division of Lan Xang into Luang Prabang, Vientiane, and Champassak by the early eighteenth century, these three Lao principalities became tributaries of Siam by 1779. The failure of King Anuvong's rebellion against Siam eventually resulted in the total destruction of Vientiane by Siamese forces by 1828. Several hundred Lao families were forcibly relocated from the east bank to the west bank of the Mekong. As a result, a number of Lao *meuang* were established on the west bank of the Mekong.

Following the treaty between Siam and France in 1893, the east side of the Mekong became a part of French Indochina, eventually forming the basis of the present national border between Thailand and Laos, while leaving the Lao population on the west side of the Mekong under the control of Siam.[2] With French colonization, as William Logan (2007, 80–81) writes, "the Mekong changed from being the main street of Lao society to become a political boundary dividing the Lao people between French and Siamese control." The French government tried to encourage people who had been relocated to the west side of the Mekong to return to the east side by asserting that the returnees would have the right to French protection. Hundreds of families suffering from Siamese repression crossed the Mekong to return to Vientiane in response to the French appeals, but many of them subsequently realized that the French officials lacked the ability to accommodate them and returned to Siamese territory (Logan 2007, 80).

Laos proclaimed its independence in 1953, but instead of forging ahead with the nation-building process, the country was drawn into a civil war between the communist Pathet Lao and the Lao Royal Government. In this period, Vientiane experienced an unprecedented economic boom, and transportation infrastructure was improved by USAID, which supported the Lao Royal Government army. The most important transport infrastructure improvement in this period was the construction of port facilities at Thadeua and ferry service to Nong Khai in 1958.[3]

Vientiane seemed to be developing, but the growth was nothing but a wartime economic boom that depended on U.S. war expenditures (Stuart-Fox 1997). The Cold War period changed Vientiane into a town of casinos, nightclubs, and prostitutes. Many landless peasants from as far away as northeast Thailand and refugees from airstrike bombing sites in northern Laos also moved to Vientiane. As a result, the population of Vientiane increased from 45,000 in 1945 to 175,000 in 1973 (Stuart-Fox 1997).

In 1975, following the fall of Phnom Penh and the arrival of the People's Army of Vietnam in Saigon, the Pathet Lao took over Vientiane. Thailand quickly imposed a trade embargo and the border was virtually closed, although many people crossed the Mekong to flee to Thailand. From May 1975 to June 1985, 309,694 people fled Laos and entered Thailand, and as many as 60,000 more people are estimated to have been absorbed into Thai society without registration (Stuart-Fox 1986, 52). The patrol at the Mekong riverbank became stricter than ever; for the Thai government, the Mekong River became the frontier against socialism, while for the Lao government it was a border that cut off the anti-socialist influence of Thai imports. The tension between the two countries did not resolve until 1990, when Thailand lifted trade restrictions.

Even after this brief look at the history of Laos, it is still unclear in what sense we should understand the notion that Thailand is "not a foreign country," as stated by my Lao friend. Is it that my friend lives in a world composed of *meaung*? If not, is it that he is in some way feeling the effects of the recent regional integration resulting from the GMS Program? In what follows, I set out to explore ways to understand how people in Vientiane relate to the border itself from the standpoint of their experiences with the Mekong River as the edge of the nation.

The Mekong River as a porous border

Examining the history in this region through *meuang* and the unstable structure of mandala polities—a non-nation-oriented historical view—leads us to do more than just heed *transnational* interactions between the nations. However, Andrew Walker (1999; 2009) and Holly High (2009) caution against overemphasis on the notion of the strong centers and vague peripheries or untamed frontiers characterizing the pre-colonial mandala structures, because such a simple center–periphery approach tends to underestimate the number of ways people experience borders and relate to state power.

The approach Werner and Zimmermann (2006) call *histoire croisée* is helpful here going along with Walker and High's attempt to reinvigorate the predominant account of the history of the Mekong frontiers. *Histoire croisée* explicitly addresses the manner of crossing of an analytical boundary, such as the bounded distinctions between local, national, and international. It calls for an exploration of multiple viewpoints derived from focusing on the entanglements intrinsic to the objects of study and from the researcher's reflection on their use of categories, scales, terms, or analytical frames. Thus, "[in] contrast to the mere restitution of an 'already there,' *histoire croisée* places emphasis on what, in a self-reflexive process, can be generative of meaning" (Werner and Zimmermann 2006, 32). If we only *replace* the position that nation-states used to hold in the classic account of transnational history with *meaung* and mandala, we will fall into the pitfall of another categorical immobilization.

With the notion of *histoire croisée*, it is not difficult to see that the issue goes beyond an open–closed or center–periphery dichotomy. The opening of the Friendship Bridge has enhanced the accessibility between Vientiane and Thailand, but it also regulates cross-border mobility by funneling the flow of people and goods over the bridge and making them pass through two checkpoints. However, that does not eliminate the possibility of productive interference in the state's attempts to control the mobility. Thus, what matters here is not focusing on whether the border is open or closed, liberated or regulated, but on how people experience the border.

Here it is useful to explain the current situation regarding border crossings from Vientiane to Thailand. There is a public bus service that regularly crosses the Friendship Bridge, charging 6,000 Lao Kip ($0.80 USD) to take passengers from the Vientiane Central Bus Terminal to the nearby border town of Nong Khai in Thailand. Buses are usually crowded with Vientiane residents on their way to shop in Nong Khai, while the *nouveaux riches* drive their registered vehicles across the bridge.[4] Many Vientiane residents possess a blue Lao–Thai border pass booklet (*bpeum pahn dehn*), which is much easier to acquire than a passport.[5] The border checkpoint in Thanaleng is located 30 minutes by bus from the city center. Immediately after crossing the border most people take a share taxi or a *jumbo* directly to a nearby shopping mall in Nong Khai that has a Tesco. There, they can have lunch at KFC or a Japanese restaurant chain, and then shop for a huge range of consumer goods, including canned foods, bottles of Coca-cola, plastic plates, and toilet paper, none of which is produced in Laos. After the shopping spree, travelers transport their vast array of shopping bags by hiring a *tuk tuk* (a three-wheeled vehicle) to take them back to the bridge. On the way back, they usually stop for some last-minute shopping at the local market where they buy fresh seafood, most commonly small red clams and squid, which are valued by people who live in a landlocked country with no access to the sea. People with too many shopping bags stop at a bus station and take an international bus to the Lao side of the border, where they negotiate with porters to help load their items onto a bus or a *jumbo*, which takes them back home.

Travelers spontaneously connect and use the recently available means of transport including *tuk tuks*, buses, and informal border porters to bring goods across the border without declaring them at customs. Many of these travelers are women who own small shops in Vientiane and are bringing back goods from Thailand to sell in their shops. Due to the lack of manufacturing industries in Laos, most of the goods found in shops are imported from neighboring countries, particularly Thailand, by these shop owners.[6] Such shops bolster the livelihoods of people in Laos, including those who cannot afford to travel to Thailand.

Economists have praised the opening of the Friendship Bridge, as it has clearly increased the amount of imported goods in Vientiane compared to the time when there was only boat service.[7] However, the actual amount of goods does not match the official data recorded by Thai customs. Even though goods purchased in Thailand valued at 1,000 THB (about $26 USD in September 2015) or more must be declared, this rule is not implemented strictly (Keola 2013, 177–178). Lax customs officers at the Friendship Bridge contribute to the free flow of goods; they rarely check travelers' shopping bags and boxes. Thus, the bridge enables even more goods to flow across the border, but in a way that was not quite envisioned in the GMS Program's discourse outlining economic integration—nor intended by the state.

Thinking through infrastructure

Having first portrayed the border crossing from Vientiane to Nong Khai, we should now consider the borderline itself—the Mekong River—as it is experienced by the people of Vientiane. This requires a slight detour to look at recent trends in the exploration of infrastructure in anthropology.

Infrastructure is conventionally understood as the material technological systems that allow for the interconnection and movement of people, things, and ideas. The English word "infrastructure" has proliferated through development discourse via the work of influential economists like Albert O. Hirschman (1958), who argued that the provision of core infrastructure such as electricity, roads and transport was the state's responsibility (Anwar 2014). Such an account legitimized the notion that international institutions and governmental organizations of developed countries should provide support for developing countries unable to afford the construction of such core infrastructure.

In this capacity, infrastructure is deeply intertwined with the Enlightenment ideals of modernity and progress (Edwards 2003). This is why, given their spectacular materiality, infrastructures have come to symbolize colonial superiority and even to embody what it means to be modern in colonial settings (Mrázek 2002). Infrastructure is certainly built to be functional, but it is also a spectacle in the sense that it serves as a conduit for a state's promise of modernization and its fantasies of linear progress. In other words, infrastructure operates at both the material and poetic levels (Larkin 2013). Recently, many anthropologists have started exploring the ways that infrastructure can operate as a fantasy object (e.g. Dalakoglou 2010; Khan 2006; Larkin 2008). For example, Dalakoglou (2010) describes a situation in Albania in which paved roads were built even though people were not allowed by the socialist state to own vehicles. The point was less to facilitate transportation than to emit signals of modernization via the imposing materiality of the new roads.

In this vein, the development of a border facility is not only brought about through an economically functional and rational decision to enhance and control cross-border interactions; rather, the facilitation of border infrastructure can be also understood as a representation of the nation-state's control of population and mobility.

Aside from the opening of the Friendship Bridge, the landscape along the riverbank in the city center also started changing in 2009 as the result of a riverbank management project, which was intended to deal with concerns about erosion. Financed with a soft loan from the Korean government, the aim of this project was to build a 12-kilometer concrete embankment as well as a paved road and a park between the river and the busiest area in the city center, the latter catering especially to foreign tourists. This concrete embankment was the first full-scale river embankment to be constructed in Laos. Before it was built, central Vientiane's waterfront still had the feel of a semi-natural landscape, where local restaurants and food stalls sat along an unpaved riverbank. At that time, there was not an obvious boundary between the river and residential areas. Currently, however, the landscape has given way to a tourist night market built on an asphalt foundation. Residents who lived along the river in the city center used to catch fish and grow vegetables on the riverbank. However, as one woman in her forties told me during my stay in Vientiane in 2013, "now the river is far away" (see Figure 14.2)

Construction of a shopping center and residential building complex called Vientiane New World has also been launched at the end of the concrete embankment by a Lao–Chinese joint venture. As the company announced in the English-language newspaper the *Vientiane Times* in 2012, Vientiane New World aimed to offer people "a modern lifestyle," which was thought to be "one of the main reasons that Lao people like to cross the border to Thailand." Thus, the Lao–Chinese joint venture was intended to stop the flow of people crossing the border and "prevent the outflow of foreign currency" (*Vientiane Times*, 22 May 2012).

The Mekong River has been transforming into a national border that must be facilitated through the development of infrastructure because it is a part of the "economic corridor" leading to "regional integration," according to the development discourse of the GMS Program. At the same time, the semi-natural riverbank was intentionally concreted over in order to create a clear spatial distinction between the river border and the population, and to represent the state's ability to exert control. As a result, the rapid and recent development, which has been limited exclusively to the central waterfront area in Vientiane, is sometimes seen by foreign development experts and educated Laos as nothing other than bravado and a vain attempt to "show off" to Thailand.

Figure 14.2 Vientiane Mekong promenade.

Source: Author.

The Mekong River as natural infrastructure

However, it is still questionable whether the Mekong has effectively become a boundary dividing two peoples. It seems relevant to take into account the variability of the river as a natural object. It flows and is continuously moving. As the Mekong River Commission points out, the Mekong River in this region has long been regarded as the "people's highway" (Starr 2003), transportation infrastructure that connects distant regions rather than a border whose main function is to divide countries.

When I occasionally visited a village next to the Mekong River, about 20 kilometers from the center of Vientiane, my Lao friends told me how their relatives, who now live in the US, fled Laos by swimming and taking a boat to Thailand to avoid official checkpoints.[8] I learned how easy it was in the old days to exchange goods with Thai traders, who would appear late at night in a boat on the river. I was also told other stories: for example, how the villagers helped members of the Hmong ethnic group escape by crossing the river, necessary because the Hmong people cannot swim because they come from the mountains; they also recounted that some "bad people" purposely let the Hmong refugees drown, in order to rob them of their money and belongings. During the dry season, certain parts of the Mekong become easily accessible by land, and it has been reported that a large number of people escaped Laos by swimming and walking to the Thai side, especially during the revolution (cf. Scott 1989, 161–262). The irresistible characteristic of the flowing river as a natural object that surrounds the Mekong gives people the sense that the river is not an insurmountable border.

Here, the notion of *histoire croisée* can be expanded to reconsider the idea of the human–non-human boundary with help from actor network theory, in which humans and non-humans are treated as analytically equivalent (Latour 1993). Anthropologists are increasingly following this call in their ethnographic explorations of non-humans and their entanglements with humans. This represents a reconsideration of the human-oriented view of anthropology, in which the relation between humans and non-humans holds that the former always act upon the latter. Recent trends in anthropological studies on infrastructure concur with such reflections.

Here we have the Mekong and the embankment. The construction of the embankment was part of the state's attempt to control nature, and thus the decision was made to construct a concrete embankment. As Penny Harvey (2010) argues, concrete has long been a strong symbol of modernity, associated with industry and state power. However, as a substance it is also vulnerable, fragile, even relational, and in need of constant care. In relation to this reality, the concrete embankment in Vientiane projects a strong image of modernization and represents the state's capacity to produce reliable and predictable physical structures. On the other hand, due to the vulnerability of concrete and its intrinsic resistance to the flow of water, there is always the risk that the Mekong could lose its function of effectively representing the stability of the state. It is hard to describe the Mekong embankment as a technological conquest over nature, as the management of riverbank infrastructure requires constant labor and capital in order to maintain and expand its function (see also Carse 2014; Jokinen 2006).

Infrastructure can be regarded as a physical form of a network that allows goods, ideas, people, and funds to travel through space. However, due to the nature of infrastructure's innumerable interactions with other materials and people who use, imagine, and interpret it, these networks may have effects that are different than what the planners or developers had expected. As many anthropologists who are interested in infrastructure have shown, pre-existing infrastructure is not discarded or erased by newly built infrastructure; it continues to have unpredictable effects on the new infrastructure (Star 1999). Newly built infrastructure projects, as Brian Larkin observes, referring to Henri Lefebvre's (1991) account of the city as historical layers of many social spaces, "do not eradicate earlier ones but are superimposed on top of them, creating a historical layering over time" (2008, 6).

Conclusion

In the borderland between Nong Khai and Vientiane lies a natural infrastructure that has always been present—the Mekong River. From a main street or a highway to a borderline between nations, from a fishing and farming spot to a night market and leisure area, the Mekong is constantly transforming in relation to human activities. Each feature of the river does not function in isolation; rather, these features act upon each other in concert.

However, at this point, what we need to reconsider is the utopian view of a world in which technology gives everyone free and equal access to mobility, opportunity, and information. For *nouveaux riches* in Vientiane who obtain a passport or a border pass and have some money to spend in Thailand, the bridge and other means of transport are useful structures, but infrastructure is a relational concept. As Star (1999, 380) puts it, "One person's infrastructure is another's topic, or difficulty." She elaborates with an example: "The cook considers the water system as working infrastructure integral to making dinner. For the city planner or the plumber, it is a variable in a complex planning process or a target for repair." The relationality of infrastructure becomes crucial when looking at those who cannot afford to use it. To the same extent that infrastructure tends to betray the expected functions that are imagined by planners, engineers, and donors, certain transportation infrastructure might immobilize certain people. Holly High writes of how the residents at her field site in southern Laos experience the Lao–Thai border:

> A "borderless" lifestyle is one accessed only by the few, and particularly by the rich. It seems significant that on Don Khiaw, jet planes flew regularly overhead, because flyover rights are a major money earner for the Lao government, but residents themselves rarely if ever boarded these icons of globalisation.
>
> (High 2009, 78)

Thus, it must be emphasized that "discussing infrastructure is a categorical act" (Larkin 2013, 330). Nevertheless, focusing on infrastructure with a notion of its relationality invites more human and non-human actors—conceivably disordered and ill-mannered in a productive way—into our research on borderlands. To do so is not to neglect historical or ethnic background, but rather to illuminate them in different ways, giving us an opportunity to work on more decentered ethnographies.

Notes

1 *Meuang* or muang, can be translated in various ways: it could mean "city state" or "urban area," or it could refer to a country, as in "Meuang Lao" or "Meuang Thai."

2 When the process of colonization started, more Lao people lived on the west side of the Mekong than in French Indochina. Some estimate that half of the population of Siam was comprised of Lao people (Toye 1968, 48).

3 The service had "two pusher tugs, two barges for cargo and passengers, a ramp and customs facilities on the Lao side of the river, all provided by U.S. aid" (Long and Askew 2007, 127).

4 Additionally, a 3.5-km-length international railway between Thanaleng and Nong Khai was built on the Friendship Bridge in 2009.

5 According to an explanation from a Lao civil servant, this border pass allows a traveler to go as far as the provinces in Thailand adjacent to the border province over the course

of three days. Thus, people from Vientiane can travel to Nong Khai and all the provinces bordering Nong Khai, such as Udon Thani and Sakon Nakhon. However, it is common to stay longer and journey further than the regulations allow. In fact, the civil servant himself said he once stayed two years in Khon Kaen and even went to school there.

6 While people in Vientiane do not see Thailand as a foreign country, they certainly recognize Thai products as distinctive. There are affinities here with research on transnational movement focusing on the material culture of post-socialist Polish migrants to Britain (Burrel 2008) and Romanian villagers' consumption practices across the Romania–Hungary border in the 1980s (Chelcea 2002).

7 The increase in exports from Vientiane to Nong Khai has been much less pronounced.

8 The Thai government had a tolerant attitude toward Lao refugees who fled to Thailand during the revolution and did not require them to offer proof of their refugee status (High 2009, 88).

References

Anwar, Nausheen H. 2014. *Infrastructure Redux: Crisis, Progress in Industrial Pakistan and Beyond.* Basingstoke, UK: Palgrave Macmillan.

Askew, Marc. 2007. "From Glory to Ruins." In *Vientiane: Transformations of a Lao Landscape*, ed. Marc Askew, Colin Long and William Logan, 43–72. Oxford: Routledge.

Burrel, Kathy. "Materializing the Border: Spaces of Mobility and Material Culture in Migration from Post-Socialist Poland." *Mobilities* 3(3): 353–373.

Carse, Ashely. 2014. *Beyond the Big Ditch: Politics, Ecology, and Infrastructure at the Panama Canal.* Cambridge, MA: The MIT Press.

Chelcea, Liviu. 2002. "The Culture of Shortage During State-socialism: Consumption Practices in a Romanian Village in the 1980s." *Cultural Studies* 16(1): 16–43.

Dalakoglou, Dimitris. 2010. "The Road: An Ethnography of the Albanian–Greek Cross-Border Motorway." *American Ethnologist* 37(1): 132–149.

Edwards, Paul N. 2003. "Infrastructure and Modernity: Force, Time, and Social Organization in the History of Soiotechnical Systems." In *Modernity and Technology*, ed. Thomas J. Misa, Philip Brey, and Andrew Feenberg, 185–225. Cambridge MA: MIT Press.

Enfield, Nick J. 1999. "Lao as a National Language." In *Laos: Culture and Society*, ed. Grant Evans, 258–90. Chiangmai, Thailand: Silkworm Books.

Evans, Grant. 1998. *The Politics of Ritual and Remembrance: Laos since 1975. Studies in Contemporary Economics.* Honolulu: University of Hawaii Press.

——. 1999. "Introduction: What is Lao Culture and Society?" In *Laos: Culture and Society*, ed. Grant Evans, 1–34. Chiangmai, Thailand: Silkworm Books.

——. 2002. *A Short History of Laos: The Land in Between.* Crows Nest, New South Wales: Allen & Unwin.

Ferguson, James. 1990. *The Anti-politics Machine: Development, Depoliticization, and Bureaucratic Power in Lesotho.* Cambridge: Cambridge University Press.

Harvey, Penelope. 2010. "Cementing Relations: The Materiality of Roads and Public Spaces in Provincial Peru." *Social Analysis* 54(2): 28–46.

High, Holly. 2009. "Dreaming Beyond Borders: The Thai/Lao Borderlands and the Mobility of the Marginal." In *On the Borders of State Power: Frontiers in the Greater Mekong Sub-Region*, ed. Martin Gainsborough, 75–100. Oxford: Routledge.

Hirshman, Albert O. 1958. *The Strategy of Economic Development.* New Haven, CT: Yale University Press.

Jerndal, Randi and Jonathan Rigg. 1999. "From Buffer State to Crossroads State: Space of Human Activity and Integration in the Lao PDR." In *Laos: Culture and Society*, ed. Grant Evans, 35–60. Chiangmai, Thailand: Silkworm Books.

Jokinen, Ari. 2006. "Standardization and Entrainment in Forest Managament." In *How Nature Speaks: The Dynamics of the Human Ecological Condition*, ed. Yrjö Halla and Chuck Dyke, 198–217. Durham, NC: Duke University Press.

Keola, Souknilanh. 2013. "Impacts of Cross-Border Infrastructure Developments: The Case of the First and Second Lao–Thai Mekong Friendship Bridges." In *Border Economies in the Greater Mekong Subregion*, ed. Masami Ishida, 163–185. Basingstoke, UK: Palgrave Macmillan.

Khan, Naveeda. 2006. "Flaws in the Flow: Roads and their Modernity in Pakistan." *Social Text* 24(4 89): 87–113.

Larkin, Brian. 2008. *Signal and Noise: Media, Infrastructure, and Urban Culture in Nigeria*. Durham, NC: Duke University Press.

——. 2013. "The Politic and Poetics of Infrastructure." *Annual Review of Anthropology* 42: 327–343.

Latour, Bruno. 1993. *We Have Never Been Modern*. New York: Harvester-Wheatsheaf.

Logan, William. 2007. "Land of the Lotus-Eaters: Vientiane Under the French." In *Vientiane: Transformations of a Lao Landscape*, ed. Marc Askew, Colin Long, and William Logan, 73–110. Oxford: Routledge.

Long, Colin and Marc Askew. 2007. "Arena of the Cold War." In *Vientiane: Transformations of a Lao Landscape*, ed. Marc Askew, Colin Long, and William Logan, 111–50. Oxford: Routledge.

Long, Colin, Marc Askew, and William Logan. 2007. "Shaping Vientiane in a Global Age." In *Vientiane: Transformations of a Lao Landscape*, ed. Marc Askew, Colin Long, and William Logan, 179–207. Oxford: Routledge.

Mrázek, Rudolf. 2002. *Engineers of Happy Land: Technology and Nationalism in a Colony*. Princeton, NJ: Princeton University Press.

Rigg, Jonathan. 2005. *Living with Transition in Laos: Market Integration in Southeast Asia*. London: Routledge.

Sage, William W. 1998. "Setting the Stage." In *New Laos, New Challenges*, ed. Jacqueline Butler-Diaz, 1–8. Temple: Arizona State University.

Scott, Joanna C. 1989. *Indochina's Refugees: Oral Histories from Laos, Cambodia and Vietnam*. Jefferson, NC: McFarland.

Star, Susan L. 1999. "The Ethnography of Infrastructure." In *American Behavioral Scientist* 43: 377–391.

Starr, Peter. 2003. *The People's Highway: Past, Present and Future Transport on the Mekong River System. Mekong development series no. 3.* Mekong River Commission.

Stuart-Fox, Martin. 1986. *Laos: Politics, Economics and Society*. London: Frances Pinter.

——. 1997. *A History of Laos*. Cambridge: Cambridge University Press.

Tambiah, Stanley J. 1977. "The Galactic Polity: The Structure of Traditional Kingdoms in Southeast Asia." *Annals of the New York Academy of Sciences* 293: 69–97.

Thongchai, Winichakul. 1994. *Siam Mapped: A History of the Geo-body of a Nation*. Honolulu: University of Hawaii Press.

Toye, Hugh. 1968. *Laos: Buffer State or Battle Ground*. London: Oxford University Press.

Vientiane Times, Vientiane, Laos. 2012. "Vientiane New World: Glory of Laos." May 22.

Walker, Andrew. 1999. *The Legend of the Golden Boat: Regulation, Trade and Traders in the Borderlands of Laos, Thailand, China and Burma*. Honolulu: University of Hawaii Press.

——. 2009. "Conclusion: Are the Mekong Frontiers Sites of Exception?" In *On the Borders of State Power: Frontiers in the Greater Mekong Sub-Region*, ed. Martin Gainsborough, 100–111. Oxford: Routledge.

Werner, Michael and Bénédicte Zimmermann. 2006. "Beyond Comparison: Histoire Croisée and the Challenge of Reflexivity." In *History and Theory* 45(1): 30–50.

Wolters, Oliver W. 1982. *History, Culture and Region in Southeast Asia Perspectives*. Singapore: Institute for Southeast Asian Studies.

Part VIII

Diasporas

15 The frontier of belonging

Repatriation and citizenship of the overseas Chinese in colonial Malaya

Low Choo Chin

Introduction

This chapter draws upon Anderson, Gibney, and Paoletti's (2011) contextualization of deportation, citizenship, and the boundaries of belonging. According to these authors, deportation serves to highlight how nation-states "conceptualize both who is a member and who has the right to judge who belongs" (2011, 547). They further define deportation as "an act of membership definition" (2011, 556). Legal vulnerability to deportation means the loss of the right to stay and the severing of the relationship between the state and the individual. Citizenship, then, is a frontier of sorts, determining both belonging and membership. Examining deportation is crucial to understand the legal conception of citizenship and how states draw a dividing line between inclusion and exclusion. In many cases, deportation "encourages contestation over the question of membership" (Anderson et al. 2011, 556).

The chapter examines the intersections between deportation, citizenship, and the frontier of belonging, in a geographical rather than metaphorical sense, using the case study of the Overseas Chinese in British Malaya. The Chinese problem of divided loyalty contradicted the growth of local nationalism and complicated the process of nation building, integration, and citizenship construction in Southeast Asia. The state's closure of citizenship highlighted the complex negotiation underwent in the process of defining inclusion and exclusion (Rigg 2003, 97). Exclusion from the space of local citizenship further rendered Chinese people vulnerable to deportation and they were "forced to flee," as evidenced in the cases of Indonesia, the Philippines, Burma, Siam, and Malaya (Peterson 2012, 103). Peterson showed (2012, 102) that the mass movement to socialist China was "one of the most understudied aspects of the Chinese diasporas." In De Genova's words, deportability turned the overseas Chinese into a "disposable commodity" (2002, 438). As a state practice, deportation "provides an instructive occasion and vital impetus for the critical re-examination of dominant conceptions and conceits regarding the privileges and practices of citizenship, and the constitution of state sovereignty itself through the universal distinction between alien and citizen" (Peutz and De Genova 2010, 4).

Prior to 1939, banishment had been used against the alien population in colonial Malaya. The Banishment Ordinance had been used mostly against Chinese secret societies and exercise of the banishment power enabled the government to maintain law and order in Malaya. This weapon could be used effectively, as there was free movement of Chinese between Malaya and China. Mass Chinese influx from South China into Malaya was tolerated by Malays, since the latter were convinced that these immigrants were "a floating population, whose undesirable members could be returned to China at any time."[1] The operation of the Chinese Nationality Law offered the British great flexibility in deporting unwanted Chinese dissidents. Banishment presented a convenient tool for the British to get rid of the Chinese if they misbehaved or became burdensome (Ong 1995, 131).

During the era of the Cold War, repatriation of the overseas Chinese became the central preoccupation of Malaya's counter-insurgency policy in response to the communist uprising. The "colonial emergency" of 1948 witnessed an unprecedented forced removal of tens of thousands of resident alien Chinese as supporters of the Malayan Communist Party (MCP) (Bonner 2007, 147–150). Their story gives indication of the existence of a political frontier that drew upon the notion of belonging. Hara captures the significance of Malayan deportation in the following ways: "Most of these people had lived for decades or generations in Malaya. Not only did they have no economic ties with China, but many did not even have relatives there" (Hara 2003: 64). Peutz and De Genova (2010, 22) remind us of "a person's vulnerability, historically, to banishment despite having been a long-established legal resident and ostensible citizen." This chapter discusses the development of repatriation and citizenship policies in British Malaya in three stages: (1) repatriation during economic slumps (1932–1945); (2) banishment of undesirable aliens to suppress post-war lawlessness (1945–1948); and (3) mass deportation during the communist insurrection (1948–1952). The Malayan discourse on deportation highlights the tensions between security and the right to belong. Throughout the chapter, the terms banishment, repatriation, and deportation will be used interchangeably.

Transnational migration and redefinition of aliens in pre-war Malaya

Transnational migration of the Chinese diaspora to Malaya began in the early nineteenth century. Until the government introduced the first immigration restriction scheme in January 1933, Chinese immigration to Malaya grew substantially in response to the development in Malaya's rubber and tin industries. The flourishing of the coolie trade was facilitated by the British government and the Malay Rulers' open door policy. Unrestricted entry into Malaya until 1933 created a few generations of local-born Chinese populations. In the eyes of the British, the Chinese were resident aliens and did not have a claim to local citizenship. Many Chinese immigrants also did not nurture the idea of becoming permanent residents of Malaya. It was during the economic slumps that the British colonial authorities found it necessary to limit Chinese immigration (Means 1976, 26–27). This restriction was soon followed by the introduction of the controversial Aliens Bill which justified the banishment of those who did not belong to Malaya.

Control by banishment was deemed necessary on two grounds: security and the economy. The need for control of aliens who spread subversive political ideas, coupled with widespread unemployment among the alien population (and the Malayan-born population) prompted the government to introduce the Aliens Bill in the Legislative Council in 1932. The Aliens Bill drew a sharp line between citizens and aliens. According to the Secretary for Chinese Affairs, A. M. Goodman, an alien was defined as a person not being a British subject or the subject of a Malay Ruler (*Straits Times*, August 9, 1932). As "nearly every alien in this country is a Chinese," the Chinese members of the Legislative Council regarded the Bill as a discriminatory act. It was regarded as "part of an anti-Chinese policy" (*Straits Times*, October 27, 1932).

Following the passing of the Aliens Ordinance of 1932 in the Colony of Singapore, a similar piece of legislation was introduced to the Federated Malay States (FMS). The Aliens Bill, though it aroused opposition among the Chinese population of the FMS, was passed on January 24, 1933 by the Federal Council after a long heated debate. Since the large majority of aliens in Malaya were Chinese nationals, the Chinese community was the category most affected by the legislation. According to the definition of "alien" in the Bill, the Chinese (who were neither British subjects nor British protected persons) were to be subject to stricter immigration rules. As the FMS Government already had vast powers for removing undesirable elements from the country, there was strong opposition against imposing further restrictions. Banishment alone was a deadly weapon wielded by the government against the Chinese (*Straits Times*, January 25, 1933).

In the Federal Council debate on the Aliens Enactment, A. M. Goodman justified the need for a control measure to keep out of Malaya a source of unrest and sedition. The spread of communist propaganda in Malaya was palpable. According to Goodman, 99 percent of it was the work of communist agitators from China and 99 percent of the propaganda was in the Chinese language. Communist propaganda was widely circulated among Chinese alien residents (particularly aliens from the island of Hainan). This propaganda had to be checked, Goodman added, pointing out also that the underclass was particularly vulnerable to its lure: "Our greatest danger in the last four or five years has been the spread of this propaganda among the unemployed or discharged labourers" (*Straits Times*, January 25, 1933). This comment highlights the justification of repatriation policies on economic grounds. During the slump of the early 1930s, mass unemployment witnessed the repatriation of unemployed Chinese, especially the Chinese mining coolies in Malaya. During the first seven months of the tin restriction period, from March to September 1931, 21,365 Chinese were repatriated at a cost of just over $15 a head (*Straits Times*, November 18, 1931).

The debate on the Aliens Enactment aroused strong opposition from Chinese members, who were speaking on behalf of the China-born population. A *Straits Times* correspondent reminded the public that the China-born Chinese had no right to object to this legislation: "He is an alien, coming to a foreign country by permission and under certain conditions. He comes on our terms or not at all." On the positive side, the Chinese who were British subjects or British protected persons were excluded from the operation of the new enactment (*Straits Times*, January 31, 1933). In other words, the Alien Enactment recognized the existence of Malayan Chinese who were to be exempted from this regulation and protected them against economic and political dangers. It gave the Malayan-born or locally domiciled Chinese a distinct "Malayan" status. A *Straits Times* editorial applauded the British administration in the colony and the Malay states for making an important step forward with the Aliens Enactment, saying:

> the local born Chinese can now legitimately put up the argument that as the government has thought it fair and safe to grant concessions to them in respect of immigration, and has for the first time written into a legislation a clear cut demarcation between Malaya-born and China-born Chinese.
>
> (*Straits Times*, February 6, 1933)

The Chinese community posed a serious problem for the British in terms of maintenance of law and order. As the migrant Chinese brought in their existing social and political organizations, security problems caused by secret societies, gang robberies, and extortion were prevalent. Incidences of unrest swelled when Chinese secret societies associated themselves with Chinese–oriented political parties (the Kuomintang and the MCP) (Means 1976, 28–29). Political strife worsened after years of political-military mobilization during Japan's occupation of Malaya. According to Means (1976, 47), the Chinese became "more accustomed to violence, intimidation and extortion" and "a source of much political turmoil in the postwar years." This is what I will turn to in the next section.

Lawlessness and the banishment of undesirable aliens in the post-war period

When the British returned to Malaya, they had to deal with a high level of political unrest. On the one hand, there were waves of gangsterism. Over 600 murders were reported during the period of the British Military Administrative from September 1945 to April 1946. Kidnapping and extortion were common throughout the peninsula as well as was piracy. Episodes of racial clashes, originated as a result of Japanese policies, had claimed the lives of innocent civilians. Suffering the brutality and hardships of the Japanese occupation, Chinese guerrillas took the law into their own hands by persecuting the alleged Japanese collaborators, whom were the Malays. The military government was widely criticized for its inability to restore law and order and to protect innocent people (Stubbs 1989, 15–16; Ramakhrisna 2002, 8–9).

In the aftermath of the war, a more efficient banishment order against undesirable aliens was urged upon the government. The new Malayan Union Government was facing an unprecedented crime wave, especially in Johore, Selangor, and Perak. Many people lived under a constant sense of fear and intimidation as it was regularly reported in the press that foreign criminals—thieves, bandits, gangsters, extortionists, intimidators, murderers—were "free to roam" in the countryside. In September 1947, the English-language newspaper, *Straits Times*, criticized the Malayan Union Government on the virtual suspension of banishment: "Far too many foreign criminals are free to roam and ravage the countryside as they will, not because they cannot be captured, but because, when rounded up, known criminals, associates of criminals and members of secret societies can no longer be expeditiously deported." Whitehall was held responsible for the "ineffectual way" in which lawlessness was being dealt with. According to the editorial, the police were handicapped by their inability to use their pre-war banishment powers to deal with undesirable foreign nationals:

> It is an open secret that within recent months high police officers have expressed the opinion that the immediate banishment of two hundred persons in the Malayan Union would result in the saving of six lives a month and a fifty per cent reduction in the amount of crime.
>
> (*Straits Times*, September 15, 1947)

The *Straits Times* editorial comment on the virtual suspension of banishment was supported by the *Malay Mail* editorial: "the banishment law is not being made effective use of in the Malayan Union and that if that were done the post-war crime wave would very soon be put an end to" (*Straits Times*, September 18(a), 1947). The *Singapore Free Press* (September 15, 1947) stated that the police in Singapore received rather fewer than 50 reports of extortion a month, although there should be hundreds of cases daily. The activities of armed violence went unchecked as the public dare not make reports to the police for fear of reprisals. In many cases, the police had difficulty finding witnesses to give evidence in court. The cooperation of the public could be easily obtained if they were sure that the convicted alien criminals would be subject to banishment. The press lamented the "strange reluctance" of the government to utilize the banishment laws, as it had done in the pre-war period.

The Governor of the Malayan Union, Sir Edward Gent, was pushed to "insist upon the most ruthless application of the [Banishment] Ordinance" (*Straits Times*, September 18 (b), 1947). A very large percentage of criminals in Malaya were aliens and expulsion was necessary. Among other suggestions, the *Straits Times* editorial strongly advised for an amendment to the Banishment Ordinance. First, the newspaper said, the ordinance should be renamed the Deportation of Undesirable Aliens Ordinance as the original name "banishment" implied that people could be removed from their place of birth or citizenship. Second, there should be specific mandatory provisions to deport aliens who committed crimes punished with 10 years' imprisonment or more (*Straits Times*, September 18 (b), 1947). Third, an expedited banishment procedure was urgently needed as only 40 out of 200 banishment orders had been approved in the previous months. Fourth, the lists of banishees ought to be made public, their crimes with large photographs posted throughout the country. Publicity then served as the only way to check lawlessness. If the banishment ordinance was not functioning, the editorial even recommended a new ordinance carrying a mandatory death penalty for major crimes, namely subversion, shooting policemen and civilians, and extortion (*Sunday Times*, September 21, 1947). A shared consensus among the public on the effective method of dealing with lawlessness was deporting alien criminals: "Alien criminals fear banishment as they fear no other penalty, for their own governments are far less finicky in their handling and treatment of such people. Banishment is the one deterrent that means anything to them" (*Straits Times*, September 26, 1947).

Following the intense circulation in the media, similar debates were taking place in the Malayan Union Advisory Council raising the pressing need to speed up banishment through administrative and police channels, to apply the existing law more rigorously, and to use corporal punishment wherever possible for crimes of violence. Before January 1, 1947, no banishment orders were either approved or executed. Since then, 127 orders had been made, of which 49 had been approved and 19 actually carried out (*Singapore Free Press*, October 4, 1947) (see Table 15.1).

The operation of the Banishment Ordinance in Malaya raised similar concerns in the House of Commons: whether the Colonial Secretary was satisfied with the Malayan Union Government's use of the Banishment Ordinance to break the present wave of crime (*Straits Times*, October 29, 1947). During the House of Commons debates in November 1947, David Rees-Williams, the Under-Secretary of State for the Colonies, expressed his satisfaction with banishment enactments in force in Malaya. They were effectively used to combat robbery and murder as in the past (*Straits Times*, November 14, 1947). The application of banishment on certain categories of persons was questionable. Besides maintaining law and order, deportation was also used as a political weapon to deport politically undesirable persons such as trade unionists and members of the Malayan Peoples' Anti-Japanese Army (MPAJA). Furthermore, it was debatable how many detainees had been acquitted on criminal charges before being banished. Likewise, it was also debatable how many would have qualified for Federation citizenship had they not been banished. Deportation, nevertheless, was justified as follows: "Deportation is not, after all, a punishment. It amounts to no more than a free passage to the banishee's own country" (*Singapore Free Press*, February 13, 1948). As Barber (1971, 65) remarked, "the threat of deportation to starving China from prosperous Malaya had always been one of the greatest deterrents to crime."

Table 15.1 Approved banishment orders in Malayan Union, March–September 1947

Month	*Jan*	*Feb*	*March*	*April*	*May*	*June*	*July*	*Aug*	*Sept*
Banishment orders	0	0	3	2	3	8	21	2	10

Source: "Action." *Sunday Times*, October 5, 1947.

Note: 42 police applications for banishment orders were still under considerations

Another equally complex issue facing the British was the political status of the non-Malay communities. The concern over this issue brought the introduction of the Malayan Union citizenship, which granted automatic citizenship to non-Malays who could claim to belong to the country through birth, or residence (Stubbs 1989, 24). Due to Malay political pressure, the Malayan Union citizenship scheme was replaced with a restricted version of citizenship based on the concept of double *jus soli*: a local-born child of an alien was entitled to citizenship if both of their parents were born and had resided in the Federation for 15 years. In other words, only the second generation of local-born children was eligible for Federal citizenship through the operation of law. The first generation was eligible through registration upon fulfilling certain residential and language criteria (Ratnam 1965, 76).

When the Federation of Malaya Agreement came into force on February 1, 1948, 3,120,000 persons became Federal citizens automatically through the operation of law (2,500,000 Malays, 350,000 Chinese, and 225,000 Indians, Pakistani and Ceylonese). The Chinese were given the means of acquiring citizenship, though in a much restricted form compared to the Malayan Union scheme. The Malays and local-born British subjects, on the other hand, enjoyed the privileged status of automatic citizenship (Ratnam 1965, 84). The qualifications for citizenship by operation of law were restricted because "It is by no means certain that birth in Malaya necessarily implies either permanent residence in or loyalty to Malaya. The answer lies largely in a more open profession by the Chinese of their single minded interest here and nowhere else."[2] Although the Malays recognized that there were some Malayan-minded Chinese, it was difficult to distinguish between those who gave undivided loyalty to Malaya and those who gave allegiance to both governments.[3]

With the declaration of the Malayan Emergency, deportation as a practice took on a more sinister complexion. Its application was not only used on criminal aliens, but widely extended to include those suspected of supporting the MCP. Emergency Regulation 17D was passed on January 10, 1949, and granted unprecedented power to the High Commissioner to order the mass detention and banishment, without right of appeal, of the aliens suspected of aiding, abetting, consorting, or harboring communist agitators.[4] Returned deportees would be guilty of an offense and be liable to imprisonment for a term not exceeding three years.[5] The orders were not applicable to (a) any Federal citizens or (b) any British subject born in the Federation or in the Colony.[6] During the Emergency, citizenship was crucial as non-belonging would mean deportation. The following section deals with two aspects of the Malayan counter-insurgency: the repatriation of those unassimilated communist-indoctrinated Chinese aliens, and the granting of citizenship to those who were willing to regard Malaya as their object of loyalty.

Communist insurrection and mass deportation in the Federation of Malaya

In June 1948, the security condition in the Federation of Malaya worsened with the declaration of the nationwide Emergency. Creech Jones, the Secretary of State of the Colonies, fully appreciated the situation and set up a clear line of action to Malcolm MacDonald, the Commissioner-general:

> I will deal first with the Federation, and with those who cannot (repeat cannot) be regarded as belonging to the Federation. I have decided to give High Commissioner forthwith authority to use the Banishment Ordinance without prior reference to me, against any persons who are not Federal Citizens and who are implicated in acts of violence, or in organising or inciting persons to take part in strikes, disturbances or demonstrations in which violence or threat of violence is used.[7]

Creech Jones was pressured by the European business community in Malaya to take drastic actions—banishment and flogging—to deal with the lawlessness sweeping Malaya. The subversive elements (communists) were mainly Chinese nationals but they hired Indians (who were British nationals and could not be deported) to carry out subversive activities.[8] Members of the European delegation demanded the expansion of the Banishment Ordinance to include British subjects as well.[9] Time and again, the *Straits Times* editorial was compelled to urge the government for expedited deportation of the "undesirable element" at the early stage of the Malayan Emergency: "It must be clear to the Government that no real, final and lasting control of the situation can be gained unless the illegal immigrant is apprehended and deported with all possible speed" (*Straits Times*, September 20, 1948). They made their point clear that "it is a tragedy that by granting freedom to a small, lawless alien clique, the people of this country have been denied what they have a right to expect—individual freedom and the rule of law" (*Straits Times*, September 20, 1948).

A clear message was made public by the government in November 1948: "No alien can expect to continue living in the Federation if he assists the Communists with money, food, information or in any other way" (*Straits Times*, November 26, 1948). Those assisting the communists in their campaign of lawlessness either willingly or under pressure were to be deported. These aliens were disrupting the life of citizens and preventing citizens from enjoying the full privilege of citizenship (*Straits Times*, November 26, 1948).[10] Finally in December 1948, the High Commissioner, Henry Gurney took a firm stance against terrorism and the alien Chinese squatters, who were supporting the communist bandits: "the only answer to this problem is to require dangerous alien elements to leave the Federation."[11] In his telegram to Creech Jones, Gurney asked for the power to repatriate these aliens rapidly. Repatriation was urgently needed on the followings grounds: (1) squatters were providing the insurgents with food; (2) resettlement in several areas would be impractical and undesirable because it would merely disperse bad elements; (3) the government would be failing in its duty to the people if it failed to restore law and order; (4) detentions would increase the strain on Security Forces, who could be more usefully employed; (5) wealthy Chinese were forced to contribute to bandits' funds or suffered severe consequences; and (6) a prolonged terrorist war would involve huge financial cost.[12]

In a desperate attempt to "separate the Communist 'fishes' from the 'sea' of squatters" Emergency Regulation (ER) was subsequently promulgated. ER 17 (1) provided for the issuance of detention orders for individual persons, with the right to appeal. ER 17C authorized the High Commissioner in Council to deport any person, who was neither a Federal citizens nor a British subject born in Malaya and was amended on 22 January 1949 to include the deportee's dependents as well (Ramakrishna 2002, 65). On 10 January 1949, a more sweeping Emergency Regulation provided for collective detention and deportation. The controversial ER 17 D empowered the High Commissioner in Council to order the detention and deportation of all or any of the inhabitants of a particular area, who were suspected of helping the bandits. According to Ramakrishna, ER 17D resembled the principle of "collective responsibility" (Ramakrishna 2002, 66).

Immediate deportation was essential as collective detention created serious political, administrative, financial, and security difficulties. By June 1950, the total number of detainees in the Federation and Singapore totaled 9,000 persons (including dependents). According to one official estimation, the figure increased by 525 a month (including dependents). The detention of some persons without trial for two years, it was believed, would strengthen their communist faith and subject the government to human rights criticism.[13] Besides the maintenance cost of the detention camps, the Federation Government bore the expenses, at $125 per head, to enable the repatriates to reach their final destinations in China. The repatriates did not arrive destitute at Swatow, the Chinese port of entry. On disembarkation, each adult and each child received $25 (Straits currency) and $10 as a pocket money allowance respectively.[14] From the administrative perspective, the operation of the detention camps required about 1,500 guards. Most importantly, the government's ability to deport undesirable aliens was crucial to the Malays as the British were responsible for bringing the Chinese coolies into the Malay States since the nineteenth century. Mass deportation had been carried out since January 1949 and the High Commissioner aimed to repatriate 2,000 Chinese detainees a month to the mainland through Swatow.[15]

In 1949, the Federation deported 5,994 aliens and 106 British subjects (*Straits Times*, March 8, 1950) (see Table 15.2). In 1950, a total of 8,900 aliens

Table 15.2 Total banishments under Emergency Regulation 17C, January–August 1949

Month	*Jan*	*Feb*	*March*	*April*	*May*	*June*	*July*	*Aug*	*Total*
Aliens	344	289	526	490	376	988	458	1,074	4,545
British subjects	5	5	21	11	4	10	2	8	66
Total	349	294	547	501	380	998	460	1,082	4,611

Source: "4,611 Deportations in 8 Months: Over Thousand in August." *Singapore Free Press*, September 7, 1949.

Note: The total number of persons detained under the Emergency Regulations at the end of August was 6,203 compared with 6,437 in July.

and 202 British subjects were deported under Emergency Regulation 17C, under which dependents of deported persons were also liable to deportation (*Straits Times*, January 11, 1951) (see Table 15.3).

The problem of repatriation in Malaya went far beyond the security and political issues. Though the detainees were legally Chinese nationals, they were born in Malaya. The repatriation policy brought up the whole issue of belonging and citizenship. On the one hand, the mass deportation of the Chinese detainees was justified on the grounds that they were not citizens. Yet, many of the deportees were born in Malaya, had never been to China, and were looked upon as natives of Malaya by outsiders too. Forced repatriation also caused the splitting up of families, many wives not knowing where their husbands were.[16] At the landing place in Swatow, it was remarked that many of the deportees were not natives of the district—some were from Kwangtung, Kwangsi, and Hainan—and the local authorities were facing difficulties forwarding them to distant villages. Moreover, many of the deportees arrived without sufficient funds to enable them to reach their "native" village.[17]

As far as the government in Malaya was concerned, the detainees were a communist-indoctrinated alien population, trying to overthrow the legitimate government. As Henry Gurney put it: "They are all Communists, of degrees varying with the length of time that they have inevitably been under indoctrination in the camps, and are not capable of re-absorption into Malayan society, with the exception of those selected for the Taiping Rehabilitation Camp."[18] The Chinese-dominated MCP could not be regarded as a "genuine Malayan Nationalist movement" according to Griffiths. The Malays were strongly anti-communist and the British did not anticipate the spreading of communism beyond the Chinese community. In the Federation of Malaya, there were 3,000 to 5,000 permanent armed communist terrorists, who relied heavily on many thousands of supporters for food, money, propaganda, and information. A major problem facing the government was galvanizing the support of the Chinese against the MCP.[19]

In restoring the confidence of the Chinese, the government's task was complicated by internal and external circumstances. Externally, the success of the Chinese Communist Party in China created a stimulus to Chinese national pride. It was necessary to counter this sentiment by giving the Chinese people of Malaya a "constructive and Malayan nationalism."[20] Internally, the government recognized the need to give the Chinese a greater share in the political life of the country. By widening the door of citizenship, it was hoped that the Chinese would be

Table 15.3 Total banishments under Emergency Regulation 17C, January–May 1950

	Jan	*Feb*	*March*	*April*	*May*	*Total since the Emergency*
Aliens	0	188	113	122	222	6,639
British subjects	0	10	5	6	8	135
Total	0	198	118	128	230	6,774

Source: "230 Were Deported in May." *Straits Times*, June 6, 1950.

"brought to renounce their political ties with their Chinese homeland and accept undivided loyalty to Malaya."[21]

There were positive signs of acceptance from the Malay community to liberalizing the citizenship provisions as a result of the discussions taking place in the multi-ethnic Communities Liaison Committee (CLC) (Heng 1988, 149–150). The two-year inter-ethnic deliberation at the CLC resulted in the introduction of the principle of delayed *jus soli.* Delayed *jus soli* signified the unqualified right of the second-generation aliens to Federal citizenship. Local-born children were granted access to automatic citizenship if one of their parents had been born in the Federation. This principle was regarded as fair as the second generation could show a higher level of assimilation compared to the first generation: "a non-Malay of the first generation of local birth will not be assimilated to the Federation's way of life" (Federation of Malaya 1951, 15).

It was clear that assimilation was the benchmark of citizenship criteria. Only those who had identified themselves with "the Federation's way of life" enjoyed the privileged status of citizenship. Based on the established benchmark, unconditional *jus soli* was only applicable to the Malays since "it is reasonably certain that these will be readily assimilated to the Federation's way of life" (Federation of Malaya 1951, 15). Anderson, Gibney, and Paoletti (2011, 559) are right to point out that belonging was no longer measured by the length of residence. A more relevant criterion was the "demonstration of an allegiance" to Malay culture. On September 15, 1952, the Federation of Malaya Agreement (amendment) Ordinance 1952 and the State Nationality Bill came into effect (Stubbs 1989, 185). On the midnight of the 14th, 1.1 million Chinese and 180,000 Indians became Federal citizens by operation of law, whereas those eligible by registration were queuing at the government's offices nationwide to show their proof of eligibility (Barber 1971, 162).

Conclusions

The Malayan Emergency witnessed the controversial measure of mass deportation, but it also opened up the gateway to citizenship for the Malayan Chinese. The government policy on the problem involved two different attitudes: (1) the communist supporters and sympathizers were subject to the full force of law and (2) the neutral and the fence-sitters were to be weaned to the government's side and to be transformed into loyal citizenry. The first measure was drastic to the outside world when it involved detention, expulsion, and distraught families landing in China in destitution. While, as it was believed in the press and government, the state of emergency in Malaya necessitated the repatriation of the dangerous element, its application to the Malayan-born Chinese—who were aliens in the eyes of the Federal citizenship law—proved to be challenging. On the other hand, the amendment to the Federation citizenship law gave law-abiding residents a stake in belonging. As a conclusion to this chapter, it could be argued that citizenship and deportation practices in colonial Malaya complement each other in ways that were not always intended or anticipated.

Notes

1 Henry Gurney, Memorandum on repatriation to China, 31 May 1950, the United Kingdom National Archives (TNA), Foreign Office (hereafter FO), FO 371/83542, FC 1822/22. All FO sources used in this article from the TNA were accessed online through Archives Direct: http://www.archivesdirect.amdigital.co.uk.ezp.lib.unimelb.edu.au.
2 Letter from Sir H Gurney to Sir T Lloyd on the problems and methods of winning Chinese support, 20 December 1948, CO 537/3758 (in Stockwell, 1995, 90).
3 Ibid.,
4 Emergency Regulations (Federation of Malaya No. 10 of 1948), Regulation 17D FO371/75934, F462/1584/10.
5 Ibid.
6 Ibid.
7 Telegram No. 70 from Mr. Creech Jones to Mr. M. J. MacDonald on the use of deportation and detention, 12 June 1948, CO 717/167/52849/2/1948 (in Stockwell 1995, 17).
8 Notes by W. G. Sullivan of a meeting between the CO and a delegation representing European business interests on Malayan lawlessness and Sir E Gent's counter-measures, 22 June 1948, CO 717/172/52849/9/1948 (in Stockwell, 1995, 22).
9 Ibid., 28.
10 Ibid.
11 Telegram No. 1326 from Henry Gurney to Mr. Creech Jones on measures to deal with alien Chinese squatters, 25 October 1948, CO717/167/52849/2/1948 (in Stockwell 1995, 78).
12 Ibid.
13 The repatriation of Chinese nationals under orders of banishment or detention virtually ceased in October, 1949 when the Swatow port was closed due to the communist takeover of the mainland. J. C. Sterndale Bennett, "Report on the disposal of Chinese Detainees and Banishees at present in Singapore or the Federation," 1 June 1950, FO371/83542, FC 1822/24.
14 Memorandum prepared by the Colonial Office, "Malaya: Detention, Repatriation and Resettlement of Chinese," 1951. FO 371/92371, FC182/23.
15 "Report on the disposal of Chinese Detainees and Banishees at present in Singapore or the Federation," 1 June 1950. FO371/83542, FC 1822/24.
16 Report on visit to Hong Kong by Mr. E. B. David, from 6th to 8th April 1949, FO 371/75935, F 7219/1583/10.
17 Telegram from Sir. R. Stevenson, Nanking to Foreign Office, 7 April 1949. Difficulties made by the authorities at Swatow over the landing of repatriated Chinese undesirables from Malaya. FO 371/75934 F 5161/1583/10.
18 Henry Gurney, "Memorandum on repatriation to China," 31 May 1950, FO 371/83542, FC 1822/22.
19 Memorandum prepared by James Griffiths, "Political and economic background to the situation in Malaya" for Cabinet Defence Committee, PREM 8/1406/2, DO (50) 94, 15 November 1950 (in Stockwell 1995, 261–262).
20 Ibid., 263.
21 Ibid., 261

References

Anderson, Bridget, Matthew J. Gibney, and Emanuela Paoletti. 2011. "Citizenship, deportation and the boundaries of belonging." *Citizenship Studies* 15(5): 547–563.

Barber, Noel. 1971. *The War of the Running Dogs: How Malaya Defeated the Communist Guerrillas 1948–60*. London: Collins.

Bonner, David. 2007. *Executive Measures, Terrorism and National Security: Have the Rules of the Game Changed?* Aldershot: Ashgate.
De Genova, Nicholas. 2002. "Migrant 'Illegality' and Deportability in Everyday Life." *Annual Review of Anthropology* 31: 419–447.
Federation of Malaya. 1951. *A Bill Intituled an Ordinance to Re-Enact with Amendments PART XII of the Federation of Malaya Agreement 1948*. Kuala Lumpur: Government Press.
Hara, Fujio. 2003. *Malayan Chinese and China: Conversion in Identity Consciousness, 1945–1957*. Singapore: Singapore University Press.
Heng, Pek Khoon. 1988. *Chinese Politics in Malaysia: A History of the Malaysian Chinese Association*. Singapore: Oxford University Press.
Means, Gordon P. 1976. *Malaysian Politics*. London: Hodder and Stoughton.
Ong, Pamela Siew Im. 1995. *Blood and the Soil: A Portrait of Dr. Ong Chong Keng*. Singapore: Times Books International.
Peterson, Glen. 2012. *Overseas Chinese in the People's Republic of China*. Oxon: Routledge.
Peutz, Nathalie and Nicholas De Genova. 2010. "Introduction." In *The Deportation Regime: Sovereignty, Space and the Freedom of Movement*, edited by Nathalie Peutz and Nicholas De Genova, 1–30. Durham: Duke University Press.
Ramakrishna, Kumar. 2002. *Emergency Propaganda: The Winning of Malayan Hearts and Minds 1948–1958*. Richmond: Curzon.
Ratnam, K. J. 1965. *Communalism and the Political Process in Malaya*. Singapore: University of Malaya Press.
Rigg, Jonathan. 2003. "Exclusion and Embeddedness: The Chinese in Thailand and Vietnam." In *The Chinese Diaspora: Space, Place, Mobility and Identities*, edited by Laurence J. C. Ma and Carolyn L. Cartier. Lanham: Rowman & Littlefield Publisher.
Stockwell, Anthony J. 1995. *British Documents on the End of Empire, Series B, Vol. 3. Malaya, Part II: The Communist Insurrection, 1948–1953*. London: HMSO.
Stubbs, Richard. 1989. *Hearts and Minds in Guerrilla Warfare: The Malayan Emergency 1948–1960*. Singapore: Oxford University Press.
Ucko, David H. 2010. "The Malayan Emergency: The Legacy and Relevance of a Counter-Insurgency Success Story." *Defence Studies* 10(1–2): 13–39.

Newspaper articles

Singapore Free Press

"Power to Fight the Gangster." September 15, 1947.
"Banishment Speed-up Demanded: Council Plea for Drive on Crime." October 4, 1947.
"The Banishment Laws." February 13, 1948.
"4,611 Deportations in 8 Months: Over Thousand in August." September 7, 1949.

Straits Times

"The Chinese Coolie in the Crisis: Federal Council Praise." November 18, 1931.
"Chinese Topics in Malaya." January 21, 1932.
"Alien Population to Be Registered." August 9, 1932.
"Chinese Topics in Malaya." October 27, 1932.
"Aliens Bill and Chinese in the F.M.S: Federal Council Debate." January 25, 1933.
"Malaya's Aliens." January 31, 1933.

"Divided Allegiance." February 6, 1933.
"Handcuffing the Police." September 15, 1947.
"A Union View of Banishment." September 18, 1947.
"Banishment Humbug." *Straits Times*, September 18 (b), 1947.
"This Is More Like It." September 26, 1947.
"Commons Interest in Banishing." October 29, 1947.
"Police Winning War on Gangs." November 14, 1947.
"Undesirables Must Go." September 20, 1948.
"Communists' Friends Must Go." November 26, 1948.
"198 Expelled In Feb." March 8, 1950.
"230 Were Deported in May." June 6, 1950.
"716 People Deported in Two Months." January 11, 1951.

Sunday Times

"Get Rid of Them." September 21, 1947.
"Action." October 5, 1947.

16 James Tigner and the Okinawan emigration program to Latin America

Pedro Iacobelli D.

Introduction

In the course of modern history, exchanges across the Pacific Ocean have been frequent and substantial. Nevertheless, while the flows tethering both sides of the ocean (and the countries within) are extremely important for millions of people, the academic studies of the phenomena have been rather scarce. As Immanuel Wallerstein has succinctly explained, the emergence during the Cold War of academic divisions under the rubric of "area of studies" encouraged the study of specific regions, and not necessarily the interconnections between two or more of them (Wallerstein 1997, 195). Far Eastern Studies, Latin American Studies and African Studies were the regional capsules for which social scientists could find generous financial support to conduct their research at home; thus, the scholarly division of areas, initially developed in the US and later adopted by other nations such as Japan, became a firewall to the examination of movements of ideas, goods and people within an inter-area context (Karashima 2015). The historical exchanges of people, goods and ideas between Asia and the Americas—the Pacific as a place of encounters and as a realm of transnational flows—was substantially obscured during most of the Cold War era.

In the last two decades, the conceptualization of the Pacific as a field of study in the humanities has acquired new life. For example, the seminal work of authors such as Arif Dirlik, Rob Wilson, Miyoshi Masao and Wang Hui illuminate the discussion around the power relations at play within and without the "Pacific Rim" (Wilson and Dirlik 1995; Wang 2007). More recently, the scholarship of Matt Matsuda; the May 2014 special journal issue dealing with transpacific history in *Pacific Historical Review*; and the valuable contributions of the authors of *Transpacific Studies: Framing an Emerging Field* have added more theoretical content to the examination of the relationships within and without the Pacific (Matsuda 2012; Kurashige et al. 2014; Hoskins and Nguyen 2014). In much of this literature, the US and its policies in Asia are central in articulating both cooperation and resistance across the region.

This chapter examines one of the rare cases of inter-Area Studies conducted in the 1950s, by tracking the work of James Lawrence Tigner (1917–2007), then a doctorate candidate at Stanford University, who was commissioned by the U.S. military to survey the Japanese communities in Latin America. Even though Tigner pioneered Japanese migration studies, the epistemological basis and motivation of his research have not been considered by most scholars on the field (cf. Amemiya 1999). His work is fascinating inasmuch as it links Cold War-period Area Studies to migration studies, bringing about a transpacific perspective to the geopolitical considerations of his time. This chapter also takes a fresh look at the origin of the state-led emigration program from Okinawa, then occupied by the U.S. military, to Latin America, by considering migration from the sending state perspective. The noteworthy role of James Tigner, and the influence of the U.S.-based Area Studies in the formative years of the Okinawan emigration program to Bolivia, Argentina and Brazil, are at the center of this chapter's inquiry. In this sense, this paper contributes to the growing literature of postwar Japanese migration.

Indeed, studies on migration history, as one of the forefront topics in transnational studies, have incorporated inter-area research connecting Asia and the Americas (a critical review of the Japanese "transnational" historiography can be found in Iacobelli et al. 2015). Japanese migration is one of the most meaningful movements of people across the ocean—and with the people, the movement of their material culture and kinship relationships. The prewar migration movement, connecting the lives of Japanese imperial subjects and local communities in several countries, gave birth to large Japanese presences in California, Sao Paulo and Lima, to name the biggest communities in the Americas. While the Japanese migration literature has benefited from the works of scholars on both sides of the Pacific Ocean, there persist a number of shortcomings when it comes to analysis of the role of the sending state during the Cold War. For example, the postwar migration from Japan to the Americas has been mainly studied as a continuation of the prewar Japanese movement (Adachi 2006; Befu 2010; Stanlaw 2006). Also, in these narratives, the sending state either has no influence in the organization of the migration flow or plays a minor role in it (Masterson and Funada-Classen 2004; Yanaguida and Rodriguez 1992). Furthermore, the distinction between migrants from mainland Japan and from Okinawa is not always acknowledged in this literature (Suzuki 1992). Finally, the much fewer works written by Latin American authors tend to rely excessively on oral sources and published material, and they focus more on the consolidation of a Japanese (Okinawan) community in specific regions, rather than on the border-crossing process that preceded it (Takeda 2006; Sanmiguel 2006; Morimoto 1999). Furthermore, since the United States played such an important role in the political life of many states within the "Western bloc," the question of to what extent the American position of power in the Pacific region shaped patterns of mobility within the same bloc during the Cold War needs to be asked.

By examining James Tigner´s involvement in the postwar migration program to South America, this chapter contributes to elucidate and observe the postwar political developments that connected Asia and the Americas. Building on sources in English, Spanish and Japanese, this chapter proposes to see borders between Japan and the Americas as a much more fluid reality than they normally they have been depicted. The chapter opens with a brief description of the background of the U.S. position of power in Okinawa and the limitations to border crossing, then explores the context, funding and mechanics of James Tigner's project to study Japanese communities in Latin America. It concludes with a final reflection on the nature of the frontier and the role of the dominant hemispheric power in bridging the gap between continents.

Mobility and the American interests in Okinawa in the early 1950s

The first and last battle fought on Japanese soil during the Pacific War (1941–1945) was in Okinawa from March to June 1945. While some of the main Japanese cities were air bombed before and after June 1945, Okinawa prefecture was the only one in the country to experience on-the-ground fighting. As a result of this battle, more than 200,000 Japanese people—half of them Okinawan civilians—perished, and the victorious U.S. military occupied and began administering the territory (Arasaki 2005, 1–31). When Japan surrendered on August 15, Okinawa prefecture was devastated, towns were completely ruined and the local population, fearful of the horrors of the American invaders—as told to them by the Japanese soldiers—were psychologically in tatters (Dower 1999). Following the declaration of Potsdam, Japan was divided in two administrative territories: the mainland islands of Japan were managed by a U.S.-led Allied Powers administration and Okinawa prefecture, also known as Ryukyu Islands, was entrusted in exclusivity, to the U.S. army. General Douglas MacArthur, as Chief Commander of the U.S. military in the Far East, led both occupations.

The U.S. military government that followed the end of the hostilities was sluggish in initiating the reconstruction of the territory and in improving living conditions for the Okinawan people. The Okinawa prefecture attracted more resources and the attention of American authorities only later, when the early events of the Cold War such as the Communist triumph in China in 1949 and the Korean War in 1950, began to unfold. The National Security Council decisively pushed for the retention of the Ryukyu Islands in a peace treaty with Japan. The National Security Council's documents such as the NSC 49 "Current Strategic Evaluation of U.S. Security Needs in Japan" of June 15, 1949 and the NSC 60/1 "Japanese Peace Treaty" of September 8, 1950 expressed the view that any future treaty with Japan must guarantee the U.S. "exclusive strategic control of the Ryukyu."[1] These documents echoed the widely-spread idea that the United States had to protect its areas of interest from the Soviet-led spread of Communism. The NSC 68 "Objectives and Program for National Security" of April 14, 1950 (approved by U.S. President Harry Truman in September the same year) incorporated the principles of the containment policy into a single document.[2] If in the late 1940s the U.S. Department of State considered it important to secure certain industrial and military centers in Asia which were thought to be of vital importance for the national security, in the early 1950s all points in the West Pacific region were equally vital to the United States (Gaddis 2005, 89–121). In short, the NSC 68 expressed the idea that the balance of power between Washington and Moscow was at stake constantly, everywhere in the world.

As a result of the increasing Communist threat in East Asia the US sought in the peace treaty with Japan to attain permanent control over Okinawa. The Ryukyu Islands represented a strategic location from which the continuity of the American defense perimeter in the Pacific could be secured. Thus, the State Department sought to obtain political and legal sanction for American rule of Okinawa through the peace treaty. In the peace treaty with Japan, the American government, heavily influenced by the NSC 68 and NSC 60/1, granted extraordinary power to the U.S. army to maintain its position of power in the islands, even though the rest of Japan was rendered independent. Article 3 of the peace treaty, granted "all power of administration" of the Ryukyu Islands to the United States, including the government of the local people (Eldridge 2001, 301–314).[3]

Okinawa's landscape was transformed radically in the 1950s. U.S. bases were established throughout the archipelago and U.S. military personnel were ubiquitous in Okinawa's main towns and villages. Okinawa played such an important role in the U.S. security perimeter that journalists and military officials alike began to call it the "U.S. keystone in the Pacific." However, as the number of bases increased, the area of arable land and the perimeter available for housing developments were both severely reduced. Moreover, the local population experienced a rapid increase due to the repatriation of Japanese citizens from the former colonies and a "baby boom" threatened to intensify the shortage of food and jobs in U.S.-occupied Okinawa. For the U.S. military authorities in the island, the increase of the population proved to be a problem that was difficult to solve. One of the early measures considered by the military command in Okinawa was to allow migration as a means to soothe the social pressures that resulted from the "overpopulation problem." Resettlement, particularly overseas, was a very popular idea in Okinawa in the 1950s; indeed, migration touched every aspect of the U.S. administration in the Ryukyu Islands.

Mobility in postwar Asia was heavily determined by the political contingencies during the Cold War. The conflict between the US and the Soviet bloc hindered border-crossing movements between these blocs (cf. Morris-Suzuki 2015). Okinawan people—living in a militarized region—could not leave the island without a permit from the U.S. authorities and from authorities in the hosting state. Thus, in order to begin a migration program the sending state had to orchestrate a transnational effort to coordinate a controlled emigration program. The United States Civil Administration of the Ryukyu Islands (hereafter USCAR), was established to improve the administration of the islands in 1950, but had to evaluate the feasibility and convenience of leading such an emigration program.

As the idea of allowing Okinawan migration gained support in the American officialdom, new potential benefits for the U.S. governments were formulated. The American authorities trusted that out-migration could reduce the population problem, help the local economy recover and help the USCAR to reduce social tension caused by the foreign occupation, and that it would ultimately improve the U.S. image overseas. In 1950 Major General Sheetz, Military Governor of the

Ryukyu Islands (1949–1950), considered migration "[t]he most practical means of alleviating the effects of population pressure in Ryukyus." Hence, he requested that the Department of State explore "with other countries the possible facilitation of such emigration by establishment of large quotas, minimum prerequisites, and perhaps even financial assistance."[4] Brigadier General James M. Lewis, USCAR Civil Administrator (1951–1953), regarded the matter of establishing an effective emigration program as being of the "utmost importance [that] cannot be overemphasized." Brigadier General Lewis, recalling the abovementioned problems of overpopulation, the war-ravaged economy and the pressure of the U.S. base-building process upon the Okinawan people, saw in an extensive emigration program "a major contribution towards the solution of existing and potential problems affecting [U.S.] interests in this area."[5] Similarly, the headquarters of the Far East Command also believed the "matter of establishing an effective emigration program to be of utmost importance." Moreover, migration was expected to increase the remittances coming from overseas Ryukyuan communities. Okinawans living overseas had generously supported their countrymen in the islands in the past. As Tamashiro Migorō, one of the organizers of the emigration program in Okinawa, has pointed out, remittances were "one of the most important reasons for the promotion of migration by the Japanese government before the war" (Tamashiro 1979, 95). In the early postwar years, Okinawan associations abroad contributed financially to the reconstruction efforts in their homeland. As a result, the American plan for Ryukyu emigration considered the remittances as another favorable outcome (95–96). In the same vein, John A. Swezey, Chief of Customs and Immigration, USCAR, considered that an out-migration program could have a positive impact on the U.S.–Ryukyu relations and thus on the U.S. image abroad.[6] Along this line, the USCAR organized several friendship associations such as Naha Ryukyuan-American Friendship Association in order to enhance understanding between the military forces in Okinawa and the local population.[7] Migration was seen as a contribution to the betterment of the relationship between occupiers and local subjects. As Swezey put it, one of the emigration program's goals was the "creation of better American-Ryukyuan relations."[8]

The next step was to coordinate a survey to research the feasibility of a mass out-migration program. Since the rest of Asia was closed to Japanese migration due to the horrors and atrocities committed by Japanese troops a few years earlier during the Pacific War, Latin America emerged as a possible destination.

Area Studies and "Tigner's Report"

Area Studies was reframed in the early years of the Cold War. It sought to present, as mentioned in the introduction of this chapter, practical information about specific regions coming from a various disciplines within the social sciences. The spatial framework of understanding—the image of "area"—had as principal objective, as Latin Americanist Julian Steward described it in 1950, to understand "the nations in foreign areas so thoroughly that we could know what to expect of them," and this exercise "required the data of the social sciences and humanities"

(Steward 1950, xiii). Inter-area research was extremely rare, without funds and lacking the interest of most scholars at the time. The partition of the world in areas was achieved through a large infusion of money coming from private foundations, and as an outcome of the intermediation between the private sector and the government by scholarly institutions such as the Social Science Research Council (Harootunian 2000, 29).

Okinawa as an "area" was a latecomer to this partition of the world. While China and Japan gathered scholars' attention, and it was not uncommon for many American universities to have a Chinese or Japanese Studies department, Okinawa—as an independent area of research—became conceivable only during the postwar period. The embryonic "Ryukyuan Studies" of the 1940s were encouraged to expand at the onset of the USCAR in 1950. To this end, the Department of the Army contracted the main Area Studies research organization in the US, the Pacific Science Board of the National Research Council of the U.S. National Academy of Science, to conduct mainly ethnographic and botanical research in the Ryukyu Islands in 1951 (Sensui 2010, 150). Harold J. Coolidge, Director of the Pacific Science Board during the period of 1946–1970, was a well-known international conservationist directly engaged with conservation projects in Africa and the Pacific (Coolidge 1981; Talbot 1982). Coolidge became an important interlocutor between the U.S. army and the scholarly circles during his career in the Pacific Board.

The "Scientific Investigations in the Ryukyu Islands" (hereafter SIRI) was a series of research activities about the archipelago. Among the many projects conducted, the Pacific Science Board commissioned two studies of history: one by George Kerr, whose research for the SIRI became the basis of his *Okinawa: The History of an Island People* (1958), and another by James L. Tigner, a doctoral degree student at the Hoover Institution, Stanford University, who was commissioned to write a report on the history of Okinawan migration to Latin America (Kerr 1958; Sensui 2010, 151). Kerr's report filled the gap in the general historical knowledge on Okinawa needed by the U.S. administration, for there were no major history textbooks on Okinawan history written in English. Tigner's report on the Okinawan communities in Latin America was a result of specific circumstances such as the need to encourage overseas migration to reduce the population pressure at home. Brigadier General James Lewis, Civil Administrator USCAR, explicitly requested a report on Japanese migration from the Department of the Army in July 1951 (Tigner 1954, iii).[9] The demographic and socio-economic problems of the Ryukyu Islands were accompanied by an extraordinary interest in emigration among Okinawans. Brigadier General James Lewis was, according to Tigner's progress report, "especially interested in learning why Okinawans wish to out-migrate to South America."[10] The final goal for USCAR, as Lewis hoped, was that an agreement with Latin American republics would be made on a diplomatic level before promoting emigration. Tigner's mission was given great importance in the USCAR as it promised to provide a better understanding on Okinawan emigration culture and to assist in the advancement of diplomatic negotiations with Latin American governments.

James Tigner, after a successful but short career in the counterintelligence section of the U.S. Air Force, began his postgraduate studies in history at Stanford University in 1950. He had retired from the military with the rank of Major, having received the following awards: the Army Occupation Medal, the Asiatic Pacific Service Medal, the World War Two Victory Medal and the American Campaign Medal.[11] It was during the early stage of his doctorate investigation, in his early forties, when the Pacific Science Board first contacted him. Tigner assumed responsibilities for conducting the study in July 1951 and in August he was in Japan and Okinawa locating source materials on Ryukyu emigration.[12] From Naha, Tigner continued to Hawaii and then to Washington where he met Harold Coolidge, Director of the Pacific Science Board, and officials of the Department of the Army and the Department of State who helped him to organize his fieldwork in Latin America. The nine months following his visit to Washington were spent surveying Okinawan communities in Latin America.

Tigner's field trip to Latin America was thorough in the number of communities visited and systematic in the way these communities were examined. Indeed, Tigner's itinerary followed the location of Okinawan communities and the time spent in each of them varied depending on the size of the community and the prospects for Ryukyuan emigration and settlement. Tigner, in each of the 12 countries visited, followed a similar pattern: his first stop was in the local American diplomatic mission where he consulted with the highest U.S. official, and only then proceeded to meet with the local Okinawan community.[13] Tigner relied heavily on interviews with first and second generation migrants to write his report, due to the lack of published information about the Japanese communities in South America, let alone the Okinawan ones. The medium of communication was Spanish and Portuguese. Neither Japanese nor any of the Okinawan languages was used—after all, Tigner was a Latin Americanist rather than an Asian studies expert.

Brazil, Argentina and Bolivia emerged as the key locations during Tigner's field trip. He stayed several months in Brazil, particularly in the Western regions of Campo Grande and Mato Grosso. Even though Okinawan communities in those regions and the local authorities expressed their desire to receive immigrants from the Ryukyu Islands, Tigner considered that there were important social hurdles that had to be overcome first. In particular, there were Japanese groups that did not recognize the Japanese defeat during the war, the so-called *kachigumi*; these exerted pressure against those who acknowledged the American victory. Even though Japanese and Okinawan migration to Brazil resumed in the late 1950s, the phenomenon of quasi-religious groups supporting the defeated Japanese empire was a source of constant consideration by the American planners of postwar Okinawan migration.

In Argentina, as in Brazil, the local community as a whole enthusiastically embraced the idea of a new Okinawan settlement (*colonia*) (Tigner 1967, 216–220). However, unlike the Brazil case, the Argentinean authorities were not supportive. The Peronist government's anti-American stance during the Cold War further complicated a U.S.-organized settlement in Argentina. Tigner had to be more careful here than in other countries, "in view of President Peron's continuing unfriendly pronouncements against the United States, and the police state character of his regime."[14] The American officials at the Embassy in Buenos Aires went further, to recommend avoiding contacts with the Argentinean government officials, for that that type of exposure "might react adversely in the survey."[15] Nevertheless, Tigner found out that Japanese (Okinawan) migrants would be allowed to immigrate. While the Argentinean preferred European migration, some concessions could be expected because of the nation's desperate need for farm labor.

James Tigner found in Bolivia both a government willing to receive immigrants and a small but very active local Okinawan community. As a result of the revolution in 1952 the Movimiento Nacionalista Revolucionario (Nationalist Revolutionary Movement or MNR for short) had attained power in Bolivia and was keen to populate the region of Santa Cruz, which had a strong potential to become Bolivia's key agricultural outlet. Tigner noted "Bolivia is the only Latin American country where Japanese may enter as free immigrants without being subject to rigid conditions for admission and settlement."[16] Moreover, the local prewar Bolivian Okinawan community, a result of secondary migration from Peru and Brazil during the first half of the twentieth century, was working at the time of Tigner's visit on a plan to support migrants coming from their motherland (Tigner 1963, 203–213).[17] The Okinawan association in Riberalta founded the Uruma Agricultural and Industrial Association (or Uruma Association) in 1949.[18] This association was instrumental in developing a plan for an Okinawan colony in Santa Cruz.[19] In fact, the Uruma Association had already purchased 2,500 hectares of land in Santa Cruz and held advanced conversations with the local prefect and the central government of Bolivia for the establishment of an Okinawan agrarian settlement by early 1952.

Tigner believed that conditions in Bolivia were extremely promising. The Bolivian authorities expressed eagerness to bring in Japanese or Okinawan migrants; there was vast land available in Santa Cruz department for new immigrants; the US had a well-established technical assistance office in the country; and more importantly, there was a local Okinawan community that was developing their own immigration plan. Furthermore, the location of the Uruma colony, even though it was distant from the main roads, had great potential since it lay near a railroad project being constructed to link Santa Cruz with the Brazilian city of Corumba and from there to the capital of Rio de Janeiro.[20] In this sense, Tigner's fieldwork was a success. Not only had he surveyed the continent and found immigration opportunities in various countries, but he also had "discovered" a mature immigration project that was ready to be implemented.

Together with Paul H. Skuse, USCAR Chief of Public Safety, Tigner authored a policy report for the military government in Okinawa which advised beginning the emigration program to Bolivia immediately. For Skuse and Tigner, the emigration plan was almost complete at the Bolivian end since the Uruma plan had received authorization by the Bolivian government. Moreover, after visiting the settlement's location, the U.S. foreign aid officers in Bolivia (dubbed Point IV) had favorably described the agricultural properties of the land. The military authorities in the islands successfully dispatched the first such groups of migrants in 1954, resuming thus the prewar transpacific flow. By the end of the decade Brazil and Argentina had followed as destinations for Okinawa's migrants. The "Ryukyu Emigration Program," as it was dubbed, was a successful program that dispatched over 17,000 people within 10 years. Though far from the optimistic initial estimates of the U.S. officials, it offered an opportunity for Okinawans to leave the islands and seek better economic and social opportunities in South America.

Conclusions

The U.S. involvement in Okinawan migration is a phenomenon worthy of research, as it shows that the mechanics of transpacific mobility were reliant on the actions taken by the then hegemonic power. Tigner's report, in this sense, is a symptom and a consequence of the domination of American politics in the Asia-Pacific region. The bridge between Asia and Latin America was built on Cold War considerations and had as its raison d'être the control and dominion of a stronger military empire in the West coast of the Pacific. Heonik Kwon has noted that the Cold War in the Asia-Pacific region had a distinct postcolonial dimension—and for him it was not only a conflict between Soviet and Western blocs (Kwon 2014, 64–84). But when we look at the case of the Okinawan migration to South America, we notice the opposite: that is, mobility during the Cold War highlights the (neo)colonial dimension of newly acquired territories by a hegemonic power (cf. Iacobelli 2013).

The Ryukyu Emigration Program serves as an example of a Cold War policy conducted in order to improve the governability of an U.S. military administered area, but it also shows the transnational connections that allowed bridging the gap between two distant regions. It should be noted that this migration program was planned using the skills of a U.S.-based academic, who joined a bigger group of social scientists in shaping the American knowledge about the Ryukyu Islands. Indeed, the origin of the state-led emigration program to Brazil, Argentina and Bolivia cannot be disconnected from its Cold War time contingencies. In a world divided by geopolitical considerations, Tigner's survey is a remarkable case of bloc-wide solidarity. Along these lines, international migration was framed, at least in the Okinawan case, within the realm of the U.S.-dominated world.

Notes

1 Quoted in Yoshida, 2001: 45. NSC 60/1 was a revised version of NSC 60 of 27 December 1949. NSC 49 was revised in NSC 49/1 of 4 October, 1949.

2 The NSC 68 erased the early distinction made by George Kennan between peripheral and vital interest zones. There is much discussion as to what was "peripheral" and what was "vital." For Kennan, Japan and Western Europe were part of the vital zone of

interest. For the NSC 68's authors, "the assault on free institutions is worldwide now, and in the context of the present polarization of power a defeat of free institutions anywhere is a defeat everywhere" (National Security Council 1975). The policy planning group behind the NSC 68 was led by Paul H. Nitze.

3 The Japanese negotiators in many occasions expressed the view that severing the Ryūkyū from Japan would be a mistake (Department of State 1977, 811, 33, 960–961, 1163). The emperor, among other key political figures, supported the exchange of Okinawa for U.S. security. For comments on the so-called "Tenno message" see Ōta 1990, 314–329; Matsuoka et al. 2003, 84–85.

4 National Archives and Record Administration (hereafter NARA), RG.59. 1950–54. box.5689, folder 1.

5 "Ryukyuan Emigration to Brazil and Bolivia," 9 June 1953. NARA RG.319.270.18. ex.60 box.30.

6 "Ryukyuan Emigration, An outline," 3 June 1954. NARA RG.319.270.18. ex.60 box.30.

7 In 1953 the "Friendship Week of May" was established for the same purpose. See "Report of Government and Political Developments—May 1953," 4 June 1953. NARA RG.260.B190.2/1 ex.2169. FRCs603. Folder 14.1.

8 "Ryukyuan Emigration, An outline," 3 June 1954. NARA RG.319.270.18. ex.60 box.30.

9 Tigner's report was translated into Japanese in two volumes in 1957 and 1959.

10 James Tigner, "SIRI: Progress Report on Survey of Okinawan Communities in Latin America with Exploration of Settlement Possibilities," 15 October 1951. N/A, "United States Administration Materials, RYUDAI (the University of the Ryukyus), Sengo Shiryō Shitsu." Vol. No. S312 UN 17(2).

11 See University of Nevada, Reno "Remembering Friends," accessed October 1, 2014, http://www.unr.edu/silverandblue/online/winter2008/readmore/friends.html.

12 James Tigner, "SIRI: The Okinawans in Latin America: Investigation of Okinawan Communities in Latin America with Exploration of Settlement Possibilities," August 1954. N/A, "United States Administration Materials, RYUDAI (the University of the Ryukyus), Sengo Shiryō Shitsu." Vol. No. S312 UN 17(6), iii.

13 Tigner, decades after completing his field trip, published some of his conclusions in an academic journal (Tigner 1981).

14 Tigner, "Progress report 31 May 1952," pp. 2/3. N/A, "United States Administration Materials, RYUDAI (the University of the Ryukyus), Sengo Shiryō Shitsu." Vol. No. S312 UN 17(3).

15 Comment made by the First Secretary Mr. Lester Hallory, in ibid., 3.

16 Tigner, "The Okinawans in Latin America," SIRI August 1954. N/A, "United States Administration Materials, RYUDAI (the University of the Ryukyus), Sengo Shiryō Shitsu." Vol. No. S312 UN 17. P480.

17 The local Japanese Association, though few in numbers, was established in the city of La Paz in the 1920s and served the local Japanese and Okinawan community until it was forced to close during World War Two since Bolivia supported the United States during the war. The president of the La Paz Japanese Association was interned in Crystal Palace, Texas in May 1943.

18 Uruma is the name of a small region in central Okinawa (Tigner 1963, 213–214).

19 James Tigner, "SIRI: Progress Report on Survey of Okinawan Communities in Latin America with Exploration of Settlement Possibilities," 31 May 1952, 28. Also see N/A, "United States Administration Materials, RYUDAI, Sengo Shiryō Shitsu." Vol. No. S312 UN 17(3).

20 James Tigner, "SIRI: Progress Report on Survey of Okinawan Communities in Latin America with Exploration of Settlement Possibilities," 31 May 1952, 30. N/A, "United States Administration Materials, RYUDAI, Sengo Shiryō Shitsu." Vol. No. S312 UN 17(3).

References

Adachi, Nobuko. 2006. "Introduction: Theorizing Japanese Diaspora." In *Japanese Diasporas: Unsung Pasts, Conflicting Presents, and Uncertain Futures*, edited by Nobuko Adachi, 1–22. Oxon: Routledge.

Amemiya, Kozy. 1999. "The Bolivian Connection: U.S. Bases and Okinawa Emigration." In *Okinawa: Cold War Island*, edited by Chalmers Johnson, 53–70. Cardiff: Japan Policy Research Institute.

Arasaki, Moriteru. 2005. *Okinawa gendaishi*. Tokyo: Iwanami Shinsho.

Befu, Harumi. 2010. "Japanese Transnational Migration in Time and Space: An Historical Overview." In *Japanese and Nikkei at Home and Abroad: Negotiating Identities in a Global World*, edited by Nobuko Adachi, 31–46. Amhers: Cambria Press.

Coolidge, Harold J. 1981. "Profile of Harold Jefferson Coolidge," *The Environmentalist* 1.

Department of State. 1977. *Foreign Relations of the United States, 1951. Vol. VI, Asia and the Pacific (in Two Parts) Part 1*. Washington, D.C.: Government Printing Office.

Dower, John W. 1999. *Embracing Defeat: Japan in the Wake of World War II*. New York: W. W. Norton & Company, Inc.

Eldridge, Robert D. 2001. *The Origins of the Bilateral Okinawa Problem: Okinawa in Postwar US–Japan Relations, 1945–1952*. New York: Garland.

Gaddis, John Lewis. 2005. *Strategies of Containment: A Critical Appraisal of American National Security*. Oxford: Oxford University Press.

Harootunian, Harry. 2000. *History's Disquiet: Modernity, Cultural Practices, and the Question of Everyday Life*. New York: Columbia University Press.

Hoskins, Janet and Viet Thanh Nguyen. 2014. *Transpacific Studies: Framing an Emerging Field*. Honolulu: University of Hawaii Press.

Iacobelli, Pedro. 2013. "The Limits of Sovereignty and Post-War Okinawan Migrants in Bolivia," *The Asia-Pacific Journal* 11(34). http://www.japanfocus.org/-Pedro-Iacobelli/3989/article.html, accessed May 28, 2016.

Iacobelli, Pedro, Danton Leary and Shinnosuke Takahashi. 2015. *Transnational Japan as History: Empire, Migration and Social Movements*. New York: Palgrave.

Karashima, Masato. 2015. *Teikoku Nihon no Ajia kenkyū: sōryokusen taisei/keizai riarizumu/minshu shakaishugi*. Tokyo: Akashi Shoten.

Kerr, George H. 1958. *Okinawa, the History of an Island People*. Rutland, VT: Charles E. Tuttle Co.

Kurashige, Lon, Madeline Y. Hsu and Yujin Yaguchi. 2014. "Introduction: Conversation on Transpacific History," *Pacific Historical Review* 83(2): 183–188.

Kwon, Heonik. 2014. "The Transpacific Cold War." In *Transpacific Studies: Framing an Emerging Field*, edited by Janet Hoskins and Viet Thanh Nguyen, 64–84. Honolulu: University of Hawaii Press.

Masterson, Daniel and Sayaka Funada-Classen. 2004. *The Japanese in Latin America*. Urbana and Chicago: University of Illinois Press.

Matsuda, Matt. 2012. *Pacific Worlds: A History of Seas, Peoples, and Cultures*. Cambridge: Cambridge University Press.

Matsuoka, Hiroshi, Yoshikazu Hirose, and Yorohiko Takenaka. 2003. *Reisenshi: sono kigen, tenkai, shūen to Nihon*. Tokyo: Dobunkan.

Morimoto, Amelia. 1999. "Inmigración y transformación cultural. Los Japoneses y sus descendientes en el Perú," *Política Internacional* 56(Abril/Junio): 15–24.

Morris-Suzuki, Tessa. 2015. "Prisoner Number 600,001: Rethinking Japan, China, and the Korean War 1950–1953." *The Journal of Asian Studies* 74(2): 411–432.

National Security Council. 1975. "NSC-68: A Report to the National Security Council," *Naval War College Review* XXVII(May–June).
Ōta, Masahide. 1990. *Kenshō: Showa no Okinawa.* Naha: Naha shuppansha.
Sanmiguel, Ines. 2006. "Japoneses en Colombia. Historia de inmigración, sus descendientes en Japón," *Revista de Estudios Sociales* 23(Abril): 81–96.
Sensui, Hidekazu. 2010. "Okinawa no chishikenkyū: senryōki amerikajin shugaku no saikentō kara." In *Teikoku no shikaku / shikaku*, edited by Tooru Sakano and Changon Shin, 147–176. Tokyo: Seikyusha.
Stanlaw, James. 2006. "Japanese Emigration and Immigration: From Meiji to the Modern." In *Japanese Diasporas: Unsung Pasts, Conflicting Presents, and Uncertain Futures*, edited by Nobuko Adachi, 35–51. Oxon: Routledge.
Steward, Julian H. 1950. *Area Research, Theory and Practice.* New York: Social Science Research Council.
Suzuki, Jōji. 1992. *Nihonjin dekasegi imin.* Tokyo: Heibonsha.
Takeda Mena, Ariel. 2006. *Anecdotario histórico—japoneses chilenos—primera mitad del siglo XX.* Santiago: Margarita Hudolin.
Talbot, Lee M. 1982. "Dedication to Dr. Harold J. Coolidge," *The Environmentalist* 2(4): 281–282.
Tamashiro, Migorō. 1979. "Okinawa kaigai ijū kankei kiroku," *Ijū kenkyu* 16: 81–115.
Tigner, James Lawrence. 1954. "Scientific Investigations in the Ryukyu Islands (SIRI): The Okinawans in Latin America: Investigation of Okinawan Communities in Latin America with Exploration of Settlement Possibilities." Washington, D.C.: National Research Council.
Tigner, James. 1957. *Tigunaa hōkokusho [burajiru hen].* Naha: Ryukyu Seifu.
——. 1959. *Tigunaa hōkokusho [kōhen].* Naha: Ryukyu Seifu.
——. 1963. "The Ryukyuans in Bolivia," *The Hispanic American Historical Review* 43(2): 203–213.
——. 1967. "The Ryukyuan in Argentina," *The Hispanic American Historical Review* 47(2): 216–220.
——. 1981. "Japanese Immigration into Latin America: A Survey," *Journal of Interamerican Studies and World Affairs* 23(4): 457–482.
Wallerstein, Immanuel. 1997. "The Unintended Consequences of Cold War Area Studies." In *The Cold War and the University: Toward an Intellectual History of the Postwar Years*, edited by Noam Chomsky et al., 195–232. New York: The New Press.
Wang, Hui. 2007. "The Politics of Imagining Asia: Empires, Nations, Regional and Global Orders," *Inter-Asia Cultural Studies* 8(1): 1–34.
Wilson, Rob and Arif Dirlik. 1995. "Introduction: Asia/Pacific as Space of Cultural Production." In *Asia/Pacific as Space of Cultural Production*, edited by Bob Wilson and Arif Dirlik, 1–14. Durham: Duke University Press.
Yanaguida, Toshio and María Dolores Rodriguez del Alisal. 1992. *Japoneses en América.* Madrid: Mapfre.
Yoshida, Kensei. 2001. *Democracy Betrayed: Okinawa under U.S. Occupation.* Bellingham: Western Washington University.

Index